Advanced *p*H Measurement and Control

Digital Twin Synergy and Advances in Technology

Fourth Edition

Advanced *p*H Measurement and Control

Digital Twin Synergy and Advances in Technology

Fourth Edition

Gregory K. McMillan
Christopher Stuart
Dean Cook
Zachary Sample

Notice

The information presented in this publication is for the general education of the reader. Because neither the author nor the publisher has any control over the use of the information by the reader, both the author and the publisher disclaim any and all liability of any kind arising out of such use. The reader is expected to exercise sound and professional judgment in using any of the information presented in a particular application.

Additionally, neither the author nor the publisher has investigated or considered the effect of any patents on the ability of the reader to use any of the information in a particular application. The reader is responsible for reviewing any possible patents that may affect any particular use of the information presented.

Any references to commercial products in the work are cited as examples only. Neither the author nor the publisher endorses any referenced commercial product. Any trademarks or trade names referenced in this publication, even without specific indication thereof, belong to the respective owner of the mark or name and are protected by law. Neither the author nor the publisher makes any representation regarding the availability of any referenced commercial product at any time. The manufacturer's instructions on the use of any commercial product must be followed at all times, even if in conflict with the information in this publication.

The material and information contained in this book are for general information purposes only. Views and opinions expressed by the author(s) are solely their own and do not necessarily represent those of ISA.

Printed in the United States of America

ISBN-13: 978-1-64331-232-3 (print)
ISBN-13: 978-1-64331-233-0 (ePub)
ISBN-13: 978-1-64331-234-7 (Kindle)

ISA
3252 South Miami Blvd, Suite 102
Durham, North Carolina, USA 27703
www.isa.org

Library of Congress Cataloging-in-Publication Data
McMillan, Gregory.
Advanced ph measurement and control: digital twin synergy and advances in technology / by Gregory McMillan, Christopher Stuart, Dean Cook, Zachary Sample, fourth edition
p. cm.
Durham: International Society of Automation, 2024.
Includes bibliographical references and index.
ISBN: 9781643312323 (paperback)
LCCN: 2024931010
1. Automation and control. 2. pH measurement. 3. Digital twin. 4. Electrode technology.
https://lccn.loc.gov/2024931010

This book is dedicated to everyone working in the process industry who is seeking to advance process performance by gaining a greater understanding of first-principle process relationships and the opportunities to improve automation system dynamics and strategies. The authors especially appreciate practitioners who are willing to invest the time to use the digital twin to develop a deeper knowledge and breed innovation to get the most out of the largely untapped capability of today's measurement and control systems to achieve consistent and reliable process control improvements.

Acknowledgments

I am extremely fortunate to have been in a uniquely positive creative environment for my entire 54-year career. I have worked with exceptionally talented professionals who were industry leaders in particular areas of expertise, many at the forefront of using modeling to better understand and develop process control opportunities. I was also given the freedom to explore, discover, detail, implement, and document advancements. The subsequent publication of what I had learned was encouraged, which is increasingly rare for users compared to producers of automation systems. In my tenure in Monsanto's Engineering Technology (ET) department, this culture was nurtured by its director and creator, the late James Fair, professor emeritus at the University of Texas. Steady-state modeling was the focus for process and mechanical design improvement and dynamic modeling was the focus for process control improvement for gaining deeper knowledge and fast, flexible testing. As far as the sources of my accomplishments, I estimate 30% are from models; 30% from conversations with experts, most notably those on *p*H electrodes with Jim Gray; and 40% from field experience. I also credit the books *Determination of p***H***, Theory and Practice,* second edition, by Roger G. Bates and *p***H** *and pION Control in Process and Waste Streams* by F. Greg Shinskey for providing deeper knowledge of complex process relationships.

Upon retiring from Solutia, the spinoff of Monsanto's chemical business, I was fortunate to continue to explore the synergy of modeling and control while working for Mark Nixon, director of DeltaV Research and Development (R&D) at Emerson Process Systems and Solutions, one of the great minds in developing new control system capabilities.

Working part-time in developing and demonstrating dynamic models for key unit operations and control system enhancements, particularly for the bioprocesses, has been a great finale to my career thanks to Nobin William, vice president of Simulation R&D at Emerson Process Systems and Solutions, who provides extraordinary support and knowledge on dynamic simulation. This is complemented by a positive work environment with creative energy and talent, particularly exemplified by the coauthors and vice presidents Dave Sextro and Mart Berutti.

I also thank the cofounder of the ISA Mentor Program, Hunter Vegas, and the program resources, Mark Darby, Bart Probst, Michel Ruel, and Hector Torres, for excellent technical reviews. Finally, I thank my spouse, Carol, who has supported all my endeavors for over 35 years.

Gregory K. McMillan

Contents

About the Authors

Gregory K. McMillan is a retired senior fellow from Solutia and an ISA Fellow. He has most recently retired from working part-time as a senior principal software engineer in Process Simulation Development at Emerson Process Systems and Solutions. McMillan was an adjunct professor in the Washington University Saint Louis Chemical Engineering department from 2001 to 2004. He received the ISA Kermit Fischer Environmental Award for *p*H control in 1991, received the *Control* magazine Engineer of the Year Award for the Process Industry in 1994, was inducted into the *Control* magazine Process Automation Hall of Fame in 2001, was honored by *InTech Magazine* in 2003 as one of the most influential innovators in automation, and received the ISA Life Achievement Award in 2010. He is the author of numerous ISA books on process control, his most recent being *Advances in Reactor Measurement and Control* and *New Directions in Bioprocess Modeling and Control,* second edition. McMillan has been the monthly Control Talk columnist for *Control* magazine since 2002. McMillan is the founder and co-leader with Hunter Vegas of the ISA Mentor Program for industry practitioners and started and guided the ISA- Standards and Practices committee on ISA-TR5.9-2023 PID Algorithms and Performance Technical Report. McMillan received the ISA Mentoring Excellence award in 2020. He has a bachelor of science degree in engineering physics from Kansas University and a master of science degree in control theory from Missouri University of Science and Technology.

Christopher Stuart is a senior software engineer in the Process Simulation Development group at Emerson Process Systems and Solutions. He holds a bachelor of science degree in chemical engineering with a biochemical emphasis from Missouri

University of Science and Technology. Stuart's research interests are in efficient algorithms for thermodynamic calculations and dynamic simulations.

Dean Cook is the Mimic product manager in the Process Simulation Development group at Emerson Process Systems and Solutions. He holds a bachelor of science degree in physics from Western Illinois University, has 30 years of experience developing dynamic simulation software, and continues to drive toward improving training and process optimization for a wide range of industries.

Zachary Sample is a digital enterprise consultant in Marketing/Business Development for Emerson Process Systems and Solutions. He has over 10 years of experience delivering cross-industry simulation solutions in various roles ranging from technical implementation and engineering to project management and consulting. In these roles, he has had a passion for helping process companies better leverage simulation to improve automation and operational performance. He has a bachelor of science degree in chemical engineering from Missouri University of Science and Technology and a master of science degree in chemical engineering from North Carolina State University.

Preface

There have been many advances in the technology and the understanding of the trade-off in the capability of a wide spectrum of electrode technology for different process operating conditions. This knowledge can be used to enable *p***H** measurements to achieve the incredible precision and rangeability of the hydrogen ion concentration that exceeds that of any other composition measurement by many orders of magnitude. This book provides the insights and guidance needed for electrode selection, installation, calibration, and troubleshooting. The remarkable advantages of middle signal selection are detailed to achieve the best performance, reliability, and sustainability.

We also offer a much greater recognition of what can affect the accuracy of the titration curve, the key to the best design of all aspects of the system, especially mixing equipment, piping, valves, flowmeters, and control strategies. Most critical is the realization that the often overlooked absorption of seemingly insignificant amounts of carbon dioxide from simple exposure to air creates carbonic acid that, like other weak acids, profoundly moderates the steepness of the titration curve in the neutral region. Theoretical titration curves of strong acids and strong bases commonly seen in the literature rarely exist in plants. The slopes of these theoretical curves are four or more orders of magnitude steeper than those experienced in actual *p***H** applications. The moderating effect of weak acids, weak bases, and conjugate salts in actual plants can be readily added to a widely inclusive charge balance to provide curves that match the reality. Because of a single valley and logarithmic relationship, the charge balance can be easily and quickly solved within eight iterations by a simple and reliable interval halving search technique. The literature tends to focus on direction solutions that require extensive narrowing and simplifying equations.

There are no limits to the scope or complexity of the charge balance with the interval halving solution. Furthermore, the charge balance can be readily modified to include the effect of operating conditions—most notably ionic strength, low water content, and nonaqueous solvents—on activity coefficients, dissociation constants, and, consequently, *p***H**. The resulting charge balance tuned to match plant actual titration curves in conjunction with the first principle models of the material and energy balances provides a powerful dynamic model for a digital twin.

This book presents the methodology to find and detail the best-performing and most economical design of equipment, piping, valves, flowmeters, and control strategies. Best practices are added to the end of each chapter to help summarize the material and focus on what is truly important. Project savings of $400K or more are possible when a plant eliminates a vessel and uses a static mixer, valves, flowmeters with much greater precision and rangeability, and feedforward or ratio control. The total lifecycle cost benefits, which include reagent and maintenance costs and better process performance, can be even greater.

Because of the many challenges posed by the incredible precision, rangeability, and associated nonlinearity of *p***H**, the digital twin is an essential tool for total system design, installation, maintenance, and continuous improvement. Chapter 9 has been added to provide extensive practical guidance on how to build, employ, and adapt dynamic simulations in a digital twin. Dynamic simulations are critical for developing and testing the design details, enabling a successful application that addresses the exceptional challenges to achieve a *p***H** system's true potential. Appendix M offers examples of how dynamic models are set up to simulate key combinations of static mixers and vessels and variations in piping and control system designs.

1 The Essentials

1.1 The Heart of the Matter

The *p*H in aqueous solutions has a major impact on many industrial processes. Although several ions are present in aqueous solutions, the H^+ (H_3O^+) and OH^- ions intervene most frequently in physical, biological, or chemical behavior.

This chapter is designed to provide everything you must know about *p*H in a couple of hours because time is a premium and edification a luxury. This introduction gets you started in the right direction. The succeeding chapters are layered with a level and degree of detail that ranges from the novice to the expert.

> Key CONCEPT
>
> The *p*H electrode offers, by far, the greatest sensitivity and rangeability of any measurement. Making the most of this capability requires incredible precision of mixing, reagent manipulation, and nonlinear control. *p*H measurement and control can be an extreme sport.

The logarithmic relationship between *p*H and hydrogen ion activity, as shown in Equation 1-1b, offers the ability to measure hydrogen ion concentration from 1 to 10^{-14} over the 0 to 14 *p*H scale range. In fact, *p*H measurements below 0 and above 14 are also possible, which extends the rangeability beyond 14 orders of magnitude.

$$a_H = 10^{-p\mathbf{H}} \tag{1-1a}$$

$$p\mathbf{H} = -\log(a_H) \tag{1-1b}$$

$$a_H = \gamma \bullet c_H \tag{1-1c}$$

where

a_H = hydrogen ion activity (gm-moles per liter)

c_H = hydrogen concentration (gm-moles per liter)

γ = activity coefficient (1 for dilute solutions)

*p***H** = negative base 10 power of hydrogen ion activity

The hydrogen ion activity is the effective concentration and is a measure of the hydrogen ion's ability to move and combine with other ions. For dilute solutions, the effective and actual concentrations are equal, and the activity coefficient is 1. For solutions with high concentrations of ions, the crowding and presence of other charges reduce the activity coefficient to less than 1, and the effective concentration is less than the actual concentration. For solutions with less than 90% by weight water or more than 5% by weight salt, the *p***H** becomes a noticeable function of water and salt besides hydrogen ion concentration. Certain salts, such as NaCl, also affect the millivolt potential developed by the measurement and reference electrodes. The ions from the dissociation of acids and bases also have activity coefficients that affect the charge balance and, therefore, the *p***H**.

The product of the hydrogen and hydroxyl ion concentrations must equal 10 raised to the minus power of the water dissociation constant (pK_w) per Equation 1-1d for water solutions. The pK_w, and thus the actual solution *p***H**, is a function of the process temperature. In Chapter 2, we will learn how other dissociation constants can cause the solution *p***H** to change. It is important to realize that the standard temperature compensator corrects for the temperature effect on the millivolt potential developed by the electrode and not for the changes in the actual *p***H** with temperature. Smart transmitters have recently added the option for the user to program for correcting the temperature's effect on the solution *p***H**. Except for dilute strong base solutions above 7 *p***H**, the exact relationship between temperature and solution *p***H** is not generally available and must be developed from lab tests.

$$c_H \bullet c_{OH} = 10 - pK_w \tag{1-1d}$$

where

c_H = hydrogen ion concentration (gm-moles per liter)

c_{OH} = hydroxyl ion concentration (gm-moles per liter)

pK_w = negative base 10 power of the water dissociation constant (14.0 at 25°C)

> **Key CONCEPT** The solution *p***H** changes despite a constant hydrogen ion concentration because of changes in dissociation constants with process temperature and activity coefficients with ionic strength and water content.

At the neutral point, the concentration of hydrogen and hydroxyl ions is, by definition, equal. If the temperature is 25°C, the pK_w is 14.0, which means the *p***H** is 7 and the hydrogen and hydroxyl ion concentrations are both 10^{-7}. Table 1-1a shows how the hydrogen and hydroxyl ion concentrations change by a factor of 10 for each *p***H** unit.

Table 1-1a illustrates the heart of the matter. No other type of commonly used measurement covers such a tremendous range. Also, the *p***H** electrode can respond to changes as small as 0.01 *p***H**, which means the *p***H** measurement can track changes as minute as 0.000000005 in hydrogen ion concentration at 7 *p***H**. No other commonly used measurement has such tremendous sensitivity. As with most things in life, you do not get something for nothing!

Table 1-1a. Hydrogen and hydroxyl ion concentrations in a water solution at 25°C.

*p*H	Hydrogen Ion Concentration	Hydroxyl Ion Concentration
0	1.0	0.00000000000001
1	0.1	0.0000000000001
2	0.01	0.000000000001
3	0.001	0.00000000001
4	0.0001	0.0000000001
5	0.00001	0.000000001
6	0.000001	0.00000001
7	0.0000001	0.0000001
8	0.00000001	0.000001
9	0.000000001	0.00001
10	0.0000000001	0.0001
11	0.00000000001	0.001
12	0.000000000001	0.01
13	0.0000000000001	0.1
14	0.00000000000001	1.0

The rangeability and sensitivity capabilities cause associated control system design problems that can seem insurmountable [1]. It is important to realize that these problems are caused by attempting a level of performance in the *p***H** process in terms of concentration control that goes well beyond the norm. For a strong acid and strong base control system, the reagent valve must have a rangeability greater than 1,000,000:1 for an incoming stream that varies between 0 and 6 *p***H** and a set point at 7 *p***H**. The stick-slip of the same control valve must be less than 0.00005% to control within 1 *p***H** of the 7 *p***H** set point. How, then, is this possible? Such strong acid and base systems are controlled by approaching the set point in stages and using successively smaller precision control valves. The multiple-stage requirement of *p***H** control can be visualized by comparing it to trying to sink a golf ball in the hole on a green. The distance between the tee and the green presents the rangeability requirement, and the size of the hole compared to the distance represents the sensitivity requirement. For the strong acid and strong base system mentioned earlier, the tee would be about a million yards from the green. A hole in one is possible. Using the same large control valve at each stage is like the joke about the gorilla who drives the green in one stroke but then uses his driver again and hits the ball the same distance when he tries to putt the ball in the hole.

The process gain at a given *p***H** set point is best visualized as proportional to the slope of the titration curve. The titration curve is a plot with *p***H** for the y-axis and the ratio of reagent to influent concentration, volume, or flow for the abscissa. For a strong acid and base system, the process gain between 6 and 7 *p***H** is six orders of magnitude larger than the process gain between 0 and 1 *p***H**. Figure 1-1a shows a titration curve for a strong acid and strong base at 25°C. The curve appears to be a vertical straight line between 2 and 12 *p***H** and symmetrical about the neutral point where the concentrations of hydrogen and hydroxyl ions are equal. However, zooming in on the straight line reveals another S-shaped titration curve. Successive zooms centered on 7 *p***H** should always yield additional S-shaped curves, as shown in Figures 1-1b and 1-1c, because the slope is continuously changing by a factor of 10 for each *p***H** unit deviation from 7 *p***H**. However, most titration curves will not show this true behavior because there are not enough data points. It is common for there to be only a couple of data points in the neutral region where the slope is steepest, even though this is the location of the set point and, therefore, the area of greatest interest for environmental and biological systems. Also, many users overlook or inadequately define the abscissa. For example, if management looks at Figure 1-1c and ignores the X scale, the curve looks relatively harmless, and they will wonder, "What is the big deal?" Graphical deception is a common problem in *p***H** systems.

A lab titration consists of adding drops to a beaker with a fixed sample volume and composition until an end point is reached. This is similar to traditional batch *p***H**

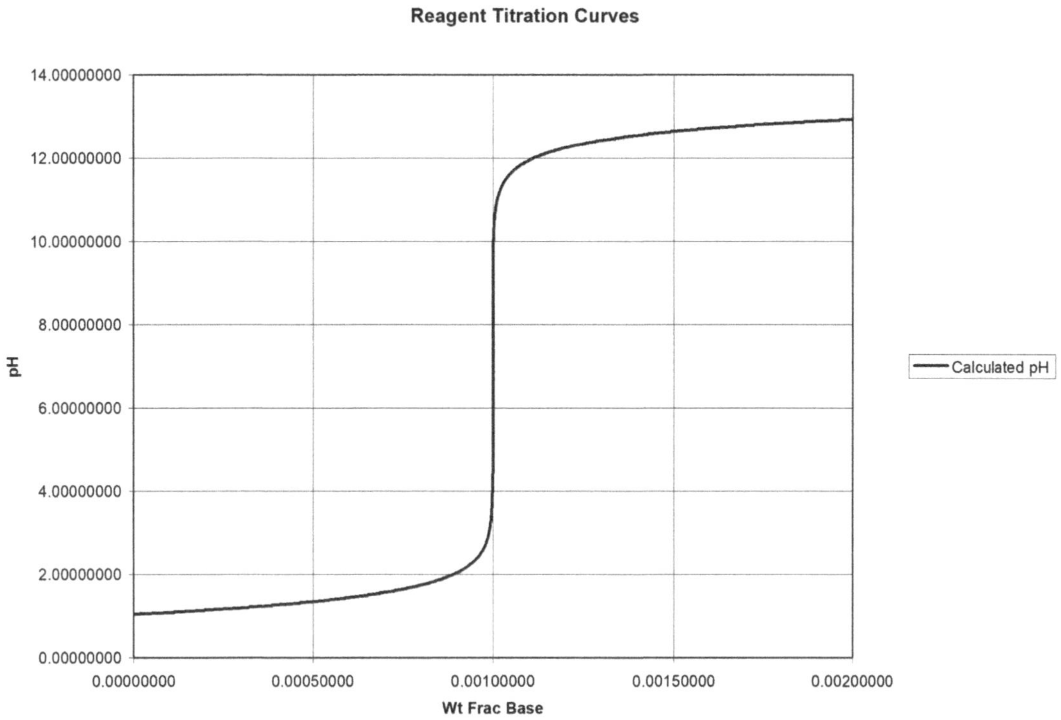

Figure 1-1a. Strong acid and base titration curve for the entire operating range.

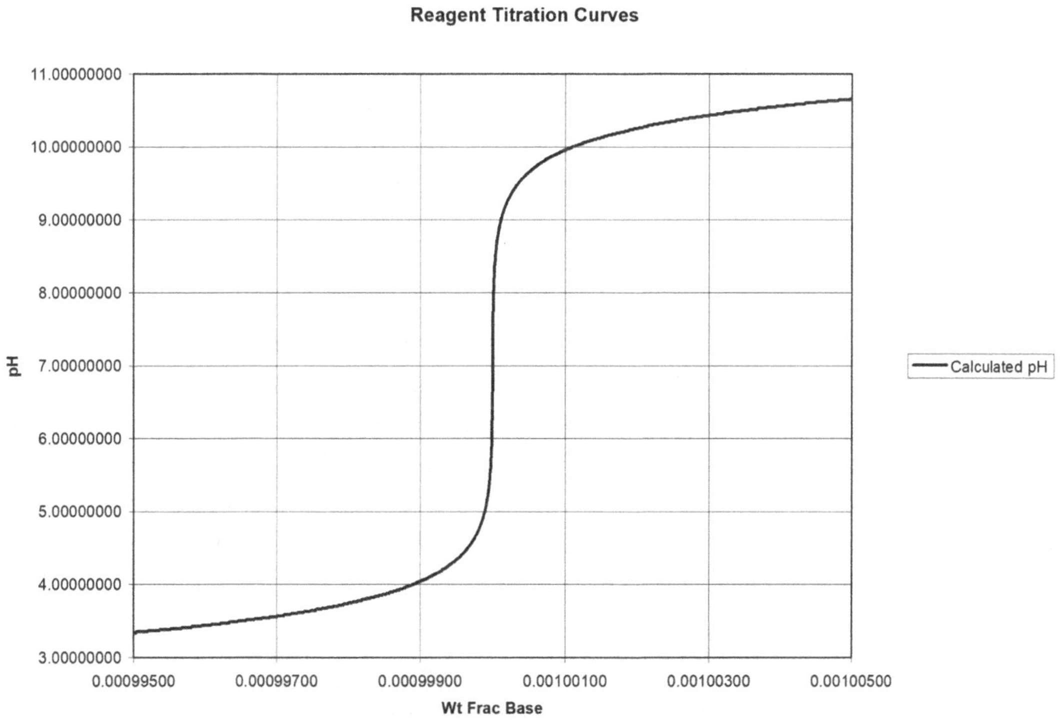

Figure 1-1b. Strong acid and base titration curve that zooms in on a 3 to 11 *p***H** range.

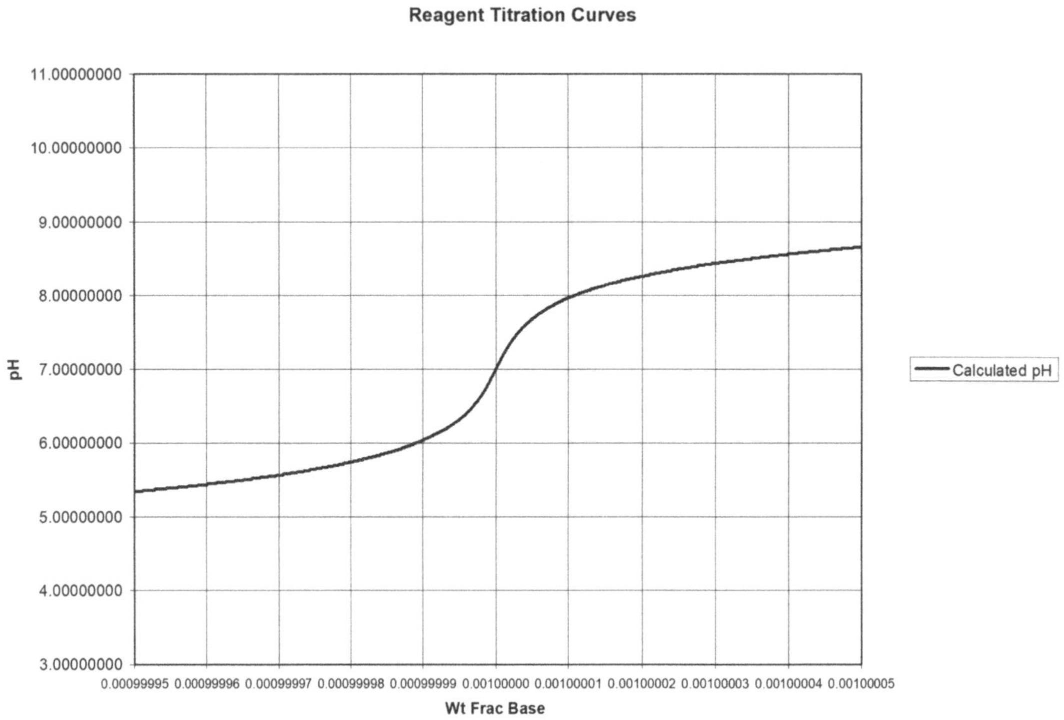

Figure 1-1c. Strong acid and base titration curve that zooms in on the neutral region.

control. If the chemist says it is too difficult to make the drops small enough during titration to add enough data points in the neutral region to see the curvature, this is a red flag. It will be even more difficult in the plant to adjust these small changes in dosage to hold a set point, particularly for upsets. In continuous *p***H** control, there is a variable feed and discharge flow and a need to keep the discharge always at a set point. This is analogous to a chemist having to vary the number/size of small drops added to a beaker with a hole in the bottom and a variable makeup flow and composition by just looking at the *p***H** meter.

For a control loop to respond equally well for all excursions along the 0 to 14 *p***H** scale, the controller gain would have to change in an equal and opposite direction to the *p***H** process gain so that the loop gain is constant. Using multiple stages to limit the *p***H** excursions to a small region around the set point reduces the change in gain seen by the controller. The set points are incremented between the influent and desired final *p***H** so that the distance in acid or base concentration between the influent and set point *p***H** is reduced for each control loop [1].

> **Key CONCEPT** There are no straight lines in a titration curve. Zooming in on any line that appears to be linear should reveal another curve if there are sufficient data points.

Fortunately, many *p***H** systems have weak acids and bases that flatten out various portions of the titration curve to provide a buffering effect. The overall difficulty is reduced, especially if the set point ends up residing on one of these plateaus, as shown in Figure 1-1d, for a weak acid and base [1]. The natural buffering of surface and ground waters from carbonates can change your mood from suicidal to merely depressed when the actual curves are compared to theoretical titration curves with pure water. Titration curves of fabricated samples will be much steeper than the titration curves of actual process samples, especially if the lab uses deionized water instead of plant water.

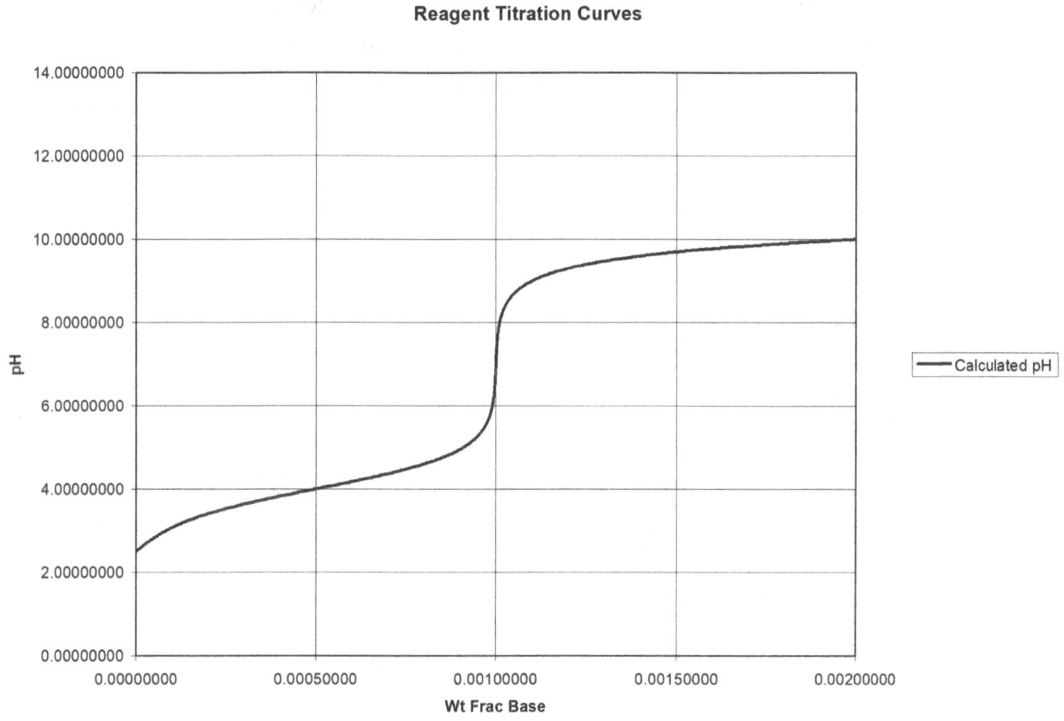

Figure 1-1d. Weak acid and base titration curve with flattened areas (plateaus).

For the control system to work, special attention must be paid to designing and installing the electrodes, transmitter, controller, control valve(s), piping, and mixing equipment. A mistake in designing or installing any component in the loop can cause the control system to not only fail but also to fail miserably. The process, mechanical, and instrument engineers must all be alert to special system requirements from the inception of the project. The top 12 mistakes are summarized in Table 1-1b. Ten of these are illustrated in Figure 1-1e, which shows the use of a horizontal tank, gravity flow reagent, and a ball valve without a positioner.

Table 1-1b. Top 12 mistakes made every day in *p***H** system design.

1.	Incorrect or missing titration curve
2.	Absence of a plan to handle failures, start-up, and shutdown
3.	Insufficient number of stages of neutralization (inadequate rangeability and sensitivity)
4.	Improper vessel geometry and agitation patterns (excessive equipment dead time)
5.	Backfilled reagent dip tube (excessive reagent delivery delay)
6.	Incorrect location of reagent injection point (short-circuiting)
7.	Gravity flow reagent (excessive reagent delivery delay)
8.	Incorrect location of reagent control valve (excessive reagent delivery delay)
9.	Control valve with excessive stick-slip (poor sensitivity and excessive variability)
10.	Electrodes submersed in a vessel (coating and maintainability problems)
11.	Electrodes located in pump suction (bubbles, clumps, and wrenches)
12.	Electrodes located too far downstream (excessive measurement delay)

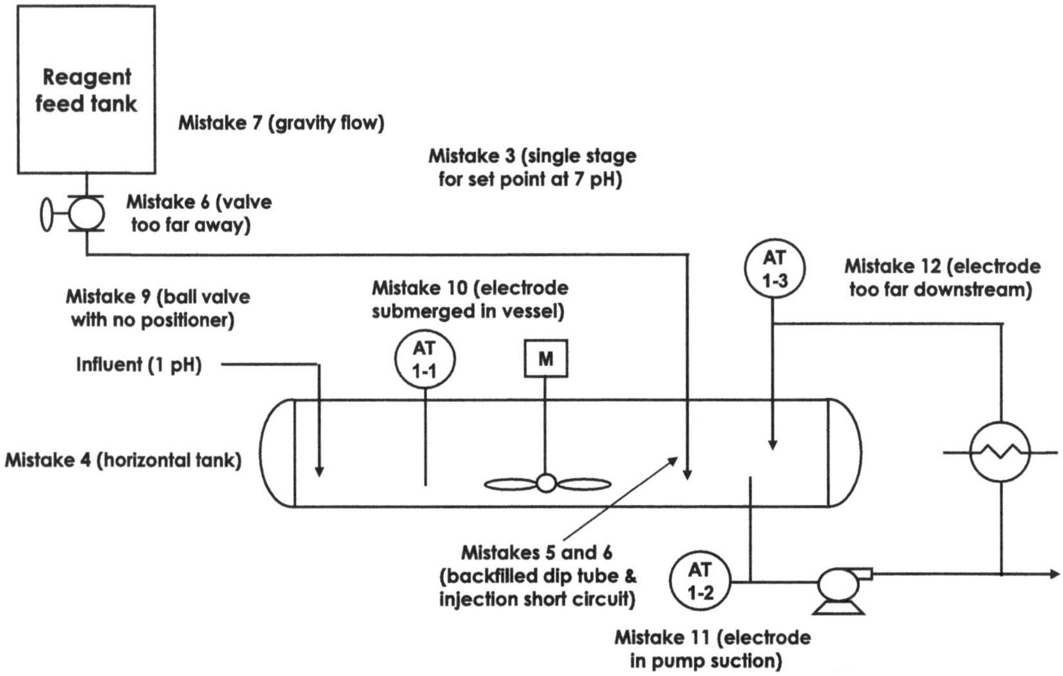

Figure 1-1e. Examples of mistakes made every day in *p***H** system design.

Key CONCEPT The flatness of the titration curve at the set point has the greatest effect on the tightness of *p***H** control. The next most important effect is the distance between the influent *p***H** and the set point.

The titration curve is the essential tool for determining the nonlinearity, sensitivity, and rangeability requirements that are important for every step, from project definition through sustaining installed performance. It determines the investment required, the design details, and the expected efficiency, operability, and maintainability. Without a titration curve, you are flying blind [1, 2].

RULE of Thumb

The first step in designing a p***H*** *system is to generate a titration curve at the process temperature with enough data points to cover the range of operation and show the curvature within the control band (absolute magnitude of the difference between the maximum and minimum allowable* p***H****).*

1.2 Stage Fright

Because of the pressure to minimize capital costs, a project may sell itself short on the number of stages required. The basic rule from the 1970s was that a well-mixed tank is needed for every two *p***H** units that the influent feed *p***H** is away from the set point. For example, if the set point was 7 *p***H**, one stage for an influent at 5 *p***H**, two stages for an influent at 3 *p***H**, and three stages for an influent at 1 *p***H** would be required. The modern-day version of this old rule would reduce the requirement by one stage if feedforward control or signal linearization could be used effectively or if the set point could be moved to a flatter portion of the titration curve. Today, three stages are rarely used. Even the most difficult systems are tackled by an inline system for the first stage, followed by a well-mixed vessel for the second stage.

Key CONCEPT

The titration curve is the essential tool for every aspect of *p***H** system design and analysis.

One of the most frequently missed opportunities in dramatically reducing the difficulty of control and saving on reagent usage is to shift the set point away from the center to the edge of a control band where the slope is flatter. While it may seem to be wrong in terms of flirting with the edge, the large oscillations from set points at the steepest point waste reagent and the sluggish tuning create a greater risk of going over the edge. For example, many environmental systems must keep the effluent between 2 and 12 *p***H** to avoid it being classified as hazardous waste. However, a set point at 7 *p***H** is the wrong choice and may lead to oscillations between 2 and 12 *p***H**. A much better choice would be an optimized set point of 4 *p***H** for acidic influent and 10 *p***H** for basic influent. You can warm up to the idea by seeing the tremendous difference at an operating point of 5 or 9 *p***H**. You can be a hero by going into a control room and simply changing the set point to a flatter part of the curve and increasing the controller gain. Of course, the loop probably was not stable to begin with at 7 *p***H**, so the better set point does not mean you can increase the gain by a factor of 10 per *p***H** unit, even though it might be a strong acid and base.

RULE of Thumb

*For a set point on the steepest part of the titration curve, one stage of neutralization is needed for every two p**H** units away from the set point. Accurate feedforward control and precise valves can eliminate one stage.*

Key CONCEPT

There is a huge opportunity to reduce reagent usage and capital investment by simply moving the set point to a flatter part of the titration curve and closer to the influent *p***H**.

In the 1980s, engineers discovered they could substitute a static mixer for the first vessel by using an appropriate measurement filter. A static mixer is a piece of pipe with baffles or tabs that subdivides the flow stream to provide radial mixing. A centrifugal pump can serve the same purpose. However, these inline mixers provide little to no back mixing or smoothing. Consequently, the process dead time is larger than the process time constant, which leads to large peak errors for fast upsets and a tendency for the loop to oscillate. Originally, engineers thought that the large oscillations meant you could only use inline mixers on the flat portions of curves or for large control bands. However, because the period and recovery time is fast for a well-designed inline system, a judiciously applied measurement filter can enable a *p***H** loop to control the average of the oscillations if there is sufficient volume downstream to smooth out the concentration fluctuations [1, 2].

Stick-slip is the most frequent source of oscillations in *p***H** control. For example, if the stick-slip is 0.2%, the reagent valve capacity is 1 gpm, and the influent flow is 1000 gpm, then the amplitude of the oscillation in the ratio of reagent to influent flow is 0.000002 along the x-axis of the titration curve. For a strong acid and base system, this could translate to an oscillation of 4 to 10 *p***H** around a 7 *p***H** set point whenever the loop is in automatic.

Stick is the percentage change in the valve signal needed to make the trim move and is caused by friction in the packing and trim-sealing surfaces. Slip is the change in position once the valve trim moves, again in the percentage of the signal span. The slip in a properly designed valve is about equal to the stick and sets the resolution capability of the valve.

For large valves, the percentage of stick-slip translates to a large error in reagent flow. Control strategies discussed in Chapters 6 and 8 have been developed to position a large reagent valve by feedforward and valve position control to keep a small valve in a good throttle range. Manipulating coarse and fine adjustment valves can eliminate a stage.

RULE of Thumb

*Using a static mixer or pump as a fast inline p**H** system can eliminate the need for a well-mixed vessel for the first stage, even for a set point on the moderately steep part of a curve, by the judicious use of a signal filter.*

For $\tau_o > T_o$:

$$A_o = A_i \bullet \left[\frac{T_o}{2 \bullet \pi \bullet \tau_o} \right] \quad [1\text{-}2a]$$

where

A_o = output amplitude (*p***H** or concentration)
A_i = input amplitude (*p***H** or concentration)
T_o = oscillation period (minutes)
τ_o = open-loop time constant (minutes)

The open-loop time constant in Equation 1-2a is either the digital signal filter time in a distributed control system (DCS) for smoothing out *p***H** oscillations and noise or the residence time for a well-mixed vessel for smoothing out concentration oscillations. Equation 1-2a is useful for estimating the vessel size needed to smooth out stick-slip and the size of a signal filter to keep short-term *p***H** fluctuations within the control band [3]. For example, if the oscillation period of *p***H** noise is 0.012 min, a digital filter of 0.04 min would reduce the noise amplitude by a factor of 20.

Filters must be used judiciously because they add delay to the response to unmeasured load upsets and give the illusion of better control by providing an attenuated version of the real world. They should only be used to help screen out transients that are so fast they will be smoothed out by a downstream volume, and any response by a controller would potentially do more harm than good. It is also important to realize that there will be an offset in the actual *p***H** of the effluent compared to the average *p***H** signal because of measurement errors and noise and any nonsymmetry of the titration curve about the set point. The effect can be estimated by translating the oscillations on the titration curve. Usually, the offset is relatively small (less than 1 *p***H**) and can be reduced by a simple bias of the upstream set point [2]. Using process variable signal characterization to embed the titration curve and create a linear reagent demand controller (described in Chapter 8) for the inline mixer can inherently eliminate the offset.

In general, the fastest *p***H** control system with the largest reagent valve should be the first stage and the first line of defense to bear the brunt of an upset so that oscillations have a smaller period and can be effectively filtered out by larger downstream volumes per Equation 1-2a. Oscillations should be translated between the axes of the titration curve as seen in Figure 1-2a for a weak acid and weak base system. The translation enables the effect of valve size and type, set point, filtering, tuning, and dead time. The recovery time is faster, and the amount of off-spec material is smaller for a fast system. Chapter 8 will show that the period of oscillation is proportional to the loop dead time, and the integrated error is proportional to the loop dead time squared.

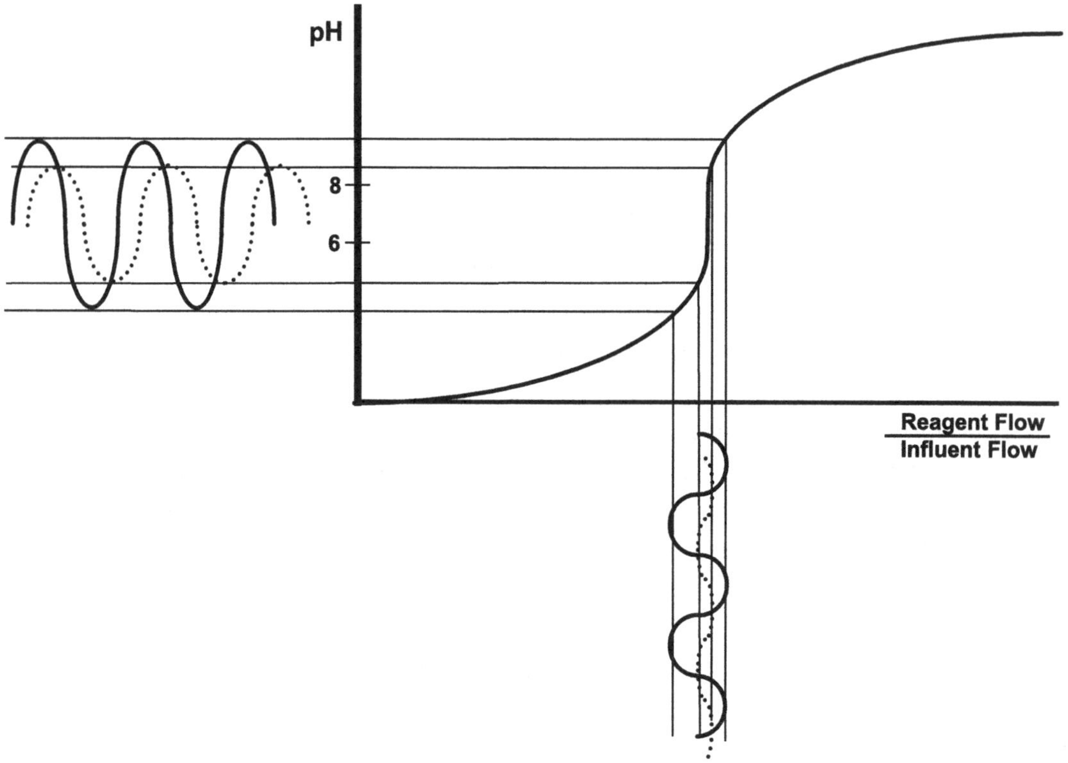

Figure 1-2a. Oscillations should be translated between axes of the titration curve.

> **Key CONCEPT** The average or filtered signal *p***H** will not equal the actual *p***H** downstream because of measurement errors, noise, and the titration curve.

Another consideration is the resonance of oscillations from two control loops in series. The first stage should have a total dead time from mixing and transportation delay that is at least five times faster than the next stage to minimize the amplification of upsets from the resonance of in-phase oscillations. Mixing and transportation time delays are proportional to volume, so this translates to a volume for the first stage that is at least five times smaller than the volume for the second stage. (Chapters 5 and 7 discuss the major design considerations to minimize dead time.) A control loop may do more harm than good if the upset has a frequency that is close to its own frequency. Consequently, inline systems with *p*H controllers should not be put in series because they have similar delays and periods.

> **RULE of Thumb** *The volume of each stage should differ by a factor of five to minimize resonance with the smallest volume first to minimize recovery time.*

A typical system design uses a static mixer with a transportation delay of 0.05 min for the first stage and a well-mixed vessel with a mixing delay (turnover time) of 0.25 min and a minimum residence time of 5 min for the second stage.

While a tank for smoothing or blending requires minimal agitation, it is critical that a vessel used for *p***H** control be well mixed with a residence time that is at least 20 times the turnover time (mixing dead time as defined in Chapter 5). Anything that falls short of this is disastrous because it creates large volumes of off-spec material and slow oscillations that persist downstream.

> **Key CONCEPT** A poorly mixed tank should never be used for *p***H** control but can be effectively used to attenuate oscillations from an upstream *p***H** loop.

For a vessel to be classified as *well mixed*, the height should be less than 150% of the diameter, and an axial pattern agitator should be used to provide a ratio of residence time to mixing turnover time that is greater than 20:1. This guidance is pertinent to any application but is especially important for *p***H** control because of the potential extreme sensitivity and nonlinearity. The turnover time should be less than 1 min, and baffles should be used to improve the mixing pattern. If conventional baffles are not permissible because of the area where polymers, particles, or microbes can accumulate, special draught tubes and contoured bottoms are used to prevent dead zones.

There is a super cheap alternative for *p***H** control. Using a large volume with an inline *p***H** controller around a pump in its recirculation line will provide an incredibly smooth *p***H** from the vessel discharge [2]. The volume does not need to be particularly well mixed. Jet mixers may be sufficient. The volume must be large, so it could be a tank salvaged from the equipment graveyard. The performance is achieved by blending a high recirculation flow with a much smaller influent flow and smoothing out the fast concentration oscillations from an inline system with a large volume per Equation 1-2a. It is important that the inline system has a properly located and precise measurement and valve per Figure 1-2b. The drawbacks to this system are the extra time required to bring the whole volume to the proper operating point on start-up or for a set point change and localized high concentrations of acids and bases in the recirculation line.

In theory, an incredibly precise valve and accurate signal characterization could enable *p***H** control from any influent *p***H** to any set point. Such a valve does not exist, but the combination of the right valve and set point can go a long way toward eliminating equipment and saving on reagent use.

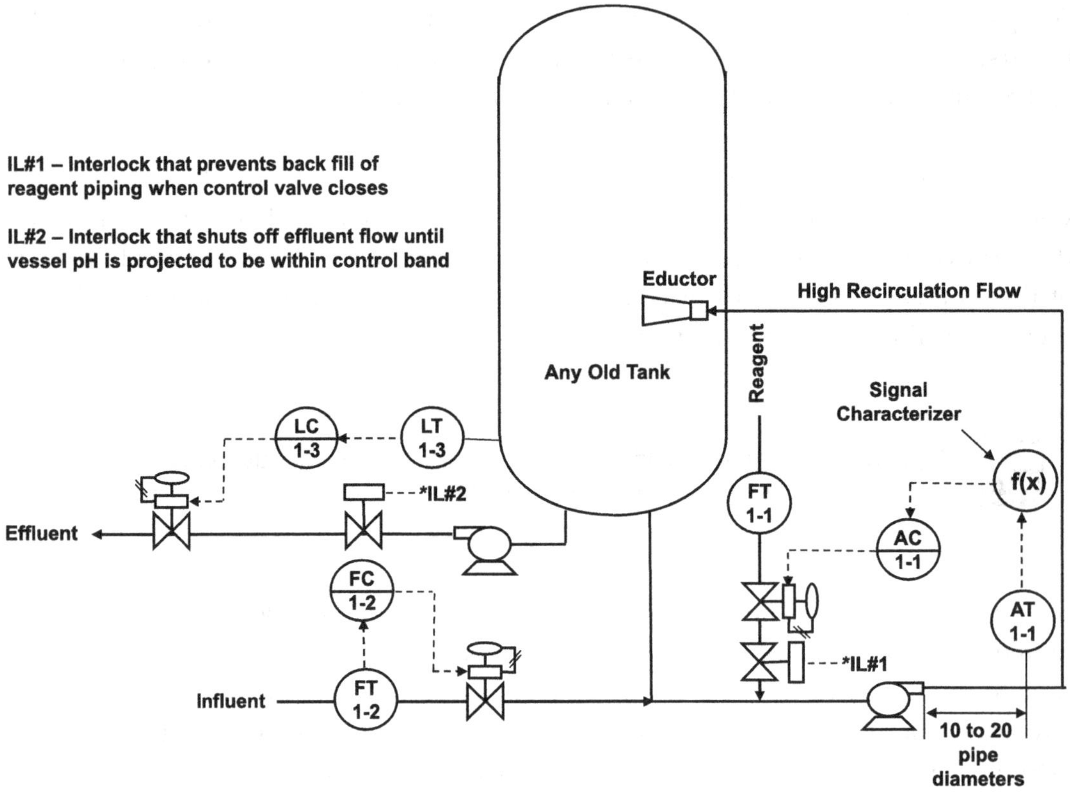

Figure 1-2b. A cheap way to achieve tight *p***H** control of influent upsets.

1.3 Size Does Matter

The small valves typically used in *p***H** control are more prone to improper sizing, irregular flow characteristics, greater stick-slip, and plugging. Here size does matter because most of these problems originate from extremely small Reynolds numbers, clearances, and stem diameters [4].

> **Key CONCEPT** Most of the problems with reagent throttling and injection stem from the extremely low flows required for *p***H** control.

For a valve flow coefficient (C_V) less than 0.01 or extremely viscous fluids, the valve may be operating in the laminar flow regime or in the region of transition from turbulent to laminar flow. Liquid flow moves from being approximately proportional to the square root of the pressure drop toward being proportional to the pressure drop as the flow goes from being fully turbulent to completely laminar. This can translate to an enormous sizing error. The result is often an oversized valve that rides the seat where the high seating friction causes excessive stick-slip. Even worse is operating in the transition region where the flow for a particular valve position has poor reproducibility because normally insignificant disturbances, such as microscopic changes in roughness or small vibrations, can trigger a switch between turbulent and laminar flow and an erratic installed flow characteristic. Some reagents, such as 98% sulfuric, have a viscosity so high that starting the injection flow is like trying to get ketchup out of a bottle.

Key CONCEPT Most reagent valves are oversized, which increases the amplitude of the oscillations from stick-slip.

RULE of Thumb *Make sure valve sizing calculations take into account any transition to laminar flow and viscosity correction.*

If you also consider the possibility of a significant distortion of the inherent flow characteristic caused by machining tolerances that are an appreciable portion of the clearances for such tiny trim sizes, the scene is set for an unknown and extreme nonlinearity. Tiny clearances can pose all sorts of problems because small particles and coatings cause plugging and sticking. The low flow velocities at the surface that are usually associated with laminar flow make the likelihood of coating much greater.

Finally, tiny stems are likely to be bent from normal handling both before and after installation. Slight deflections of the stem can cause huge amounts of stick-slip. What good is a control valve if you cannot drop or step on it?

Is the situation hopeless? Not if you go with a manufacturer that specializes in tighter machining tolerances, minimizing stick-slip, rugged stem guiding, and streamlined passages, as shown in Figure 1-3a. Stay away from stems smaller than 3/8 in and insist on getting response and flow test results. Use computer programs that properly deal with laminar flow and offer an installed characteristic curve for your piping and operating conditions. Even though the smart digital positioner may be larger and cost more than the valve, it is more important than ever that it be used and properly tuned for the small actuator volume. The challenge here is for packaged platforms that are going for the low bid not to get cheap where it hurts [4].

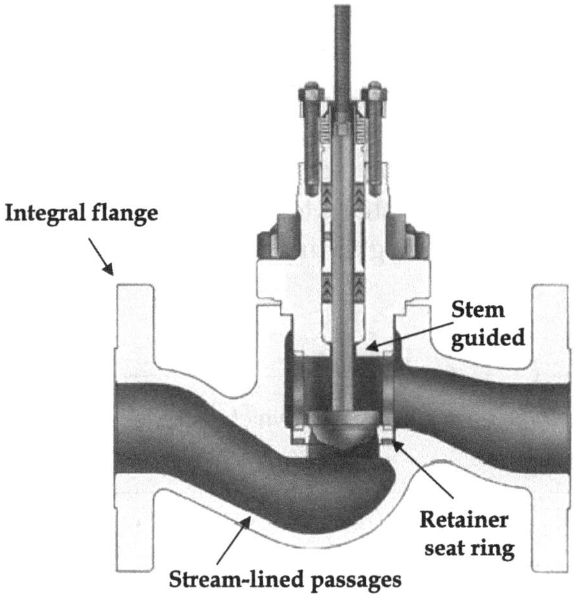

Figure 1-3a. Rugged sliding stem with streamlined passages for corrosive service.

RULE of Thumb *Use precision machined industrial sliding stem (globe) valves with digital positioners tuned for the small actuator volumes to minimize stick-slip.*

If there is a tendency for plugging, pulse width modulation is a solution when there are sufficient back-mixed volumes to attenuate the pulses. This can also provide a linear flow characteristic and a flow large enough to be turbulent. The ratio of maximum to minimum pulse width establishes the rangeability. The maximum pulse width and, therefore, cycle time determines the degree of variability that must be filtered and the additional dead time from the pulse-off time. Now stroking time can be an issue because it is desirable to have the minimum pulse be as short as possible [4].

RULE of Thumb *If the trim will plug, use pulse width modulation and set the ratio of the maximum to minimum pulse width to achieve the desired rangeability.*

Reagent dilution can eliminate valve problems associated with low flow but can introduce large reagent concentration disturbances and transportation delays if not done properly. Chapters 7 and 8 discuss methods for precise reagent concentration control.

The installed characteristic of a throttled reagent valve is the inherent characteristic because the frictional loss in the piping is so small from the low flow. The pressure drop across the reagent valve is a large percentage of the system drop except for injectors with very tiny holes. Chapter 8 on control system design illustrates that linear trims are best for vessels and equal percentage trims are best for static mixers when the *p*H controller directly manipulates a reagent valve instead of a reagent flow set point. As with any rule, there are exceptions. The flow of acid and bases consumed by reactions can be so large that the installed characteristic of an equal percentage trim starts to resemble a linear characteristic, and a linear trim starts to distort to a quick opening characteristic.

1.4 One Is the Loneliest Number

The reproducibility of a *p*H measurement on the steep part of a titration curve is rarely better than 0.2 *p*H if you include short- and long-term variability. If you move the set point to a flat part of the curve where noise is less, the reproducibility improves to 0.1 *p*H. Most of this error is caused by fluctuations in the reference electrode potential. If you use a properly designed flowing junction reference electrode, a reproducibility of 0.02 *p*H is possible. The 0.01 *p*H accuracy stated for electrodes is for the ideal laboratory environment and measurement method and is a pipe dream for plant conditions. Technicians will question your sanity if you tell them to keep a *p*H reading within 0.01 *p*H of a lab sample. Chapters 3 and 4 describe dozens of reasons why it is unreasonable to expect lab and field measurements to agree within 0.01 *p*H. For example, lab *p*H readings appear to be more consistent because the electrode is not inserted long enough for the coatings and changes in the glass surface and reference junction potential to become an issue. For nasty streams, this short-term immersion is duplicated in the field by using automatically retractable *p*H electrodes that are sequenced to be inserted for a couple of minutes and then soaked or calibrated for 10 or more minutes. However, as noted in subsequent paragraphs and detailed in Chapter 4, there are many causes of differences between lab and field measurements, including measurement electrode response time, reference electrode equilibration time, and the variation and detrimental effect of process concentrations, temperature, and phases.

Key CONCEPT: The reproducibility of commonly used electrodes is about 0.2 and 0.1 *p*H for the steep and flat portions of a titration curve, respectively, if the process is nice and the electrodes are properly selected, installed, and maintained.

RULE of Thumb: *When the reproducibility must be better than 0.1 pH, use a pressurized flowing junction reference electrode with a reference fill solution that is compatible with the process.*

RULE of Thumb: *For really nasty process fluids that seriously shorten the life expectancy of the electrode because of chemical attack, abrasion, dehydration, or coating, use an automatically retractable and sequenced assembly to minimize the electrode's process exposure and maximize its conditioning and calibration.*

The glass measurement electrode develops a potential in response to the activity of hydrogen ions at its glass surface. In theory, the hydrogen ion jumps from a hydronium ion in the aqueous solution to join with a water molecule in the hydrated gel layer to form a hydronium ion in the glass surface, as shown in Figure 1-4a [5, 6, 7]. The reproducibility of the measurement depends on the condition of the hydrated gel layer, which is extremely thin. For solutions with low water concentrations, nonaqueous solvents, or high acid concentrations, the gel layer becomes dehydrated and must be replenished by soaking the electrode in a water solution periodically. Smart transmitters that measure the electrical resistance of the measurement electrode can detect severe changes in the gel layer if the resistance is corrected for the effect of glass type and temperature.

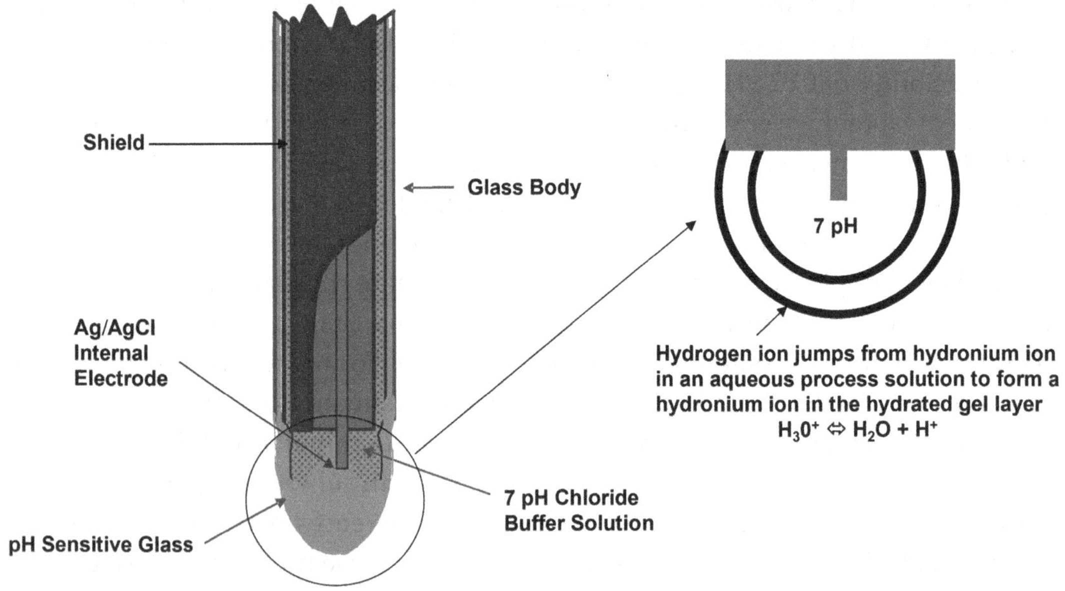

Figure 1-4a. The measurement electrode depends on a hydrated gel layer. (See Figure 4-1a for a more detailed view of the electrode.)

> **Key CONCEPT** The reproducibility of a measurement electrode depends on the condition of a thin gel layer that is easily disrupted by abrasion, dehydration, and chemical attack.

High process temperatures shorten the electrode life, as shown in Figure 1-4b, even for nice dilute aqueous solutions. High acid and base concentrations associated with operating below 2 *p*H or above 10 *p*H can chemically attack the glass surface and reduce its efficiency and life expectancy. Special glass formulations can help; in general, the higher the electrical resistance of the glass, the higher the chemical resistance. Thicker glass ("ruggedized") measurement electrodes are desirable from the standpoint of reducing breakage and resisting abrasion, but the electrical resistance and the response times of these electrodes are larger, and the active surface area of the dome shape is smaller. If these electrodes are used in an acidic stream that tends to dehydrate the gel layer, the problem is compounded. Low temperatures that exponentially increase the glass resistance also cause noisy, erratic results for these higher-resistance glasses. In general, no glass formulation provides an acceptable service life for hot (>65°C), relatively concentrated sodium hydroxide solutions (>10 *p***H**).

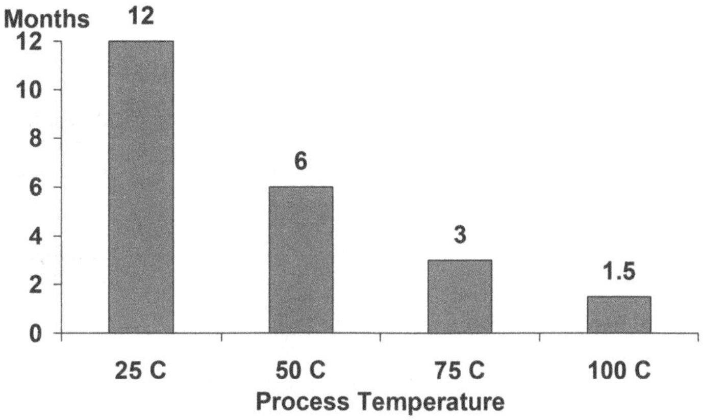

Figure 1-4b. Effect of temperature on measurement electrode life in a nice solution.

The reference electrode is designed to provide a stable reference potential and an electrical path between the process and the internal electrode. To deliver this electrical continuity, the reference fill must be in contact with the process, which figuratively and literally opens the reference electrode to all sorts of problems. The porous reference liquid junction shown in Figure 1-4c that causes the internal electrolyte ions to migrate and touch the process may enable process ions to migrate in the opposite direction, changing the reference junction potential and eventually contaminating the internals. Pulsating pressures, high ionic strength solutions, and highly mobile ions accelerate the deterioration. A second, much smaller internal junction is often used to protect the internal silver-silver chloride electrode. Particles and sticky fluids can also clog the junction. A large liquid junction surface area minimizes the clogging problem but increases the contamination problem. New solid references have large wood or plastic junctions with impregnated electrolyte ions to resist contamination. However, these junctions may be slow to reach an equilibrium potential. The most accurate and reliable alternative is a pressurized flowing junction reference. These are not popular in the United States because they require maintaining a pressurized reservoir to replenish the fill lost out through the junction. Also, a high junction flow can throw off measurements in beakers and contaminate a process. Smart transmitters can measure the junction resistance and detect clogging problems.

Most electrodes in service today are combination electrodes where the reference electrode is located in a ring around the glass measurement electrode. Manufacturers offer a combination electrode whose sleeve is plastic with immobilized electrolyte ions to form a very large reference junction area impervious to contamination. However, as discussed in Chapter 4, solid reference junctions have long equilibration times.

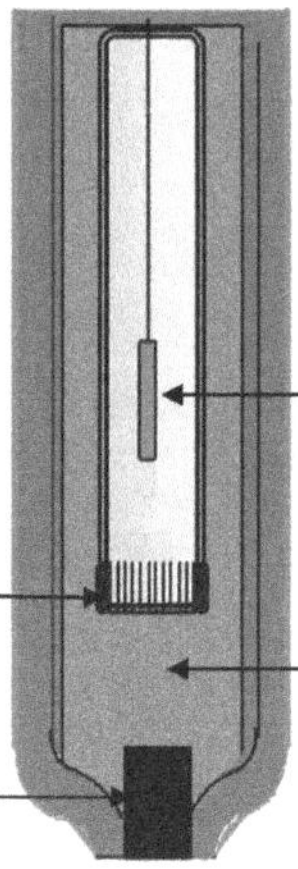

Figure 1-4c. The reference electrode depends on a porous junction.

> **Key CONCEPT** The reproducibility of the reference electrode depends on its internal electrolyte ions quickly establishing and maintaining contact with the process without the process clogging or migrating into the reference junction.

RULE of Thumb *The measurement electrode glass type and reference electrode junction and fill must be compatible with the process composition and temperature.*

RULE of Thumb *Smart transmitters should be used to continually monitor resistances to infer the condition of the glass surface and the reference junction.*

Most *p***H** systems are over calibrated while chasing sample results. If there is only one electrode, there is not much evidence to question the lab, even though time, process temperature, and carbon dioxide content could explain the disagreement.

A second electrode will raise more questions than answers because the two electrodes will never agree because of concentration gradients, velocities, and the condition of the glass surface and reference junction. As shown in Figure 1-4d, what is high today may be low tomorrow, although long-term trends may be observable (kind of like the stock market). Except for a sluggish or noisy response, it is difficult for the operator to establish which electrode to select, so suspicion and prejudice prevail. Frequent calibration adjustments often chase short-term transients and make the problem worse. The best results are often obtained by going to three electrodes and leaving them alone [5].

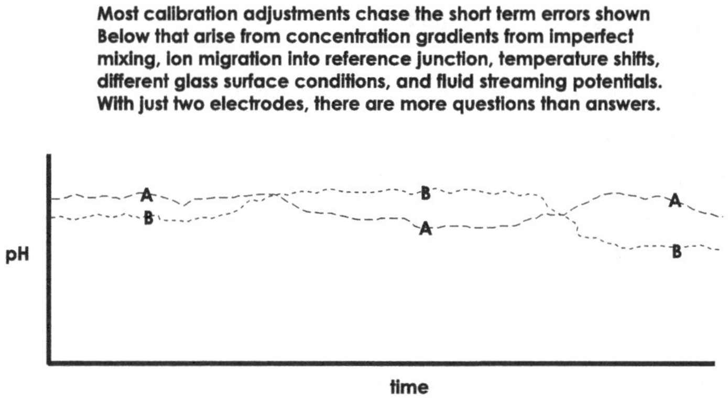

Figure 1-4d. What Is high today may be low tomorrow.

Adding a third electrode and automatically selecting the middle signal will inherently ride out a spurious signal or failure of any type from one electrode. Often, engineers think they can design a better failure strategy, but in fact, all other supposedly more sophisticated strategies for failure rejection have one or more modes of failure and can do more harm than good. The middle signal selection is often a standard function block that can be readily set up in most DCSs. Of course, if the process is truly nasty, you should not install three electrodes. You should go instead to automatically retractable and sequenced electrodes to minimize process exposure time.

RULE of Thumb *Middle signal selection will maximize the reliability and minimize the maintenance unless the process is so nasty it causes premature electrode failure.*

Finally, there is an opportunity to calculate *p***H** online for the flat part of a titration curve if there are Coriolis flowmeters to accurately measure the mass flow and density of two-component mixtures or streams with sufficiently constant acid and base concentrations. An inferential measurement or *p***H** estimator based on an online titration or charge balance model can become a third measurement that enables middle selection. In one application, the calculation was more accurate than the electrode readings because of the flatness of the titration curve. A single Coriolis flowmeter on each of the influent and reagent streams could be reliable and accurate enough to replace the use of electrodes altogether for the first stage of a neutralization system, which typically is on a very flat portion of the titration curve. Coriolis flowmeters also offer the ability to display reagent usage, process efficiency, and automatically generated titration curves online. These opportunities are explored in Chapter 8.

Key CONCEPT Coriolis flowmeters can be used to replace *p*H electrodes and provide online displays of reagent usage, process efficiency, and titration curves.

1.5 Acceleration and Stagnation

The approach to the neutral point looks like an accelerating response to the *p*H controller and is similar to what a temperature controller experiences for a runaway condition in an exothermic reactor. The tuning rule for such temperature controllers is to minimize the reset action (maximize the reset time) and maximize the rate action. Reset has no sense of direction and promotes overshoot. Rate anticipates where the process is headed by looking at the rate of change of the measurement and suppresses overshoot. However, rate will amplify noise and can generally only be used for *p*H control on back-mixed volumes where there is a large enough process lag to provide a smooth and gradual response. When operating on the flat part of the curve, the controller gain should be increased to avoid stagnation. So, for well-mixed vessels, you can become famous by Friday if you simply decrease the reset setting and increase the rate setting [8].

Key CONCEPT The approach to the neutral point looks like a runaway response.

For inline *p*H systems dominated by reagent and process transportation delays and spurious readings from concentration gradients, reset should be predominantly used to provide an averaging type of control and rate should be avoided [8].

RULE of Thumb *Tune the* p*H controller with reset action (small reset times) for inline systems and with rate action (large rate times) for well-mixed vessels.*

If the titration curve is relatively fixed and known, it can be used to provide a feedforward signal. Generally, this is flow feedforward, where the reagent flow is ratioed to the influent flow and corrected by a feedback *p*H controller [3]. It relies on a secondary reagent flow controller that receives a remote set point, which is the flow ratio multiplied by the actual influent flow corrected by the *p*H controller output. If a flowmeter is not available for either the influent or reagent flow, an inferential flow measurement can be calculated from the installed valve characteristic and pressure drop.

Flow feedforward is generally helpful as long as you under correct for set points on the steep portion of the titration curve. It is particularly important for large feed flow changes and for systems with large dead times. Using *p*H feedforward is generally not advisable if the influent measurement is on the extremes of the titration curve where the process conditions are harsh and the flatness of the titration curve increases the reagent error from measurement error. The flatness amplifies the effect of measurement error seen in the translation of an error on the *p*H y-axis to an error in the ratio of reagent to influent flow on x-axis of the titration curve. Deterioration in the glass surface and reference potential make the measurement error larger, so there is a double hit in terms of the accuracy of the feedforward signal. *p*H feedforward works best when you are going from a steep part to a flat part of the titration curve. New adaptive control algorithms can automatically identify the feedforward gain and dynamics from feed set point changes.

Key CONCEPT Flow feedforward is generally helpful but can waste reagent by an overcorrection that is more likely when a set point is on the steep portion of the titration curve.

RULE of Thumb *Do not use* p*H feedforward control if the influent* p*H is on the extremes of the titration curve.*

For batch or continuous systems, a "head start" of the reagent valve position, or reagent to influent ratio, can dramatically reduce the start-up time for low *p*H controller gains. The head start logic initializes the reagent valve or ratio to what was captured as the best value from the last run or batch. The initialization of the *p*H controller output is held until the actual *p*H reading has responded and is representative of the actual operating conditions on start-up.

Key CONCEPT A "head start" of the reagent valve position or flow ratio can dramatically reduce the start-up time for low *p*H controller gains.

RULE of Thumb *The initialization of the* p*H controller output on start-up should be held until a representative* p*H measurement is available for feedback control.*

There is a great tendency for *p*H systems to develop equal amplitude oscillations between the flat portions of the curve. People who use the ultimate oscillation method to tune a *p*H controller often choose a higher than actual ultimate gain because a wide range of gains creates equal amplitude oscillations. Higher gains tend to make the amplitude slightly larger, but this is not perceptible, and the amplitude never grows as with classic instability because the cycle is limited by the flat parts of the titration curve. The user must look for the start of damped oscillations to avoid confusion with the limit cycle [8].

Key CONCEPT A *p*H controller tends to cycle with equal amplitude oscillations.

RULE of Thumb *Do not use the ultimate oscillation method for tuning pH controllers; instead, determine the gain that causes damped oscillations.*

For a fixed and known titration curve, an opportunity exists to linearize the process gain by signal characterization of the controlled variable [2]. The x-axis of the titration curve is scaled from 0% to 100% of the operating range. Signal characterizer function blocks are then used to calculate the x-axis (reagent demand) from the y-axis (*p*H) by entering pairs of data points. The controlled variable is now 0% to 100% reagent demand. This strategy works particularly well if the set point is on the steep part of the curve. It avoids the *p*H getting hung up on the flat part of the curve. It may do more harm than good if the set point is on the bend in a curve, and you must avoid getting on the steep part at all costs because it will slow down the response on the steep part. It is better, in this case, for the controller to see the acceleration and use rate action to kick the *p*H back to the flat part.

Key CONCEPT Signal characterization of the controlled variable is a powerful tool for linearization if the set point is on the steep portion of a fixed titration curve, but it may be counterproductive to prevent excursions into the steep region for set points on the bend in the curve.

If a linear trim reagent valve is not available for vessel *p*H control, then signal characterization of the manipulated variable can reduce the nonlinearity introduced by the valve. In this case, the signal characterizer block is on the controller output.

For a variable or unknown titration curve, an adaptive controller that will identify the changes in process gain can fill in the blanks. New algorithms have been developed that can identify the process dynamics for a set point change by automatically injecting a short step into the controller output. The combination of adaptive control with signal characterization can be a powerful "one-two punch" to eliminate the nonlinearity.

RULE of Thumb *The combination of adaptive control with signal characterization can be a powerful "one-two punch" to eliminate nonlinearities.*

Figure 1-5a illustrates a myriad of advanced control techniques for a two-stage system where the first stage is a static mixer and the second stage is a well-mixed tank. It shows the use of signal characterization, *p*H and flow feedforward, adaptive control, and real-time optimization of the stage 1 set point to minimize reagent usage.

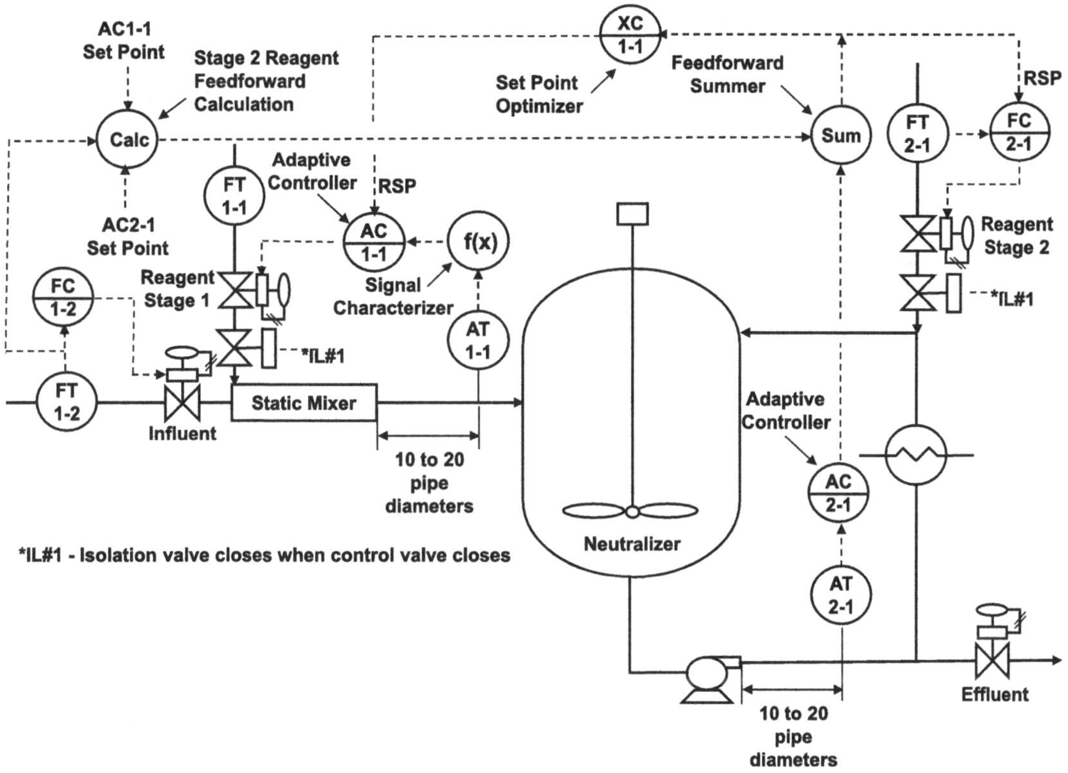

Figure 1-5a. Advanced control techniques to minimize reagent usage.

1.6 Life Is a Batch

All processes have a beginning and an end. A continuous process can be thought of as a long batch that can benefit from automating its start-up and shutdown by using sequencing techniques and standards developed for batch operations. One important distinction between batch and continuous is that during the reaction or formation of product, the vessel discharge flow is zero for batch operations. For batch reactors where the reagent is consumed or for fed batch operations where the reagent and feeds are simultaneous, the response is self-regulating and many of the design techniques developed for continuous control and optimization apply. For more traditional batch operations where the reagent and feeds are charged sequentially, the *p*H response is a ramp. This integrating response has profound implications on controller tuning and strategies. It is more critical than ever to minimize the reset action (maximize reset time) and maximize the rate action to prevent overshoot. If the *p*H goes past the end point, the only way it can be corrected is by using split-ranged acid and base reagents. Often there is a deadband between closing one valve and opening another valve to prevent cross-neutralization. However, the combination of a deadband and an integrating response results in a continuous cycle, even if there is no stick-slip. If you also consider that the split-range point for the reagent valves is the point of greatest valve gain discontinuity and stick-slip, split-ranged operation can significantly increase the cost of reagents and create excess salts. To prevent wasteful cross-neutralization, reset is turned off or reagent addition is shut off when the *p*H is within an acceptable control band.

> **Key CONCEPT** If no reagent is consumed in a traditional sequential batch operation, the *p*H response is integrating and the *p*H loop will continuously oscillate across the split-range point.

Most batch operations to date use mass flowmeters to charge a total mass of reagent specified by a recipe. The operator may then send a sample to the lab and make some manual trim adjustments. Using an online *p*H measurement and one or more of the advanced batch control strategies listed in Table 1-6a can reduce batch cycle time and improve consistency. If the starting *p*H is on a steep part of the curve, then a temperature-corrected curve can be used to calculate the charge required based on the change in mass ratio on the x-axis (item 1). However, if at the start of the reagent addition, the *p*H is on extremes where the conditions are harsh and the reagent ratio error is large, it is best to use a recipe number that was updated by the actual reagent used in the last best batch (item 2). A filtered and velocity-limited rate of change of *p*H can be multiplied by the total time delay to provide a predicted *p*H that—when compared to the desired end point—can identify an anticipation needed to prevent overshoot (item 3). Pulse width and amplitude modulation of the *p*H controller output can mimic the titration method used in the lab (item 4). Finally, inline *p*H control of a high recirculation flow can provide a smooth transition to the end point if the localized high reagent concentrations do not trigger side reactions, damage cells or crystals, or corrode recirculation piping and nozzles (item 5). The inline *p*H control set point is remotely set by a batch *p*H controller whose reset is turned off when the batch *p*H approaches the end point. What works best may be a combination of an initial charge just short of the end point based on items (1) and (2) and then a combination of items (3) and (4) or (5) to make a trim adjustment.

Table 1-6a. Advanced batch *p*H control techniques.

1.	Automatic calculation of charge from the temperature-corrected titration curve
2.	Automatic partial correction of charge based on the last best batch
3.	Automatic end point prediction and shutoff based on the rate of change of *p*H
4.	Pulse width and amplitude modulation of a proportional-only controller output
5.	Cascading batch *p*H to inline *p*H control of a high recirculation flow

1.7 Industrial Importance

In the food and beverage, chemical, petrochemical, pharmaceutical, power, pulp and paper, and textile industries, *p*H measurement and control are important for process efficiency and product quality, as well as protecting the environment. Table 1-7a summarizes the importance of *p*H for various industrial applications. The number of *p*H loops in new plants will exponentially increase as bioprocesses move from the laboratory to plant production. Suppressing alien organisms and yielding genetically engineered organisms greatly depend on tight *p*H measurement and control. The cost of a bad batch and contaminated equipment can be millions of dollars for high-value-added pharmaceuticals [9].

Table 1-7a. The importance of *p*H in the process industry.

Application	Processes and Aspects Affected by *p*H
Bacteriology	Microorganism growth and metabolism
Baking	Dough volume, texture, and color
Brewing	Yield of extract and sugar during mashing
Canning	Time and temperature for sterilization
Chemicals	Impurities and crystallization of salts
Cleaners	Effectiveness of removing paint and varnishes
Dyes	Yield and uniformity through intermediates
Electroplating	Nickel deposit hardness and brightness
Fermentation	Fermentation time and alien organism growth
Gelatin	Water absorption, solubility, and clarity
Pharmaceuticals	Effectiveness, stability, and body reaction
Pigments	Uniformity of composition in precipitation
Pulp and paper	Sizing, loading, coating, and foaming
Sewage	Digestion time, odors, and foaming
Sugar	Inversion of sugar and destruction of glucose
Textiles	Efficiency of most wet processes
Water treatment	Coagulation and softening processes

1.8 The Real Deal

Table 1-8a summarizes the facts of life from a decade of *p*H control start-ups that can be best described as a "magical mystery tour" [2].

Table 1-8a. The facts of life.

1.	Instrumentation is frequently the source of disturbances for *p***H** systems through reproducibility error, measurement noise, and valve stick-slip.
2.	Inline *p***H** loops will oscillate regardless of controller tuning if the set points are on the steep parts of titration curves.
3.	*p***H** electrode submersion assemblies with unencapsulated terminations below the liquid surface will eventually have wet terminations.
4.	Reagent control valves that are not close-coupled to the injection point will cause reagent delivery delays long enough for you to reach early retirement.
5.	You need either a flowmeter or a seer to diagnose reagent delivery problems.
6.	Flow feedforward signals must be corrected by a *p***H** feedback controller unless you like late-night calls from operations.
7.	Transportation delays to *p***H** electrodes in analyzer houses will exceed mixing time delays to such an extent that the increasing comfort of checking electrodes is offset by a decreasing comfort in checking trend recordings.
8.	Injection electrodes in recycle lines should be preferred to sample holder and submersion assemblies to reduce maintenance costs and coatings and improve response time, but not all injection electrodes are equal.
9.	Large, poorly mixed tanks are fine if you do not have to control them. Use them upstream to reduce reagent consumption or downstream to filter out oscillations. If you cannot make up your mind about where to use them, put them downstream.
10.	Two measurements without an estimator will raise more questions than give answers.

1.9 Best Practices

*p***H** measurement and control can be the most difficult and challenging application because of exceptional process sensitivity and nonlinearity; reagent precision and rangeability requirements; and the extreme vulnerability of *p***H** electrodes to noise, changes in offset and span, being stuck at 7 *p***H**, and a response that becomes so slow as to become the dominant lag in the loop providing a very attenuated view of the resulting oscillations.

The capability of *p***H** for measurement and control is extraordinary when you consider you are potentially trying to respond to changes of 14 orders of magnitude in hydrogen concentration with a precision of 10 or more decimal places for slightly basic solutions. Consequently, there is a great need to avoid many pitfalls by recognizing potential problems and realizing corresponding solutions.

I cannot overemphasize the critical importance of the digital twin. The digital twin with an actual controller configuration and a dynamic model that includes first principle process relationships, a charge balance adjusted to match titration curves, and the dynamics (gains, time constants, and dead times) of equipment, piping, mixing, and automation systems (see Chapter 9) is essential to deal with the many challenges of *p***H** measurement and control and to identify and implement the *p***H** system design needed. The digital twin can determine the best and most economical design and installation of equipment, piping, reagents, final control elements, sensors, and control strategies. A possible immediate, up-front benefit is a reduced project cost of $400,000 or more from eliminating the need for a well-mixed vessel. The process benefits in terms of better *p***H** performance and reduced lifecycle costs are even more impressive.

Once experience is gained in setting up the digital twin model, it can be readily adjusted to analyze and diagnose potential problems and develop creative, effective solutions with minimal risk. The solutions can be discovered, prototyped, tested, and used for training operators and professionals responsible for maintaining and designing the process, equipment, and automation system and for putting the best practices into action. Each chapter in this book concludes with best practices. Here is an overview of the essentials. The succeeding chapters will provide the details and understanding of what is truly needed.

1. Get lab titration curves that accurately represent the response to the reagents and the complete range of possible process concentrations and temperatures with the abscissa expressed as a ratio of reagent flow to influent flow.

2. Develop digital twin models that match the lab titration curves, paying particular attention to the moderating effect of carbon dioxide absorption and conjugate salts and the reduction in ion activity from high salt concentrations and nonaqueous solutions, and changes in dissociation constants with process temperature and ion activity.

3. From the slope of the titration curve in the control region around the set point, determine the precision of the reagent control valve or variable speed drive (VSD) required and the accuracy of the *p***H** electrodes needed in terms of the allowable reagent error.

4. From a distance along the abscissa of the titration curve, determine the rangeability of the reagent flowmeters, control valves, and VSDs needed.

5. Confirm the deadband from lost motion and the resolution limitation in reagent control valves and VSDs multiplied by the titration curve slope, that is the change in *p***H** for a change in reagent to influent flow ratio within the control region keeps the *p***H** within the control limits.

6. Total the dead time from the valve or VSD response, reagent transportation and injection delays, mixing delay, process transportation delay to the sensor,

electrode lag, signal update rate, filter time, and the controller execution rate. Minimize these sources so that the total loop dead time multiplied by the possible change in process concentration does not exceed the set point for each stage of neutralization.

7. Allocate a stage of neutralization for every 2 *p***H** units from the most extreme *p***H** (lowest and highest possible influent *p***H**) to the desired set point, where the *p***H** stage set points are rounded down or up in approximately 2 *p***H** increments.

8. Use electrodes with glass bulbs and shrouds designed to protect against direct impingement with a reference electrode whose equilibration time and offset of junction potential meet the allowable error.

9. Use three electrodes with middle signal selection to see the actual process *p***H** with minimal noise from bubbles and poor mixing and with sufficient velocity to keep electrodes clean and responsive.

10. Use precise reagent valves and signal characterization that uses the abscissa of the titration curve simply expressed as 0% to 100% reagent demand to improve control and potentially eliminate one stage of neutralization required.

11. Consider using static mixers upstream of a volume or to recirculate a volume to replace the need for a well-mixed vessel.

12. Minimize reagent injection delays by injecting reagent into a recirculation line, and minimize mixing delays by installing an axially agitated baffled vessel with a height of 1 to 2 times the diameter.

References

[1] Béla, G. Lipták, *Instrument Engineer's Handbook: Process Control,* 3rd ed. (Radnor, PA: Chilton Book Co., 1995).

[2] Gregory K. McMillan, *Process Control Case Histories – An Insightful and Humorous Perspective from the Control Room* (New York, NY: Momentum Press LLC, 2010).

[3] Terrence L. Blevins, Gregory K. McMillan, Willy K. Wojsznis, and Michael W. Brown, *Advanced Control Unleashed—Plant Performance Management for Optimum Benefits* (Pittsburgh, PA: ISA [The Instrumentation, Systems, and Automation Society], 2003).

[4] Gregory K. McMillan, "What's Your Flow Control Valve Telling You?," *Control Design* (May 2004).

[5] Gregory K. McMillan, *Process/Industrial Instruments and Controls Handbook*, 6th ed. (New York: McGraw-Hill Education, 2019).

[6] Béla, G. Lipták, *Instrument Engineer's Handbook: Process Measurement and Analysis*, 3rd ed. (Radnor, PA: Chilton Book Co., 1995).

[7] Gregory K. McMillan, "Understand Some Basic Truths of pH Measurement," *Chemical Engineering Progress* (October 1991).

[8] Gregory K. McMillan, *Good Tuning: A Pocket Guide* (Pittsburgh, PA: (ISA [The Instrumentation, Systems, and Automation Society], 2003).

[9] Gregory K. McMillan, *New Directions in Bioprocess Modeling and Control: Maximizing Process Analytical Technology Benefits*, 2nd ed. (Durham, NC: ISA [International Society of Automation], 2021).

2 The Chemistry

2.1 Nearly Normal

More than any other application, successful *p*H applications require good communication between the chemist, specialist, and plant engineer. Unfortunately, these individuals tend to speak different languages because what is normal in terms of units for one is not nearly normal for the other, and normality units, which are the most descriptive, are not typically used.

The units chemists and engineering specialists use most frequently to quantify the concentration of acids and bases in a solution are molar, molal, and normality. The concentration unit commonly used by plant engineers is weight or mass fraction. However, because process flow sheets and the specifications for control valves generally show reagent and influent flows, it is desirable to be able to convert from any of these concentration units to flow units. The remainder of the text demonstrates how valuable it will be for control valve sizing and *p*H system analysis to have the ratio of reagent to influent flow as the abscissa of the titration curve.

> **Key CONCEPT** Chemists and specialists will provide data in concentration units that should be converted to influent flow conditions and reagent flow requirements.

Chemists predominately use molar units in laboratory measurements because the calculation is based on a beaker volume. Molar concentration is the number of gram-moles per liter of solution. The number of moles is calculated by dividing the weight of the pure acid or base by its molecular weight expressed in the same weight units. The weights units used should be denoted with a dash before the word *moles* (i.e., lb-moles and gm-moles). Equations 2-1a and 2-1b show how to calculate gm-moles and the concentration in molar units. When using these equations, remember to convert all weight units to grams and all volume units to liters.

$$n = \frac{d \bullet x \bullet V}{M} \quad (2.1a)$$

$$c = \frac{n}{V} \quad (2.1b)$$

where

c = molar concentration of diluted acid or base (gm-moles per liter)
d = density of the solution (gm per liter)
M = molecular weight of pure acid or base (gm per gm-mole)
n = number of gm-moles of pure acid or base
V = volume of the solution (liters)
x = weight fraction of pure acid or base in the solution

Molal units are used predominately by engineers and scientists studying electrolytes because the concentration calculation is independent of density and, therefore, temperature. Molal concentration is the number of gm-moles per 1000 gm of solvent. Equations 2-1c and 2-1d show how to convert between molal and molar concentrations. Note that the molal concentration approaches infinity as the acid or base weight fraction approaches 1 (the grams of solvent approaches 0).

$$m = \frac{1000 \bullet n}{d \bullet (1 - x) \bullet V} \quad (2\text{-}1c)$$

$$c = \frac{d \bullet (1 - x) \bullet m}{1000} \quad (2\text{-}1d)$$

where

c = molar concentration of diluted acid or base (gm-moles per liter)
d = density of the solution (gm per liter)
m = molal concentration of diluted acid or base (gm-moles per kg solvent)
n = number of gm-moles of pure acid or base
V = volume of the solution (liters)
x = weight fraction of pure acid or base in the solution

Normality units are generally used in *p***H** simulations, which use charge balance equations for acid and base ions. Normality concentration is the grams-ions of replaceable hydrogen or hydroxyl groups per liter of solution. A shorter notation of gram-equivalents per liter is frequently used. Equations 2-1e and 2-1f show how to convert between normality and molar units. Table 2-1a shows the normality versus the weight percentage for some common reagents. Note that normality is not proportional to the weight percentage for a given reagent because the density of the solution also changes.

$$N = \frac{z \bullet d \bullet x}{M} \tag{2-1e}$$

$$c = \frac{N}{z} \tag{2-1f}$$

where

d = density of the solution (gm per liter)
M = molecular weight of pure acid or base (gm per gm-mole)
n = number of gm-moles of pure acid or base
x = weight fraction of pure acid or base in the solution
z = number of replaceable hydrogen or hydroxyl ions per molecule of acid or base
N = Normality (grams-ions of hydrogen or hydroxyl groups per liter)

Table 2-1a. The normality of a reagent is not proportional to the weight percentage.

Reagent	Wt%	Normality
Hydrochloric acid	32	10.17
($z = 1$)	38	12.35
Sulfuric acid	62.2	19.5
($z = 2$)	77.7	27.2
	93.2	35.2
	98.0	36.0
Sodium hydroxide	10	2.75
($z = 1$)	20	7.93
	50	19.1
Calcium hydroxide	5	1.36
($z = 2$)	10	2.78
	15	4.3

As previously mentioned, it is desirable to use a ratio of reagent to influent flow for the abscissa of the titration curve for *p***H** control system design. The method followed in this text is to convert the concentration of the acid or base from the reagent or influent streams in the effluent stream in molar, molal, and normality units to weight fraction via Equations 2-1a, 2-2a, or 2-3a (solve the equations for x). It is necessary to eliminate mixing and residence time dynamics when using the titration curve because it is a steady-state plot. For this purpose, it helps to visualize the influent and reagent streams combining in a pipeline instead of a tank so that there is an immediate translation from a change in reagent or influent flow to a change in the weight factions of the effluent stream. For pure (undiluted) acid or base reagent and influent streams, the ratio of the reagent to influent flow is equal to the ratio of the weight fraction of the acid or base in the effluent stream. Normally, the concentration of the acid or base in the influent and reagent is not 100%, so the weight fraction of the acid or base in the individual influent and reagent streams is used to calculate the diluted stream flows needed for control valve sizing and feedforward calculations via Equations 2-1g and 2-1h. When using weight fractions, it is necessary to designate which stream is the source of the acid or base and which stream is the destination to form a mixture. To reduce confusion in Equations 2-1g through 2-1l, the subscript $_{rr}$ designates the concentration of the acid or base reagent in the reagent stream; $_{re}$ designates the concentration of the acid or base reagent in the effluent; $_{ii}$ designates the concentration of the incoming acid or base in the influent stream; and $_{ie}$ designates the concentration of the incoming acid or base in the effluent throughout the text. Because the influent flow is usually known before the effluent flow, Equation 2-1h is substituted for F_e in Equation 2-1g to yield Equation 2-1i for calculating the reagent flow. The desired flow ratio for the abscissa of the titration curve is the ratio of the diluted reagent to the diluted influent mass flow. Per Equation 2-1i, this is equal to the ratio of the reagent to influent weight fractions in the effluent multiplied by the inverse of the ratio of the reagent to influent weight fractions in the incoming streams. To convert from mass flow to volumetric flow, each mass flow must be divided by the diluted stream density in consistent units. Note that most reagent concentrations are given in weight percentage and must be divided by 100 to get the weight fraction of the diluted reagent. If a laboratory titration curve states the initial sample volume, the abscissa has the volume of reagent titrated to reach the *p***H** set point specified, and the reagent concentration used in the laboratory and plant are equal, then , then Equation 2-1j can be used to calculate the volumetric flow rate.

The chemist often does not like working with the high concentrations used in the plant because they are more hazardous, and titration is more difficult because the drops required are much smaller. Because you do not normally have the luxury of a highly diluted reagent, you must multiply by the ratio of the plant to lab weight fractions per Equation 2-1k. The volumetric flows can be converted to mass flows by multiplying them by the respective stream's density. Equation 2-1l shows that effluent stream mass flow is equal to the sum of the influent and reagent stream mass flows.

$$F_r = \frac{x_{re}}{x_{rr}} \bullet F_e \qquad \text{(2-1g)}$$

$$F_i = \frac{x_{ie}}{x_{ii}} \bullet F_e \qquad \text{(2-1h)}$$

Solving 2-1h for F_e and substituting it into 2-1g:

$$F_r = \frac{x_{re} \bullet x_{ii}}{x_{rr} \bullet x_{ie}} \bullet F_i \qquad \text{(2-1i)}$$

$$Q_r = \frac{V_r}{V_i} \bullet Q_i \qquad \text{(2-1j)}$$

$$Q_r = \frac{V_r}{V_i} \bullet Q_i \bullet \frac{x'_{rr}}{x_{rr}} \bullet \frac{\rho'_{rr}}{\rho_{rr}} \qquad \text{(2-1k)}$$

$$F_e = F_r + F_i \qquad \text{(2-1l)}$$

where

F_e = effluent mass flow (kg per h)
F_i = influent mass flow (kg per h)
F_r = reagent mass flow (kg per h)
ρ_{rr} = density of reagent acid or base in the reagent stream in the plant (kg per liter)
ρ'_{rr} = density of reagent acid or base in the reagent used in the lab (kg per liter)
Q_i = influent volumetric flow (liters per min)
Q_r = reagent volumetric flow (liters per min)
V_i = influent sample volume used in the laboratory (milliliters)
V_r = reagent volume titrated to reach *p***H** set point in the laboratory (milliliters)
x_{ie} = weight fraction of influent acid or base in an effluent stream in the plant
x_{ii} = weight fraction of influent acid or base in an influent stream in the plant
x_{re} = weight fraction or reagent acid or base in an effluent stream in the plant
x_{rr} = wei.ght fraction of reagent acid or base in reagent stream in the plant
x'_{rr} = weight fraction of reagent acid or base in reagent used in the laboratory

> **Key CONCEPT** The calculations and curve from a chemist are often based on a reagent in a laboratory that is much more dilute than the reagent used in the plant.

2.2 Staying Active

Most discussions of *p***H** measurement and control assume the activity of the hydrogen ion has not been diminished. Equations 2-2a through 2-2c show the relationship between hydrogen ion activity and concentration. Activity is an indication of an ion's ability to do the normal things that ions like to do, such as combining with other ions (association) and breaking free from compounds (dissociation). The ratio of the escaping tendency of the component in the solution to that at a standard state is the official state-sanctioned definition of ion activity. Equation 2-2c shows that the ion concentration multiplied by an activity coefficient is equal to the ion activity. Activities are dimensionless. The activity coefficient decreases from unity as the ion concentration increases from zero. In dilute solutions, the ions are far enough apart that the interaction between ions is negligible. Ion activity decreases as ion interaction increases. The change in activity is greater for ions with a large number of charges. As the concentration increases, the activity for some ions goes through a minimum and then increases due to ions grouping together. The ion's activity also depends on the dielectric constant, the temperature of the solvent, and the concentration of the other ions in the solution. Table 2-2a lists the activity coefficients for some common ions in water for different ionic strengths. The ionic strength of solutions can be approximated as the sum of one-half of the product of the individual ion molal concentrations and their ion charge squared per Equation 2-2d from Reference 1. The ionic strength can then be substituted into Equation 2-2e to estimate its effect on the activity coefficient of the hydrogen (hydronium) ion. The hydrogen ion is less affected than other ions because it is more mobile. Most effluent streams in waste treatment are dilute enough that the activity coefficient for hydrogen can be considered equal to 1. However, this is not true for many reagents and some influent streams or for reactions where salts are formed. For highly concentrated streams, the *p***H** effectively becomes a function of water content. In other words, as the salt concentration increases and the water concentration decreases, the *p***H** will increase even though the concentration of acids or bases has not changed. Figure 2-2a shows how the *p***H** increases with the concentration of sodium chloride.

$$a_{\mathrm{H}} = 10^{-p\mathrm{H}} \tag{2-2a}$$

$$p\mathbf{H} = -\log\,(a_{\mathrm{H}}) \tag{2-2b}$$

$$a_H = \gamma_H \bullet c_H \tag{2-2c}$$

Table 2-2a. The activity coefficient depends on the solvent, temperature, ion size, ion valence, and the ionic strength of the solution.

Ion Type	Ion Size (Angstroms)	Z_i – Ion Valence (charge)	*I* – Ionic Strength in Water at 25°C			
			0.005	0.010	0.050	0.100
Hydronium	9	1	0.933	0.914	0.860	0.830
Lithium	6	1	0.929	0.907	0.835	0.800
Sodium	4.5	1	0.928	0.902	0.820	0.775
Hydroxyl	3.5	1	0.926	0.900	0.810	0.760
Potassium	3	1	0.925	0.899	0.805	0.755
Ammonium	2.5	1	0.924	0.898	0.800	0.750
Magnesium	8	2	0.755	0.690	0.520	0.450
Calcium	6	2	0.749	0.675	0.485	0.405
Carbonate	4.5	2	0.741	0.663	0.450	0.360
Sulfate	4	2	0.740	0.660	0.445	0.355
Phosphate	4	3	0.510	0.405	0.180	0.115

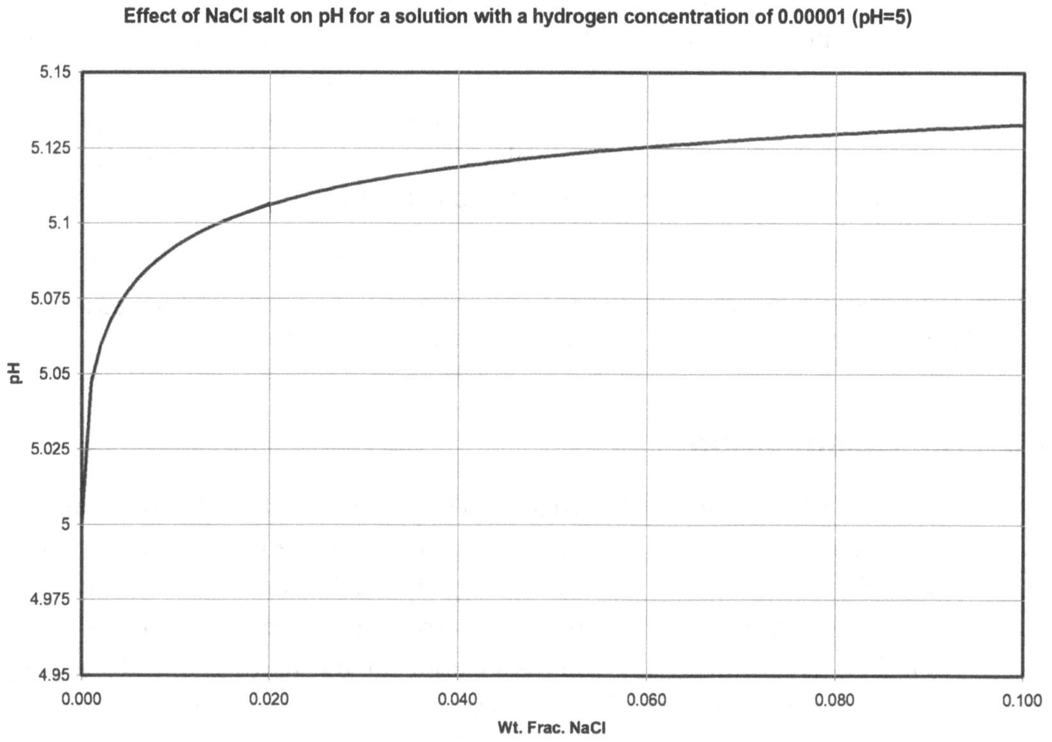

Figure 2-2a. Effect of salt concentration on *p*H.

$$I = 0.5 \bullet \sum_{i=1}^{n} (m_i \bullet Z_i^2) \qquad (2\text{-}2d)$$

$$log\ \gamma_H = \frac{-0.5 \bullet I^{0.5}}{1 + 3 \bullet I^{0.5}} \qquad (2\text{-}2e)$$

where

a_H = activity of hydrogen (hydronium) ions (gm-moles per liter)
c_H = concentration of hydrogen ions (gm-moles per liter)
γ_H = activity coefficient for hydrogen ion (1 for dilute solutions)
m_i = molal concentration of salt ion i (gm-moles per kg of solvent)
*p***H** = negative base 10 power of hydrogen ions
I = ionic strength of the stream at 25°C
Z_i = charge or valence of ion i

Equation 2-2a defines *p***H** in terms of hydrogen ion activity instead of concentration. A *p***H** measurement is a relative rather than an absolute indicator of hydrogen ion activity because the measurement is referenced to a measurement with a hydrogen electrode in a water solution (see Chapter 4, "Electrodes") that is not corrected for temperature, and the activity of a single ion is difficult to verify in practice. Also, the *p***H** measurement is generally not an indicator of hydrogen ion activity for pure acids or solvents other than water. Of greater importance is that a properly designed, installed, and maintained *p***H** measurement with a consistent reference potential and glass gel layer can provide a reproducible experimental value. For a more complete discussion of the limitations of the *p***H** measurement scale, refer to Reference 2.

> **Key CONCEPT** An increase in salt concentration or decrease in water content will increase the solution's *p***H** even though the acid and base concentration is constant.

2.3 Free Dissociation

An acid is a molecule that yields a hydrogen ion when it dissociates (breaks apart into its component ions), as shown in Equation 2-3a, and a base is a molecule that yields a hydroxyl ion when it dissociates, as shown in Equation 2-3b. Water acts as an acid and a base because it yields a hydrogen ion and a hydroxyl ion on dissociation, as shown in Equation 2-3c. Neutralization is the association of the hydrogen and hydroxyl ions to form water; it is designated by the reverse arrow in Equation 2-3c. Forward and reverse arrows are included in these equations to show that ion association and dissociation occur to maintain equilibrium between the concentrations of the species on both sides of the equation. These acid and base definitions were developed based on aqueous (water) solutions. A more general definition of an acid as a proton donor and a base as a proton acceptor is needed for nonaqueous solutions. Because visualizing these definitions is more difficult and data on nonaqueous *p***H** measurement is scarce, these definitions will not be explored further. The practical problems with nonaqueous *p***H** measurements will be discussed in Chapter 4. For more information on the proton donor and acceptor definitions, consult References 2, 3, and 4.

$$HA \Leftrightarrow H^+ + A^- \tag{2-3a}$$

$$BOH \Leftrightarrow B^+ + OH^- \tag{2-3b}$$

$$H_2O \Leftrightarrow H^+ + OH^- \tag{2-3c}$$

where

HA = acid molecule
BOH = base molecule
H_2O = water molecule
A^- = negative ion from the dissociation of an acid
B^+ = positive ion from the dissociation of a base
H^+ = hydrogen ion (proton)
OH^- = hydroxyl ion

The dissociation of some common acid and base reagents is detailed in Equations 2-3d through 2-3i. Note that sulfuric acid and calcium hydroxide dissociate twice and yield ions with a double charge.

$$HCL \Leftrightarrow H^+ + Cl^- \quad \text{(dissociation of hydrochloric acid)} \tag{2-3d}$$

$$H_2SO_4 \Leftrightarrow H^+ + HSO_4^- \quad \text{(first dissociation of sulfuric acid)} \tag{2-3e}$$

$$HSO_4^- \Leftrightarrow H^+ + SO_4^{2-} \quad \text{(second dissociation of sulfuric acid)} \tag{2-3f}$$

$$CaOH_2 \Leftrightarrow CaOH^+ + OH^- \quad \text{(first dissociation of calcium hydroxide)} \tag{2-3g}$$

$$CaOH^+ \Leftrightarrow Ca^{2+} + OH^- \quad \text{(second dissociation of calcium hydroxide)} \tag{2-3h}$$

$$NaOH \Leftrightarrow Na^+ + OH^- \quad \text{(dissociation of sodium hydroxide)} \tag{2-3i}$$

Dissociation constants are used to define the relationship between the activities of the components in equilibrium with each other. They also provide a measure of the strength of the acid or base. In most applications, the stream concentration will be assumed to be diluted enough for the charge balance equations that utilize these dissociation constants to be simplified by omitting activity coefficients. This assumption is usually valid for effluent streams, which are the streams of greatest interest in *p*H control loop performance analysis. The dissociation constant *K* typically falls numerically in the same range as the hydrogen ion concentration, so it is convenient to express it as a negative base 10 logarithm like *p*H where the small *p* designates the power function, as defined in Equation (2-3j). It is helpful to convert each pK_b to a pK_a via Equation 2-3l because the effect of the acid or base concentration on flattening the titration curve and the effect of the process temperature on the solution *p*H is greatest when the *p*H is at the pK_a.

$$pK_a = -log\ K_a \tag{2-3j}$$

$$pK_b = -log\ K_b \tag{2-3k}$$

$$pK_a = pK_w - pK_b \tag{2-3l}$$

where

pK_a = negative base 10 logarithmic acid dissociation constant
pK_b = negative base 10 logarithmic base dissociation constant
pK_w = negative base 10 logarithmic water dissociation constant

Equations 2-3r through 2-3w show the relationship between the species concentrations and the dissociation constant for the dissociation of the common reagents shown in Equations 2-3d through 2-3i. Square brackets around each species denote that the quantity is a concentration in normality units. The numerator is the product of the ion concentrations; the denominator is the unionized acid or base concentration. The double subscript is used to designate the source and order of dissociation. A lowercase *a* subscript designates that the source is an acid, and a lowercase *b* subscript designates that the source is a base. In references, the small *a* and *b* subscripts frequently are omitted. All such dissociation constants are for the hydrogen ion, even though the source may be a base. Table 2-3a lists the pK_a for some acids and bases at 25°C in a water solution of zero ionic strength ($pK_w = 14$). For bases, older references give a pK_b that must be converted to a pK_a for charge balance models and *p*H system analysis. The molecular weight is shown in the tables because it is used in molar, molal, and normality concentration calculations. For a more complete listing of dissociation constants, refer to References 5, 6, and 7.

Table 2-3a. pK_a for dilute aqueous solutions of acids and bases at 25°C [3, 7].

Acid or Base	pK_{a1}	pK_{a2}	Molecular Weight
Acetic acid	4.76	–	60.05
Acrylic acid	4.26	–	72.06
Ammonium	9.24	–	18.04
Calcium hydroxide	12.80*	12.67	74.09
Carbon dioxide	6.35	10.33	44.01
Formic acid	3.75	–	46.03
Fumaric acid	3.10	4.60	116.1
Glutaric acid	3.77	6.08	132.12
Hydrogen chloride	–6.2	–	36.46
Hydrogen cyanate	3.46	–	43.03
Hydrogen cyanide	9.21	–	27.06
Hydrogen fluoride	3.17	–	20.01
Hydrogen iodide	–8.26	–	127.92
Hydrogen peroxide	11.64	–	34.02
Hydrogen sulfide	6.97	12.9*	34.08
Lactic acid	3.86	–	90.08
Magnesium hydroxide	11.41	–	58.33
Maleic acid	1.91	6.33	116.1
Nitric acid	1.44*	–	63.02
Phosphorous acid	2.00	6.40	82.00
Phthalic acid	2.95	5.41	166.1
Sodium hydroxide	14.67	–	40.01
Sorbic acid	4.77	–	112.1
Sulfuric acid	–3.0*	1.99	98.08
Uric acid	5.40	5.53	168.11

* Present value is disputed and should be adjusted for application per lab data

It is important to realize that dissociation constants vary with temperature, and therefore the *p*H of the solution varies with temperature. A common misconception is that the temperature compensator in a *p*H measurement circuit corrects for this variation. Such compensators correct for the change in millivolts per *p*H unit relationship for the glass electrode per the Nernst equation. Smart transmitters have added the ability to specify a process *p*H versus temperature relationship. Fieldbus-based control systems have signal characterizers to calculate the correction to the set point or measurement.

The effect on temperature is greatest for *p*H values near a pK_a or a pK_w. Unfortunately, the data on the effect of temperature on a pK_a or pK_b is sketchy. In the absence of data, the change in pK_w with temperature, which ranges from –0.01 to –0.03 per °C, as shown in Table 2-3b, can be used to estimate the variation in the pK_a above 7 *p*H at specific temperatures. For strong acids and bases, the pK_a dissociation constants are, by definition, off-scale, and the change in the pK_w with temperature dominates. Figure 2-3a shows the effects of the changes in the water dissociation constant and temperature on a strong acid and strong base solution is 0 below 7 *p*H and reaches a maximum around 10 *p*H.

Table 2-3b. pK_w for dilute aqueous solutions as a function of temperature [4, 8].

Temperature (°C)	pK_w
0	14.94
5	14.73
10	14.54
15	14.35
20	14.17
25	14.00
30	13.83
35	13.68
40	13.54
45	13.40
50	13.28
75	12.71

The acid or base strength increases as its pK_a decreases. A pK_a less than 0 means that the product of the ion activities is greater than the activity of the source molecule. Such acids and bases are called *strong* because they are, for practical purposes, completely dissociated in the 0 to 14 *p*H range. Correspondingly, acids with a pK_a greater than 1 are called *weak*. The relative weakness increases as the pK_a increases. Note that the first dissociation constant of sulfuric acid classifies it as a strong acid, while the second classifies it as a weak acid. In fact, sulfuric acid behaves like a mixture of strong and weak acids. When the *p*H equals the pK_a, the negative acid ion activity equals the acid molecule activity. The dissociation is at the midpoint. This midpoint can be spotted on the titration curve if the dissociation constants are not too close together and can be used as a flag for identifying the dissociation constants from laboratory titration curves for signal characterization and control system simulation programs. This technique will be discussed in greater detail in Chapter 3, "Titration Curves," and Chapter 7, "Reagent."

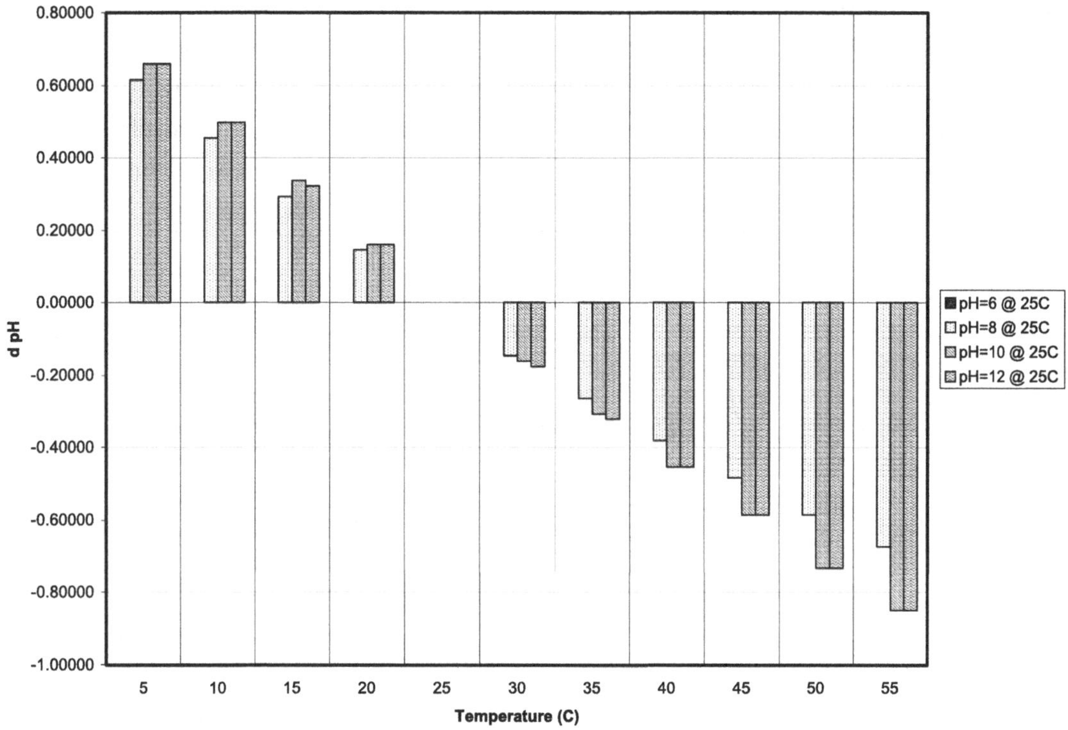

Figure 2-3a. Effect of temperature on a strong acid and base solution.

Key CONCEPT The pK_b for a base should be converted to a pK_a because the effect of ion activity and solution temperature is best seen when the *p***H** approaches a pK_a.

The dissociation constants in the literature are based on concentrations, as shown in Equation 2-3m. In reality, they should be based on activities, as detailed in Equation 2-3n, which leads to the correction factors shown in Equations 2-3o and 2-3p.

$$K_a = \frac{[H^+]\bullet[A^-]}{[HA]} \quad \text{(concentration – based acid dissociation constant)} \quad (2\text{-}3m)$$

$$K'_a = \frac{\gamma_H \bullet [H^+]\bullet \gamma_A \bullet [A^-]}{\gamma_{HA} \bullet [HA]} \quad \text{(activity – based acid dissociation constant)} \quad (2\text{-}3n)$$

$$K'_a = \frac{[H^+]\bullet[A^-]}{[HA]} \bullet \frac{\gamma_H \bullet \gamma_A}{\gamma_{HA}} \quad \text{(activity – based acid dissociation constant)} \quad (2\text{-}3o)$$

$$pK'_a = logK_a - log\gamma_H - log\gamma_A + log\gamma_{HA} \quad (2\text{-}3p)$$

For relatively dilute solutions, $log\ \gamma_H = log\ \gamma_A$, $log\ \gamma_{HA} = 0$, and γ_H can be approximated using the Debye-Hückel theory per Reference 6, which yields Equation 2-3q for correcting a pK_a for the changes in ion activity. There are sophisticated electrolyte modeling programs that take ion activity into account. In reality, the activity coefficients are not known accurately known for concentrated or multicomponent solutions because of complex ion interactions, and the user ends up tweaking the coefficients to match plant data.

$$pK'_a = pK_a - \frac{0.5 \bullet m^{0.5}}{1+m^{0.5}} \qquad (2\text{-}3q)$$

$$K_a = \frac{[H^+]\bullet[Cl^-]}{[HCl]} \quad \text{(dissociation constant for hydrochloric acid)} \qquad (2\text{-}3r)$$

$$K_{a1} = \frac{[H^+]\bullet[HSO_4^-]}{[H_2SO_4]} \quad \text{(first dissociation constant for sulfuric acid)} \qquad (2\text{-}3s)$$

$$K_{a2} = \frac{[H^+]\bullet[SO_4^{2-}]}{[HSO_4^-]} \quad \text{(second dissociation constant for sulfuric acid)} \qquad (2\text{-}3t)$$

$$K_{b1} = \frac{[OH^-]\bullet[CaOH^+]}{[CaOH_2]} \quad \text{(first dissociation constant for calcium hydroxide)} \qquad (2\text{-}3u)$$

$$K_{b2} = \frac{[OH^-]\bullet[Ca^{2+}]}{[CaOH^+]} \quad \text{(second dissociation constant for calcium hydroxide)} \qquad (2\text{-}3v)$$

$$K_{b1} = \frac{[OH^-]\bullet[Na^+]}{[NaOH]} \quad \text{(dissociation constant for sodium hydroxide)} \qquad (2\text{-}3w)$$

$$K_w = [H_3O^+]\bullet[OH^-] \quad \text{(dissociation constant for water)} \qquad (2\text{-}3x)$$

2.4 A Question of Balance

Although mass, energy, and component balances are commonplace in simulation software, the charge balance that enables calculating *p*H is rarely employed, even though the implementation is rather straightforward and the results useful for generating titration curves for *p*H signal characterization, control valve sizing, and *p*H system analysis. While there is no direct solution to the charge balance for a complex mixture, an interval halving search for the *p*H that gives a zero net charge is fast, efficient, and foolproof. In practice, the effect of changes in ion activity can be included by adjusting the acid dissociation constants[8]. Special-purpose electrolyte modeling programs can be used to provide estimates of the activity coefficients, but some tweaking of the dissociation coefficients is inevitable.

Key CONCEPT A charge balance with its pK_a and pK_w coefficients adjusted for activity and temperature effects is a powerful tool for generating titration curves.

Because solutions are neutral, the existence of just one type of ion in a solution is impossible. Thus, p**H** measurement cannot be stated as a measurement of just the hydrogen ion. In water solutions, hydrogen ions will be accompanied by hydroxyl ions per the ionic product shown in Equation 2-3x. If no other ions are in the solution, charge neutrality demands that the hydrogen ion concentration equals the hydroxyl ion concentration. If other ions are in solution, the sum of each ion concentration multiplied by its charge (normality) must equal zero. This is fortunate because it permits using an interval halving search method to iteratively search for the p**H** for a given set of ion concentrations that makes the charge balance zero. The excess charge is calculated for the p**H** guess at the midpoint of the p**H** search interval. If the excess charge is negative, the lower p**H** search limit is increased to the midpoint p**H**. If the excess is positive, the upper p**H** search limit is decreased to the midpoint p**H**. This interval halving continues until the interval is less than a specified allowable error. If the p**H** and concentrations of all but one acid or base in a solution are known, the unknown acid or base concentration can be solved directly from the charge balance equation.

The charge balance equation is set up by summing the concentration of all the ions in the solution in normality units and setting the sum equal to zero. The acid or base concentrations, not the ion concentrations, are generally given. A strong acid or base is completely ionized so that the ion concentration is equal to the acid or base concentration. A weak acid or base is only partially ionized. The concentration of the ions can be calculated from the relationships for the dissociation constants and combined in an expression based on the acid or base concentration. Equations 2-4a through 2-4h show the expressions for single, double, and triple dissociations, respectively, of a weak acid or base. For the derivations of the expressions for single and double dissociations, refer to Reference 7. A derivation of the expression for the third dissociation can be obtained by extending the same method. Note that the sign of the parameter s determines whether the expression is for a weak acid or base. Also, each base dissociation constant pK_b must be converted to an acid dissociation constant pK_a via Equation 2-3p to use the expressions for bases. If the p**H** is two or more units larger than the pK_a

for acids or two or more units less than the pK_a for bases, Equations 2-4d through 2-4f contribute less than 0.01 when summed with 1 in.Equations 2-4a through 2-4c. The numerators and denominators are then approximately equal to 1, and Equations 2-4a through 2-4c reduce to simply s • N. Thus, weak acids can be considered completely ionized for a *p***H** greater than the pK_a by 2, and weak bases can be considered completely ionized for a *p***H** less than the pK_a by 2. Weak acids behave like strong acids at a high *p***H**, and weak bases behave like strong bases at a low *p***H**. The contribution of each weak acid or weak base to the charge balance equation is represented by one of Equations 2-4a through 2-4c. Equation 2-4g shows the charge balance equation for strong acids and bases, and Equation 2-4h shows the charge balance equation for weak acids and bases. The last two terms in both equations are for the hydrogen and hydroxyl ion concentrations.

$$N_1 = \frac{1}{1+P_1} \bullet s \bullet N \qquad \text{(2-4a)}$$

$$N_2 = \frac{(1+0.5 \bullet P_2)}{(1+P_2 \bullet (1+P_1))} \bullet s \bullet N \qquad \text{(2-4b)}$$

$$N_3 = \frac{(1+0.33 \bullet P_3 \bullet (2+P_2))}{(1+P_3 \bullet (1+P_2 \bullet (1+P_1)))} \bullet s \bullet N \qquad \text{(2-4c)}$$

$$P_1 = 10^{(s \bullet (p\mathbf{H} - pK_1))} \qquad \text{(2-4d)}$$

$$P_2 = 10^{(s \bullet (p\mathbf{H} - pK_2))} \qquad \text{(2-4e)}$$

$$P_3 = 10^{(s \bullet (p\mathbf{H} - pK_3))} \qquad \text{(2-4f)}$$

For strong acids and bases, the acids and bases are completely dissociated everywhere on the *p***H** scale, which gives the following charge balance:

$$\sum_i N_{bi} - \sum_i N_{ai} + 10^{-p\mathbf{H}} - 10^{(p\mathbf{H} - pKw)} = 0 \qquad \text{(2-4g)}$$

For weak acids and bases with a single dissociation, the concentration of ions depends on the *p***H** and is computed via Equation 2-4a, which gives the following charge balance:

$$\sum_i \left[\frac{1}{(1+P_{bi})} \bullet N_{bi} \right] - \sum_i \left[\frac{1}{(1+P_{ai})} \bullet N_{ai} \right] + 10^{-p\mathbf{H}} - 10^{(p\mathbf{H} - pKw)} = 0 \qquad \text{(2-4h)}$$

where

N = concentration of an acid or base (normality)

N_1 = concentration of ions from a single dissociation (normality)

N_2 = concentration of ions from a double dissociation (normality)

N_3 = concentration of ions from a triple dissociation (normality)
N_{ai} = concentration of an acid i (normality)
N_{bi} = concentration of a base i (normality)
pK_1 = negative base 10 logarithmic first acid dissociation constant
pK_2 = negative base 10 logarithmic second acid dissociation constant
pK_3 = negative base 10 logarithmic third acid dissociation constant
pK_w = negative base 10 logarithmic water dissociation constant
P_1 = parameter for an acid or base with one dissociation
P_2 = parameter for an acid or base with two dissociations
P_3 = parameter for an acid or base with three dissociations
P_{ai} = parameter for dissociation of an acid *i* with a single dissociation
P_{bi} = parameter for dissociation of a base *i* with a single dissociation
s = ion sign (s = -1 for acids and s = +1 for bases)

The titration curves in plants tend to be more buffered than those generated by these equations because of the carbonates in plant water. It is suggested that carbon dioxide be added as one of the components in the streams and its concentration adjusted based on its solubility at process conditions and the titration curves for process samples at normal operating conditions.

RULE of Thumb *Add carbon dioxide as a stream component and adjust the concentration to provide the buffering effect normally found in plant water.*

For weak acids and bases with double and triple dissociations, Equations 2-4b and 2-4c are used, respectively, in place of Equation 2-4a and Equation 2-4h. Appendix H has the charge balance equations for i acids and bases with single, double, or triple dissociations that are solvable by interval halving. The pK_a and pK_w coefficients and the solution density should be corrected for process temperature and composition. The effect of changes in the activity coefficients with ionic strength should also be factored in [8]. Refer to References 4 and 5 for a more detailed discussion on the effect of ion concentrations on activity coefficients. The polynomial equation for acid and sodium ion error at the end of the subroutine should be replaced with one that fits the data from the glass manufacturer.

RULE of Thumb *Adjust the pK_a and pK_w in a charge balance model to include the effects of process temperature and ionic strength.*

These equations are of increasing interest to process control engineers because online charge balance models combined with Coriolis mass flow measurements of influent and reagent flows can provide online fast and reliable *p*H estimators as discussed in Chapter 8, "Control System." The pK_a in these models closest to the *p*H operating point can be manipulated online by a model predictive controller to match plant data.

> **Key CONCEPT** An online charge balance model with its pK_a adapted by model predictive control to match plant data can provide a fast and reliable *p*H estimator.

2.5 Best Practices

Knowing the chemistry is critical when analyzing the many effects of process operating conditions on the process *p*H that are seldom detailed in the literature. This lack of information and the attempt to derive exact solutions for very narrow conditions pose serious limitations on understanding the causes and effects and the ability to model *p*H for the many possibilities in industrial applications. Including the effect of changes in activities, dissociation constants, conjugate salts, and air exposure is critical. This chapter, supplemented by Appendix H, provides the knowledge of chemistry needed that, along with the groundbreaking charge balance and its solution by simple interval halving, is the key to a *p*H model matching the titration curve and consequently projecting effects of process conditions and proactively providing solutions.

1. Use the first principle relationships and ordinary differential equations detailed in Appendix F for the material balance of every component and energy balance.

2. Use concentrations in normality units, base 10 logarithmic acid dissociation constants (pK_a), and base 10 logarithmic base dissociation constants (pK_b), taking into account how they change with operating temperatures and ion activities. Then convert pK_b to pK_a in the charge balance equations to help recognize the moderating effect on the slope for an operating *p*H near a pK_a.

3. Use base 10 logarithmic dissociation constants of solvents (pK_s), hopefully, mostly water (pK_w), so that the *p*H scale is not greatly shifted. Take into account how they change with operating composition, temperature, and ion activities.

4. Include the moderating effect of carbonic acid and conjugate salts detailed in Appendix H.

5. Set the charge balance equation equal to zero (detailed in Equations 2-4g and 2-4h) that is a summation of all ions from solvents and from acids and bases realizing the these ions are a function of the total concentration of each component and the *p*H.

6. Use interval halving as the fastest and most reliable way to converge on a solution *p*H within electrode accuracy. Use reasonable upper and lower *p*H limits to identify the *p*H that makes the charge balance close to zero where the next change in *p*H is less than the electrode accuracy.

References

[1] Clark C. Wescott, *pH Measurements* (New York: Academic Press, 1978).

[2] Roger G. Bates, *Determination of pH Theory and Practice* (New York: John Wiley and Sons, 1964).

[3] William J. Moore, *Physical Chemistry*, 3rd ed. (Englewood Cliffs, NJ: Prentice-Hall, 1962).

[4] Galster, Helmuth. *pH Measurement: Fundamentals, Methods, Applications, Instrumentation* (New York: VCH, 1991).

[5] Allen J. Bard, *Chemical Equilibrium* (New York: Harper & Row, 1966).

[6] John Dean, *Lange's Handbook of Chemistry*, 14th ed. (New York: McGraw-Hill, 1992).

[7] F. Greg Shinskey, *pH and pION Control in Process and Waste Streams* (New York: John Wiley and Sons, 1973).

[8] James E. Kipp, "Computer Series 158 – PHCALC: A Computer Program for Acid/Base Equilibrium Calculations," *Journal of Chemical Education* 71, no. 2 (February 1994).

3

Titration Curves

3.1 Slippery Slopes

The task of a conventional control system to maintain a *p***H** set point on the steep slope of a titration curve avoiding the point (knee) where the curve bends to a flat slope despite upsets is similar to a novice skier trying to stop on an icy expert slope while being buffeted by high winds.

> **Key CONCEPT** Going from a knee to the steep portion of a titration curve can be like falling off a cliff with nothing to break the fall except a knee.

Unfortunately, most set points are near the neutral or equivalence points on the titration curve, which coincide with the location of the maximum slope. After an upset or start-up, the *p***H** accelerates as the operating point approaches the steepest point on the curve. To the *p***H** controller that knows only what it sees, it appears to be a fast runaway or positive feedback response. Most of the beneficial value of the investment in a well-mixed vessel in terms of slowing down a disturbance has been lost. (The decrease in the effective process time constant seen by the *p***H** controller will be quantified in Chapter 8, "Control System.") Operators will ask whether you can do anything to slow down the response because it tends to zip right past the desired set point. In Chapter 8, on *p***H** control system design and performance, we will show how signal characterization can slow down the response and eliminate the acceleration in the neutral region. This is the inherent benefit of translating the controlled variable from *p***H** to the percentage of reagent demand (x-axis of the titration curve). The more gradual response soothes the operator's nerves and restores the process time constant.

Key CONCEPT The *p*H will zoom right past a set point on the vertical portion of the titration curve, and most of the beneficial value of the investment in a well-mixed vessel in terms of slowing down the nonlinear *p*H response appears to be lost.

The *p*H control system sensitivity and process gain increase as the slope of the titration curve increases. While the details on assessing the exact implications of this statement are covered in subsequent chapters, it is appropriate to discuss how the slope changes value and location as a function of stream composition. Figures 3-1a through 3-1d show the changes in the titration curve for four possible combinations of strong and weak acids and bases. In these curves, the strong acid pK_a and the strong base pK_b are both 0 *p*H, and the weak acid pK_a and the weak base pK_b are both 4 *p*H. In every case, the sample is a strong acid, with or without a weak acid or base added, that is titrated with a strong base.

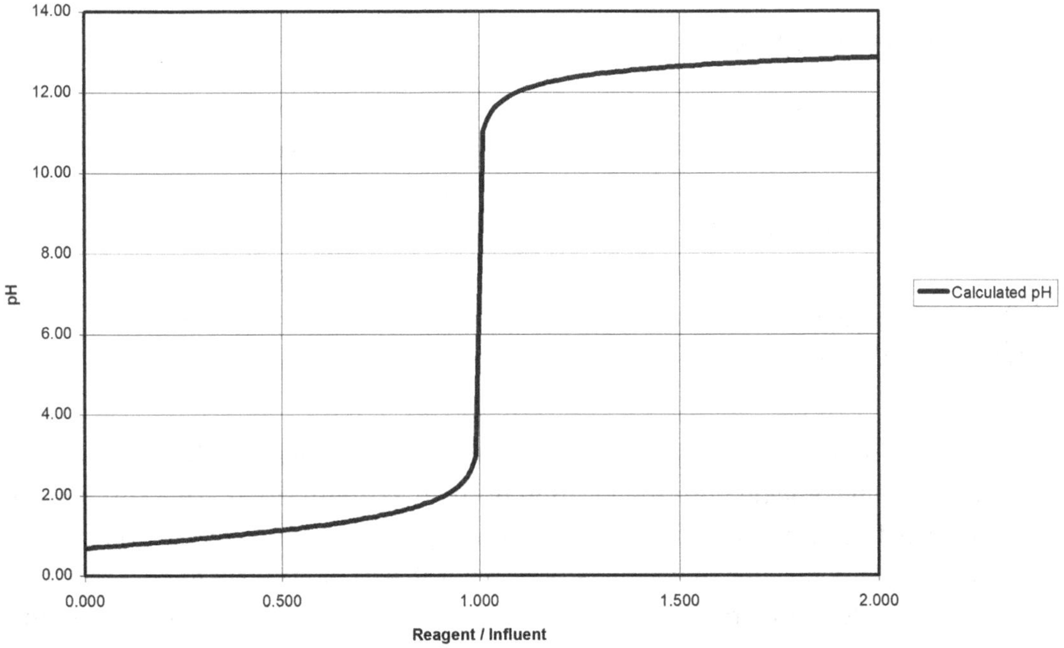

Figure 3-1a. For a strong acid and strong base, the control system sensitivity and *p*H process gain are a concern for almost all *p*H set points.

The strong acid and strong base curve shown in Figure 3-1a is distinguished by its vertical slope throughout most of the *p*H scale range. Figures 1-lb and 1-1c in Chapter 1, "The Essentials," showed that this slope is not vertical but changes by a factor of 10 for each *p*H unit deviation from 7 *p*H for a solution temperature of 25°C. True titration curves have no straight lines; the blowup of a straight line will reveal another curve if there are enough data points. A titration curve without a clearly defined abscissa is worthless because the shape will change with the abscissa range. Therefore, it is critical to have two titration curves, one that covers the whole operating range and another that details the control region.

Key CONCEPT The shape of a titration curve will drastically change with the range of the abscissa and the resolution of the data points in the knees and steeper regions.

RULE of Thumb *It is critical to have two titration curves, one that covers the whole operating range and another that details the curvature in the control region.*

The slope is steepest at the equivalence point. The equivalence point coincides with the neutral point for a strong acid and strong base system. The neutral point occurs at a *p*H equal to one-half of the pK_w, so its location depends on the solution temperature. For example, if the pK_w decreases from 14.0 to 13.54 because the solution temperature increases from 25°C to 45°C, the neutral point would decrease from 7.0 to 6.77 *p*H. The definitions of the neutral and equivalence points are as follows:

The *neutral* point on the titration curve is where the hydrogen ion concentration equals the hydroxyl ion concentration.

The *equivalence* point on the titration curve is where the acid ion concentration equals the base ion concentration.

Although the strong acid and strong base curve in this case and in most textbooks appears symmetrical about the equivalence point, the curve is only truly symmetrical if the strong acid pK_a and the strong base pK_b are equal. The midpoint of the lower flat portion of the titration curve occurs at the point where the *p*H is equal to the pK_a for the strong acid. The midpoint of the upper flat portion(s) occurs at a *p*H equal to the pK_w for water and the pK_a (not the pK_b) for the strong base. Two distinct upper flat portions occur if the pK_w and pK_a are sufficiently separated. However, the exact shape is difficult to obtain because of the large electrode error and activity coefficient error at these extremes in concentration and is unimportant because it lies outside the practical *p*H measurement and set point range.

The strong acid and strong base system has a maximum slope and range of steep slopes greater than any other *p*H system. The control system sensitivity and the *p*H process gain are concerns for any *p*H set point other than at the extremes of the *p*H scale range.

The weak acid and strong base system in Figure 3-1b is distinguished by its steep slope in the upper *p***H** scale range. The center of this steep slope is the equivalence point location which depends on the dissociation constant of the weak acid. The control system sensitivity and *p***H** process gain are a concern for *p***H** set points greater than 2 *p***H** units above the weak pK_a and 2 *p***H** units below the strong base pK_a or the water pK_w. Thus, for the pK_a used in this case, the slope is steep between 6 and 12 *p***H**. The sensitivity and gain are the least for a *p***H** set point equal to the weak acid pK_a (4 *p***H**), which is at the midpoint of the lower flat portion. The flat upper and lower ends of the curve are outside the operating *p***H** range.

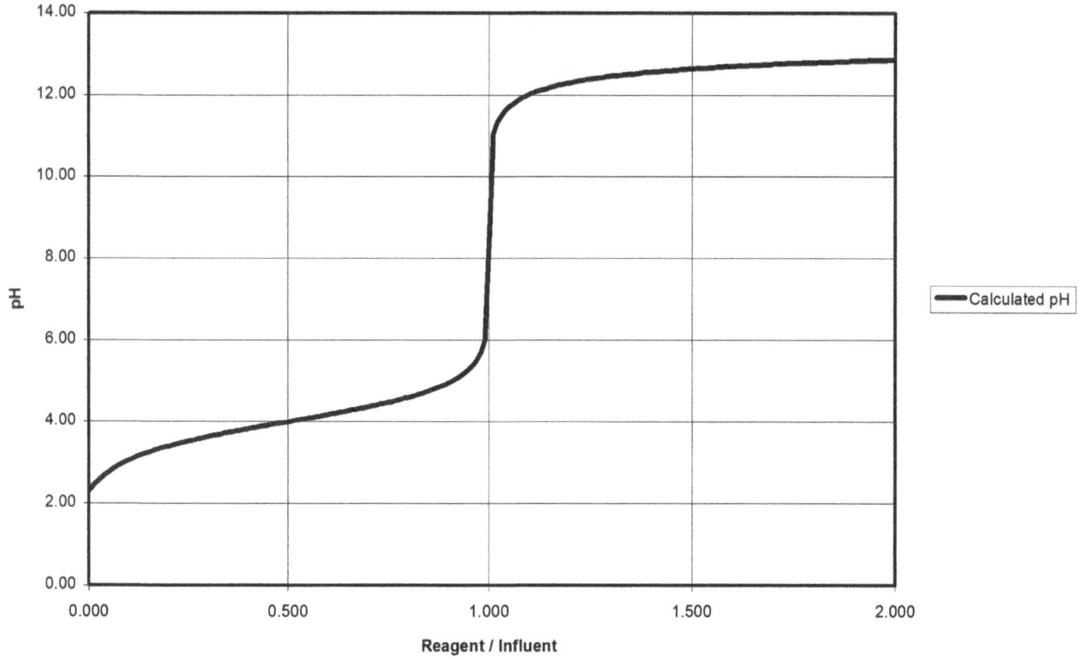

Figure 3-1b. For a weak acid and strong base, the control system sensitivity and *p***H** process gain are a concern for 6 to 12 *p***H** set points.

The strong acid and weak base system shown in Figure 3-1c is distinguished by its steep slope in the lower *p***H** scale range. The center of this steep slope is the equivalence point location which is dependent on the dissociation constant of the weak base. The control system sensitivity and *p***H** process gain cause concern for *p***H** set points more than 2 *p***H** units below the weak base pK_a and 2 *p***H** units above the strong acid pK_a. The sensitivity and gain are the least for a *p***H** set point equal to the weak base pK_a (10 *p***H**), which is at the midpoint of the upper flat portion. Again, the flat upper and lower ends of the curve are not shown because they are outside the operating *p*H range.

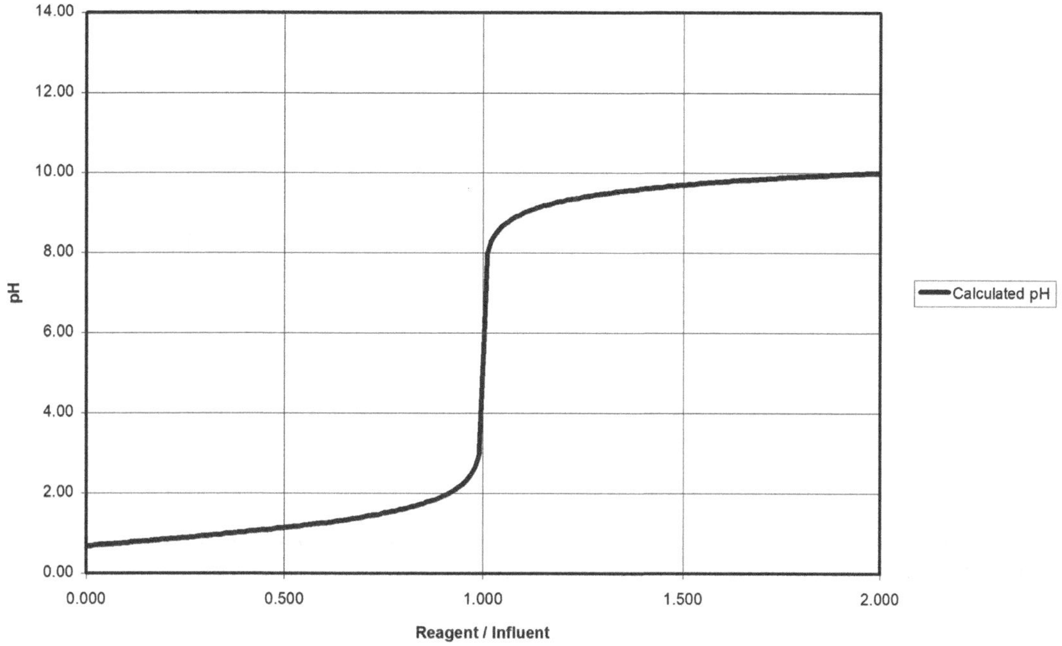

Figure 3-1c. For a strong acid and weak base, the control system sensitivity and *p*H process gain are a concern for 2 to 8 *p*H set points.

The lack of a steep slope throughout the range distinguishes the weak acid and weak base system in Figure 3-1d. The titration curve slope is greatest at the equivalence point but is still relatively small compared to the previous systems. The equivalence point ordinate depends on both the acid and base dissociation constants. Although the magnitude of the *p*H process gain is low, the gain is still nonlinear. The system should not be considered easy to control. A common misconception is that a titration curve consists of a single S-shaped curve. In Figure 3-1e, a weak acid with pK_a dissociation constants at 3 and 5 *p*H is titrated with a weak base with pK_a dissociation constants at 9 and 11 *p*H. Notice that three S-shaped titration curves are formed with an equivalence point at the point of the steepest slope for each S curve. The dissociation constants must be more than 2 *p*H units apart to create multiple S-shaped curves. The curves are symmetrical because the concentrations and the distances between the dissociation constants are equal. In actual plant applications, *p*H titration curves are not usually symmetrical because of a complex combination of acids, bases, and salts and the effect of process temperature.

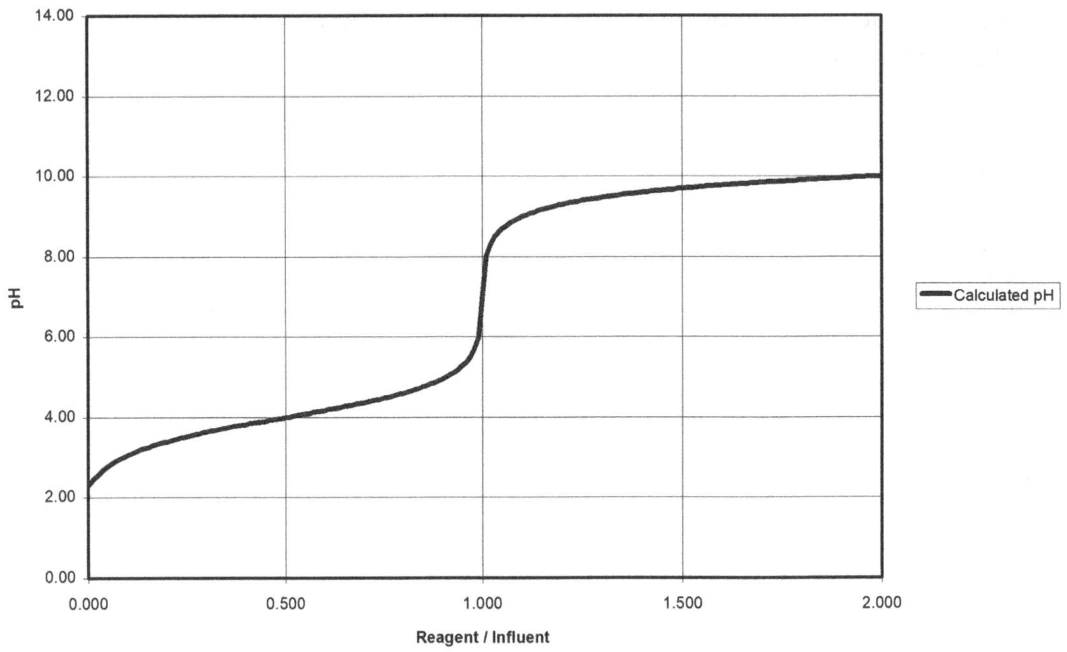

Figure 3-1d. The control system sensitivity and *p***H** process gain are moderated for a weak acid and weak base, even for set points near the equivalence point.

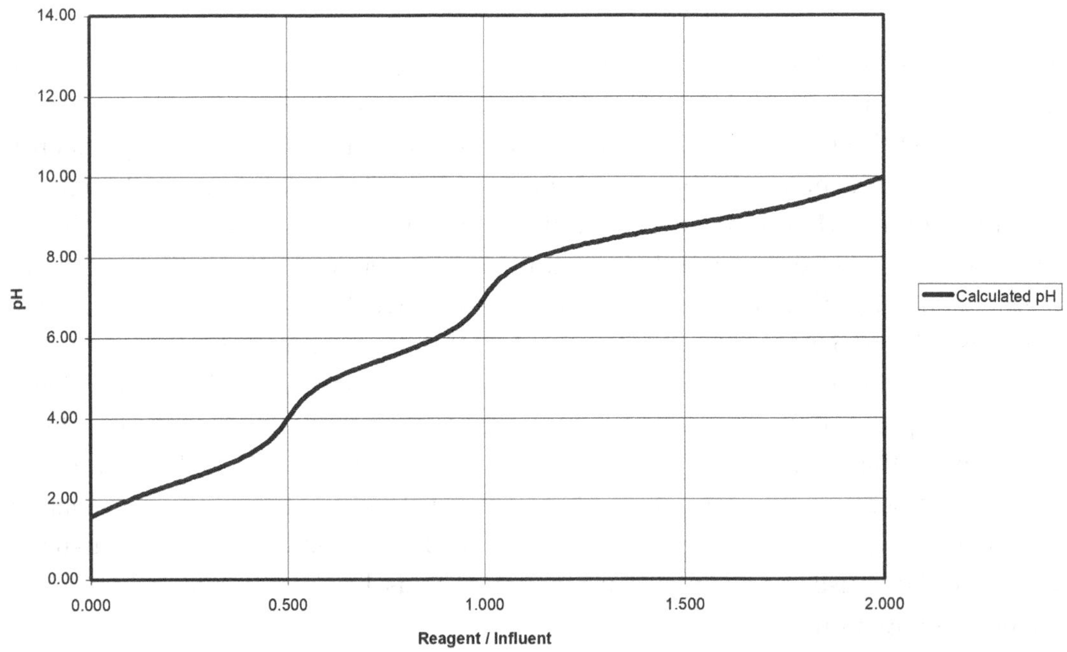

Figure 3-1e. Multiple S-shaped curves are formed from multiple weak acid and base dissociation constants if they are spread apart by more than two *p***H** units.

3.2 Laboratory Generation

Laboratory titration curves are typically created by placing a known volume of the process sample in a beaker and logging the *p***H** indication for each incremental volume of reagent added via a burette. The data points are recorded, and a smooth curve is plotted between the points. The volume of reagent added between the data points must be reduced drastically near the equivalence point. If the sample or reagent contains a strong acid or base, it is difficult to generate data points near the equivalence point. Most titrators are now automatic with interfaces to plant data historians.

If the titration curve has a long, flat tail, relatively few data points are needed until the first bend. However, the starting point representing the influent *p***H** for various operating conditions must be plotted accurately to estimate the valve rangeability and stick-slip and the mixing equipment size and agitation requirements.

If the sample or the reagent contains solids, the time required for the solids to dissolve and the ions to get into the solution to establish the solution ion balance can range from several seconds to several minutes. If the *p***H** has not reached its final value before the reagent volume is incremented, the titration curve becomes distorted. If a gaseous reagent such as ammonia is bubbled into the sample, the wait time between drops must be longer than the dissolution time. Bubbles may rise to the surface and escape from the solution before dissolving so the amount of reagent added is not equal to the amount in the solution.

If the sample volume is given and the concentrations of the reagent used in the laboratory and the control system are equal, the reagent volumetric flow can be calculated for a given influent flow and *p***H** set point per Equation 2-1j. If the concentrations are not equal, Equation 2-1k is applicable. It is important to verify the concentration of the laboratory and field reagents before using a titration curve for system design.

The laboratory temperature during titration is rarely equal to the ambient or process temperature. The sample *p***H** will change with the temperature because the dissociation constants and the water ionic product change with the temperature. This is a change in actual solution *p***H** and is not to be confused with the change in millivolts generated by the glass electrode per the Nernst equation, discussed in Chapter 4, "Electrodes." Conventional temperature compensators use a temperature sensor embedded inside the electrode to correct for the Nernst effect.

The sample temperature should be maintained equal to the process temperature at the electrodes by heating or cooling the sample. For the greatest accuracy, a separate temperature sensor should be installed at the electrodes, and the process temperature should be available to the laboratory. If the electrodes are in a recirculation line that is well insulated with no heating or cooling, the vessel temperature can be used. If the process temperature at the electrodes changes by more than a few degrees, multiple samples should be titrated at different temperatures to determine the change in actual solution *p*H with the temperature near the *p*H set point so that a correction factor can be automatically applied to the field *p*H measurement based on the deviation from a reference temperature, such as the set point. Most smart field *p*H transmitters have solution temperature compensation that is a user-configurable polynomial to match the test results of the change in actual *p*H with the temperature generated in a laboratory. Laboratory *p*H meters have the conventional electrode temperature compensation that corrects for the Nernst effect, but they tend not to have solution *p*H compensation to account for the change in dissociation constants with process temperature. Consequently, if the temperature of the sample cannot be maintained by heating or cooling at the process temperature, a temperature correction must be applied manually or by a special computer program.

Key CONCEPT The laboratory *p*H meters may not have the automatic solution *p*H temperature compensation configured into the smart field *p*H transmitter.

Every sample must be time-stamped with the exact time that the sample was taken. Some operators write the time that they mark the sample, or they round off or guess at the time. Some operators log the time of the laboratory result. The only way a laboratory *p*H measurement result can be correlated with the actual process *p*H and temperature recorded in the data historian is to have the sample time-stamped correctly.

RULE of Thumb *The sample and the curve should be marked with the exact time the sample was taken and the process temperature at the electrode.*

RULE of Thumb *The laboratory meter and the field transmitter must use the same algorithm for solution temperature compensation.*

Samples should be kept in Pyrex glass or polyethylene plastic bottles. Soft glass bottles should not be used because the glass solubility is great enough to raise the *p*H due to the sample absorbing alkali ions. Cork stoppers should not be used because cork is highly acidic. Samples exposed to the air can absorb enough carbon dioxide to lower the *p*H. This is a particular problem for nearly pure water and caustic samples. The *p*H of absolutely pure water at 25°C can change from a *p*H of 7 to a *p*H of 5.7 by exposure to air. The slope of caustic solutions will decrease in the region of 6 to 7 *p*H from the moderating effect of carbonates due to the absorption of carbon dioxide from the air. This moderating effect of carbonates helps to regulate the *p***H** of many aquatic streams and reduces the difficulty of waste and raw water *p***H** control. References 1 and 2 have information on the effect of carbon dioxide on the *p***H** of samples.

RULE of Thumb *Samples should be kept in closed Pyrex glass bottles to reduce the risk of their composition changing from evaporation and absorption.*

The errors at the high and low ends of the *p***H** scale depend on the type of *p***H** electrode, process chemicals, and the time and temperature exposure. Acidic solutions below 2 *p***H** can dehydrate the glass electrode. Hydrofluoric acid solutions below 6 *p***H** and caustic solutions above 10 *p***H** can chemically attack the glass. Both chemical attack and dehydration are greatly accelerated by temperature. In general, glass with a higher electrical resistance has a greater chemical resistance. Also, some glasses are specifically designed for low and/or for high *p***H** operation. High concentrations of relatively small alkali ions, such as sodium ions, can also penetrate the glass and suppress the *p***H** reading, although this alkalinity error is less of a problem for newer glass designs. The type of glass must be chosen to minimize the effect of the largest error.

A high ionic strength stream can also cause a slow drift in the liquid junction potential of the reference electrode. For certain types of electrodes and process conditions, it can take hours for the reference electrode to reach an equilibrium state. In cases of contamination of the internals of the reference, the drift does not stop but continues. The best method for reducing junction potential errors and contamination is a flowing junction reference, but the flow rate of the reference electrolyte must be small enough not to contaminate the sample.

Because the immersion time of laboratory electrodes is much shorter than the exposure time of field electrodes, most of the effect of the process on the glass and reference electrodes does not take place to the same degree in the laboratory as in the field. In fact, this is the key to how to deal with really nasty streams. Chapter 4 discusses using automated injection assemblies to control the time of exposure of field electrodes to the process.

Key CONCEPT The short immersion time of laboratory electrodes reduces errors from chemical attacks, dehydration of the glass of the measurement electrode, and the drift of the liquid junction potential of the reference electrode.

RULE of Thumb *The laboratory and field measurement electrodes should use the same type of glass selected to minimize the effects of process conditions.*

If the titration curve changes with time, separate samples should be gathered over a representative period and titrated individually. The samples should not be combined for titration. The *p*H controller sees individual titration curves, not a composite. This, as well as other common problems that reduce the utility of laboratory titration curves, are listed in Table 3-2a.

Table 3-2a. Common problems with laboratory titration curves.

1.	An insufficient number of data points was generated near the equivalence point.
2.	The starting *p*H (influent *p*H) data points were not plotted for all operating conditions.
3.	The curve does not cover the whole operating range, including control system overshoot.
4.	There is no separate curve zoom-in to show the curvature in the control region.
5.	There is no separate curve for each different split-ranged reagent.
6.	The effect of the sequence of the different split-ranged reagents was not analyzed.
7.	The effect of back mixing different split-ranged reagents was not considered.
8.	The effect of overshoot and oscillation at the split-ranged point was not included.
9.	The sample or reagent solids dissolution time effect on the abscissa was omitted.
10.	The gaseous reagent dissolution time and escape effect on the abscissa were omitted.
11.	The sample volume was not specified.
12.	The sample time was not specified.
13.	The reagent concentration was not specified.
14.	The sample temperature during titration was different than the process temperature.
15.	The influent sample was contaminated by absorbing carbon dioxide from the air.
16.	The influent sample was contaminated by absorbing ions from the glass beaker.
17.	The influent sample composition was altered by evaporation, reaction, or dissolution.
18.	The laboratory and field measurement electrodes had different types of glass.
19.	A composite sample was titrated instead of individual samples.
20.	The laboratory and field reagents used different compounds.

When developing titration curves for split-ranged *p*H control systems, one should take into account the sequence of reagent addition, the degree of back mixing, and the effect of overshoot and oscillation at the point of transition between split-ranged valves. Of course, if the split range consists using of a small and large valve on the same reagent, just one reagent is required in the titration and these special considerations are not applicable.

In an in-line control system such as a static mixer, there is no back mixing, so the history of reagent addition can be ignored in generating the titration curve. If there is a simple split range between an acid and a base reagent because the influent can come in acidic or basic, then a titration curve can be generated based on titrating an acid reagent into a basic influent and a base reagent into an acidic influent.

If a well-mixed vessel is used, the sample can be a mixture of influent and reagents because back mixing is part of the design. The ratio of reagent to influent is still based on the incoming stream before reagent addition, so the same curves described for the in-line mixer are still needed. However, additional curves should be generated in the control region that includes adding one reagent to overshoot the set point to the allowable control error, which is the limit of the control region, and the titration of the other reagent to bring the *p*H back to the set point.

If there are different types of basic or acid reagents, such as might occur because of the preferential use of a lower cost or waste stream as a reagent, then there should be a separate titration curve for each reagent that takes into account the sequence regardless of the degree of back mixing. A curve should be generated with the influent sample titrated with the first reagent in the split-range sequence. Additional curves should be generated by titrating the second reagent into a sample with the first reagent added according to the minimum and the maximum ratio achievable based on the maximum reagent valve capacity and the minimum and maximum influent flow and *p*H. If there is back mixing, then overshoot from one reagent and back titration of the other reagent in the control region should also be explored. Needless to say, these scenarios are rarely addressed properly or even realized, and the chances of getting it right are slim to none unless there is a detailed analysis of what the control system can see in terms of a variety of mixtures of influent and reagents.

Key CONCEPT

Creating samples with the proper influent and reagent mixtures from split-ranged reagents requires understanding the sequence of the split-ranged operation and whether back mixing creates a history of reagent addition.

RULE of Thumb

The preferential use of a low-cost reagent requires that curves be generated for titrating the low-cost reagent into an influent sample and titrating the high-cost reagent into a mixture of the influent and low-cost reagent.

RULE of Thumb

If there is back mixing, then additional curves should be generated in the control region for a mixture of the influent and the overshoot of one reagent and the titration of the other reagent to return the pH to the set point.

Automatic titrators that are electronically equipped to compute the first and second derivative of *p*H with respect to time, and therefore reagent flow, can automatically locate the equivalence point. The first derivative, which is the slope of the titration curve, reaches a maximum at the equivalence points. The second derivative—change in the slope of the titration curve—changes sign at the equivalence points. A plot of the first derivative helps in control system design because it indicates the *p*H process sensitivity. Figure 3-2a shows a plot of the first derivatives for the weak acid and weak base titration curve in Figure 3-1e. For more information on automatic titrators, consult Reference 3.

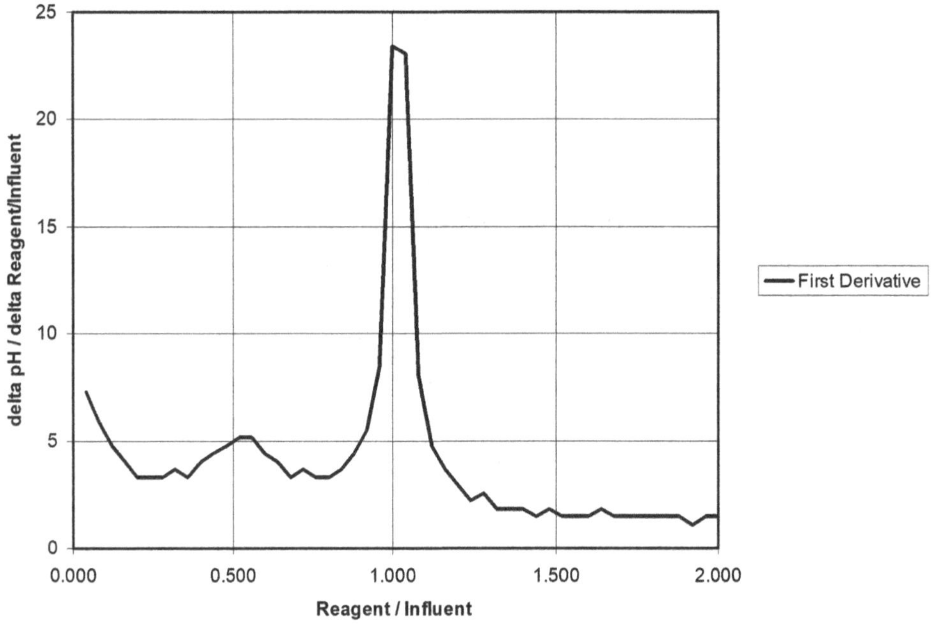

Figure 3-2a. The first derivative of the titration curve indicates the process sensitivity, which reaches a maximum at the equivalence points.

3.3 Computer Generation

The digital twin model described in Appendix M solves the charge balance equation described in Chapter 2, "The Chemistry," and computes the *p*H and ion concentrations for any number of acids and bases, each with up to three dissociation constants. The effect of ionic strength and activity coefficients on the equilibrium constants can be computed externally per Equations H-7a, b, and c. If the types and quantities of acids and bases are not known, or the effect of the activity coefficients is important, an optimization program can be used to adjust the input data, such as acid or base weight fraction or dissociation constant, to minimize the error between it and a laboratory- or an accurately computer-generated titration curve. The main advantage of the dynamic model in Appendix M is its relative computational simplicity, which enables it to be used in dynamic simulation programs without lengthy run times. The interval halving search method is simple and efficient. Although the Fibonacci method has proven to be successful, many other search methods develop instabilities because of the large concentration range covered for *p*H calculations. For more information on the problems with iterative search techniques for *p*H, refer to Reference 4.

RULE of Thumb *An interval halving search to identify the pH that satisfies the charge balance is the most efficient method of computing pH for a complex mixture.*

Dynamic simulation programs must have double-precision instead of single-precision arithmetic to show *p*H control of a strong acid and strong base; otherwise, a round-off error in the concentration calculation can cause an error of about one *p*H unit near the equivalence point. The weight fractions for in-line systems are based simply on the ratios of component mass flows to total flow, but in vessels, the net component mass flows in and out of the volume are integrated to get an accumulation of each component's mass. Variable step integration methods develop problems because of the noise from interval halving and the high sensitivity of the *p*H process. Fixed-step integration methods provide accuracies well within the limits of *p*H measurement and control in the field. Figure 3-3a shows the programming blocks and inputs needed to dynamically simulate a *p*H vessel.

RULE of Thumb *Double-precision math is needed to compute pH for a strong acid or base system.*

RULE of Thumb *Fixed-step integration techniques should be used for dynamic simulation of strong acids or bases.*

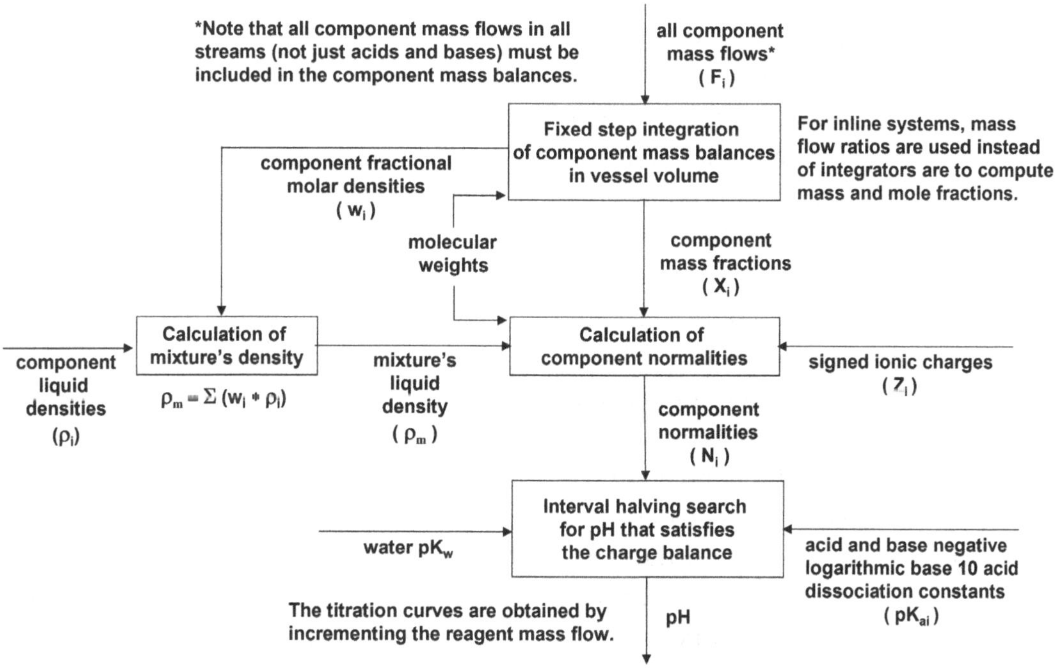

Figure 3-3a. The dynamic simulation of a *p*H vessel requires integrating all component mass balances and an interval halving search for a charge balance.

If a more accurate computer-generated titration curve is needed, the Equilibrium Composition of Electrolyte Solutions (ECES) program from OLI Systems Inc. can be used for steady-state modeling of aqueous electrolyte systems. The program first generates a set of algebraic equations describing the equilibrium conditions from the user-defined chemistry and physical properties. The program then solves the equations simultaneously for the unknown values based on a set of known values supplied by the user. The program may develop convergence problems for particularly difficult systems. While the program has a library of physical properties, such as activity coefficients, the user frequently has to identify or estimate input data for the program. For more information on the capabilities of the ECES program, refer to Reference 5.

The effect of the solution ionic strength on the acid dissociation constants can be estimated for relatively dilute solutions using Equation 2-3q and Reference 6. The pK_a dissociation constant and the A parameter are functions of the process temperature.

3.4 Field Generation

The titration curve can be field generated by measuring and keeping the influent flow constant, slowly ramping the reagent flow to an in-line mixer, and using a data historian to plot the *p*H measurement versus a delayed ratio of reagent to influent flow. The delay time applied to the ratio should equal the sum of the *p*H filter time, electrode lag time, and the process time delay, which is the reagent transportation delay through the mixer. The curve should have the proper shape if the *p*H is sufficiently filtered, the data historian compression and highway exception reporting settings are small enough, and the reagent ramp time is slower than the ratio delay time and the reagent controller's integral (reset) time that should be significantly slower than the the 63% response time of the reagent control valve. In general, if the ramp time is 1 min or larger, the reagent flow controller should be able to keep up with the ramping set point, and the distortion of the *p*H from the *p*H filter time and sensor lag time should be minimal. An in-line system is preferred because there is negligible back mixing and none of the distortion that occurs in a well-mixed vessel where there is a prior history of reagent addition and a large process time constant, which is approximately the vessel residence time. A vessel could be titrated in the fed-batch mode, where the discharge flow is shut off and the feed continues, but the ramp time must be much slower and the amount of off-spec material created from the test is much larger than would be the case for an in-line mixer. The process delay is now the sum of a reagent transportation delay and the vessel turnover time. Conventional dip tube design causes a huge reagent delivery delay. Special methods of reagent injection detailed in Chapter 5, "Mixing Equipment," must be employed or else the time shift will make a field titration curve impossible.

RULE of Thumb *The x-axis ratio should be dimensionless and should be delayed by a time equal to the sum of the process delay, electrode lag time, and* p***H*** *filter time.*

If the influent flow is shut off to the vessel so that it is in the full batch mode, and its level, weight, or initial charge can be accurately measured, a technique similar to that used in a laboratory titration can be mimicked in the field. The vessel acts as a huge beaker. The ratio is now the reagent flow divided by the initial vessel contents. Of course, the ratio should be dimensionless, so if the reagent has a mass flowmeter, the initial vessel contents should be expressed as a weight. Load cells or, even better, a mass flow totalizer for the initial charge provides the best accuracy. The tank level should be low to minimize the reagent required and the amount of off-spec material created. To show the shape of the curve for overshoot, the ramp must cause the *p***H** to go beyond the set point.

Chapters 6 through 8 illustrate the advantages of in-line systems for *p***H** control in terms of the speed of start-up and tuning and the capital investment. The chapters also show that they have an advantage for the automated online generation of titration curves. The upset created from a test to generate the titration curve via an in-line system is usually attenuated enough by a downstream volume so that the effluent or product stays within the desired control region. However, if such tests are not allowed, an in-line system can still facilitate online identification while the controller is in automatic. Data points are gathered and matched with a delayed reagent to influent flow ratio. For larger upsets and start-up, a significant portion of the titration curve can be updated. For tight control and normal operation, the region is naturally small and a wider view is not as important. Automatically updating the curve is especially useful for tuning and the signal linearization of waste treatment systems where there is a large variability in the composition of the influent and, consequently, the shape of the titration curve. Figure 3-4a shows the setup of an in-line system for automated field generation of titration curves.

Key CONCEPT An in-line system can facilitate the field generation of a titration curve by using preprogrammed tests or online identification during upsets and start-up.

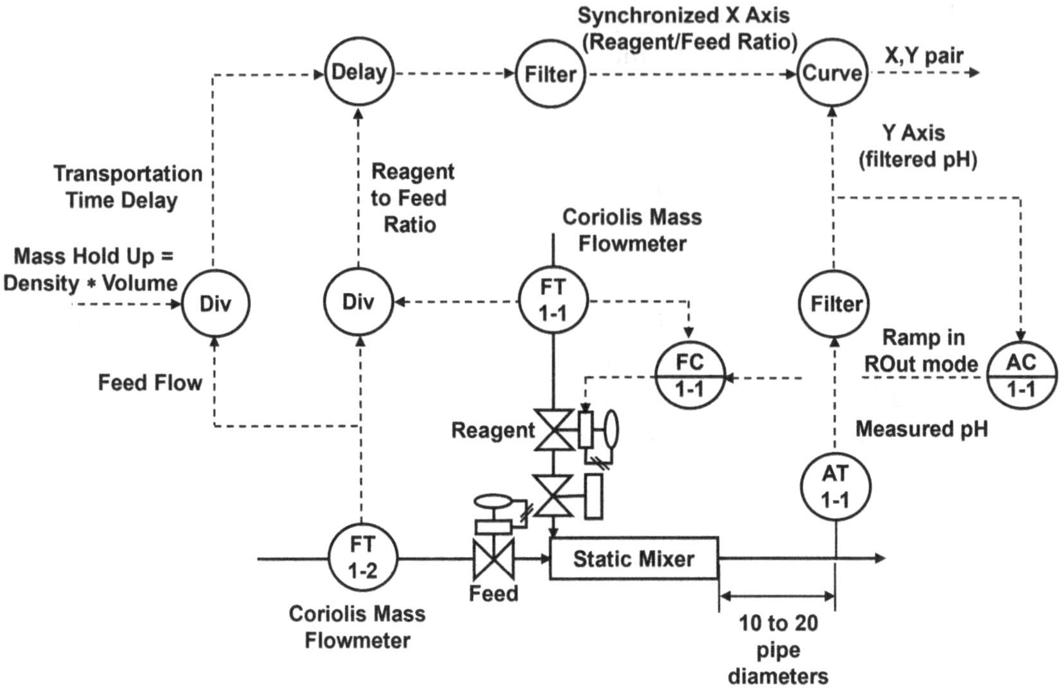

Figure 3-4a. An in-line pH control system with a short residence time and plug flow facilitates a relatively fast automated field generation of the titration curve.

3.5 Buffering

Some titration curves have a long region of relatively flat slope. Adding reagent in this portion of the titration curve has little effect on the *p***H**. The flatness of the curve can be due to buffering, which occurs in mixtures containing a weak acid or a weak base and its salt. It is also important to keep solutions used for calibration at a known *p***H** despite contamination from residue on the electrode. Some more commonly encountered buffer systems are acetic acid–acetate, carbonic acid–bicarbonates, and citric acid–citrates.

Equations 3-5a through 3-5f show the source of the buffering effect. Equation 3-5a is the dissociation of an acetic acid–acetate system solved for the hydrogen ion concentration. Taking the negative base 10 logarithm of both sides of this equation as shown in Equation 3-5b, we can convert from K_a to pK_a and from $[H^+]$ to *p***H** to yield Equation 3-5c (a unity activity coefficient is assumed here). In the buffered region, the concentrations of acetate ions $[Ac^-]$ and the unionized acetic acid $[HAc]$ are very large with respect to the hydrogen ion $[H^+]$ concentration. If hydrochloric acid residue from cleaning and rejuvenating the glass electrode gets into the solution, the large concentration of acetate ions quickly converts the extra hydrogen ions to unionized acetic acid. If the contaminant is a base such as sodium hydroxide, the extra hydroxyl ions neutralize hydrogen ions, but these hydrogen ions are quickly replaced by the ionization of acetic acid. The percentage change in the unionized acetic acid and acetate ions is small because the magnitude of these concentrations is large. The ratio of acetic acid to acetate concentration is relatively constant, so the hydrogen ion concentration and $p\mathbf{H}$ stays constant if the temperature and, thus, the pK_a are constant. In buffer solutions, the ratio of the weak acid to its salt or the weak base to its salt is nearly unity so the $p\mathbf{H}$ is approximately equal to the pK_a, which is the point of maximum buffering. Buffer solutions are used to check the accuracy of $p\mathbf{H}$ electrodes because the $p\mathbf{H}$ of the test solution remains constant even after moderate contamination. The buffer value is a measure of the buffer solution's ability to resist $p\mathbf{H}$ change and is defined as the gram-moles/liter of a strong acid or base needed to change the $p\mathbf{H}$ by one unit.

Solving the dissociation equation for the hydrogen ion concentration, we have:

$$[H^+] = \frac{[HAC]}{[AC^-]} * K_a \tag{3-5a}$$

Taking the negative base 10 logarithm of both sides provides:

$$-\log[H^+] = -\log\frac{[HAC]}{[AC^-]} - \log[K_a] \tag{3-5b}$$

Substituting in the definition of $p\mathbf{H}$ and pK_a and inverting the concentration ratio gives:

$$p\mathbf{H} = \log\frac{[AC^-]}{[HAC]} + pK_a \tag{3-5c}$$

For acid buffer solutions, this corresponds to:

$$p\mathbf{H} = \log\frac{C_{salt}}{C_{acid}} + pK_a \tag{3-5d}$$

For base buffer solutions, it can be shown:

$$p\mathbf{H} = \log\frac{C_{base}}{C_{salt}} + pK_w - pK_b \tag{3-5e}$$

A conversion from pK_b to pK_a per Equation 2-31 yields:

$$p\mathbf{H} = \log\frac{C_{base}}{C_{salt}} + pK_a \tag{3-5f}$$

where:

Ac^- = acetate salt ion concentration (normality)
HAc = unionized acetic acid concentration (normality)

H^+ = hydrogen ion concentration (normality)
K_a = acid dissociation constant
C_{acid} = molar concentration of the weak acid (gm-moles/liter)
C_{base} = molar concentration of the weak base (gm-moles/liter)
C_{salt} = molar concentration of the salt ion (gm-moles/liter)
pK_a = negative base 10 logarithm of the acid dissociation constant
pK_b = negative base 10 logarithm of the base dissociation constant
pK_w = negative base 10 logarithm of the water dissociation constant

3.6 Uses

The titration curve is the most important piece of information for designing, commissioning, and troubleshooting *p*H control systems. The uses listed in Table 3-6a will be discussed in subsequent chapters.

Table 3-6a. Titration curves have many important uses for a successful application.

1.	Assessing application difficulty
2.	Estimating the steady-state process gain
3.	Selecting the best set point for each stage
4.	Approximating the nonlinearity effect on the mixing time constant
5.	Including nonlinearity effects on control loop errors
6.	Linearizing a controlled variable by signal characterization
7.	Calculating the continuous *p*H feedforward signal
8.	Specifying the ratio factor for flow feedforward

3.7 Best Practices

Developing accurate titration curves is the key first step for nearly every aspect of a successful *p*H application. Unfortunately, getting a titration curve that shows with sufficient accuracy the slopes across the entire range of process operating conditions with enough data points in the control region, often near the neutral point with the steepest slope, is rare. Commonly given is a skeleton of a curve. The resulting lack of knowledge of the process nonlinearity and sensitivity creates a lot of dead ends in the design. The result is missing and misleading details on equipment, piping, final control elements, electrodes, ontrol strategies and the controller tuning needed to achieve the desired *p*H loop performance. The following list is a concise guide to how to get full-bodied curves with sufficient features and resolution to avoid dead ends. See Reference 7 for more details on the creation, value, and use of titration curves.

1. Get a series of lab titration curves with sufficient resolution using reagents and process samples that are identical to what are used in the control system for the complete range of process operating conditions.
2. Use lab titration curves to estimate the slope at the set point and the change in slope throughout the operating *p***H** range to gain knowledge of process gain and its effect on valve precision and rangeability, measurement accuracy, and the number of stages required.
3. Use lab titration curves to prototype a piecewise linear fit of the curve in a digital twin for signal characterization of the controlled variable and set point to improve performance. Determine whether the ability to remove most of the process nonlinearity by transitioning from *p***H** to reagent demand control (described in Chapter 8) provides significantly better *p***H** system design, tuning, and control.
4. Use lab titration curves to develop and improve dynamic models in a digital twin, described in Chapter 9, "Digital Twin," for more intelligent and effective total system design, implementation, testing, and continuous improvement and ongoing training of operators, engineers, and technicians.

References

[1] Taylor Chemicals Inc., *Modern pH and Chlorine Control* (1945): 83–86.

[2] F. Greg Shinskey, *pH and pION Control in Process and Waste Streams* (New York: John Wiley & Sons, 1973): 69–71.

[3] Douglas A. Skoog and Donald M. West, *Principles of Instrumental Analysis,* 2nd ed. (Saunders College, 1980): 573–576.

[4] J. D. Richter, et al., "Waste Neutralization Control – Digital Simulation Spots Nonlinearities," *Instrumentation Technology* 21, no. 4 (April 1974): 35–40.

[5] Steven J. Sanders, "Case Studies in Modeling Aqueous Electrolyte Solutions," *Computers & Chemical Engineering* 9, issue 3 (1985): 223–244.

[6] James E. Kipp, "Computer Series 158 – PHCALC: A Computer Program for Acid/Base Equilibrium Calculations," *Journal of Chemical Education* 71, no. 2 (February 1994): 119–121.

[7] Christopher Stuart, "Improve pH Control," *Chemical Processing* (March 2016).

4 Electrodes

4.1 A Dose of Reality

The incredible sensitivity and rangeability of the *p*H measurement are based on a series of assumptions, most of which are not sufficiently explained in the literature. Manufacturers have been doing a better job of noting some considerations, especially as they develop new electrode features that can address a particular application problem. This section presents equations that reveal underlying assumptions to help the user cut through the hype. With *p*H electrodes more than any other sensor, there is a need to "keep it real."

The accuracy and speed of response of the *p*H measurement depend on the condition of a hydrated gel layer on the surface of the glass electrode that is only 10^{-4} mm thick and the condition of a junction of the reference electrode illustrated in Figure 4-1a for a combination electrode that provides a physical contact between its internal electrolyte and the process solution. The need for the fragile gel layer and the process junction of the reference electrode to be structurally consistent, clean, and free from contamination has profound implications for properly selecting, calibrating, and maintaining *p*H measurements. The fact that *p*H sensors perform as well as they do despite exposure to a wide variety of chemicals, temperatures, and pressures is truly remarkable and a testimony to the ingenuity of various electrode designs.

> Key CONCEPT
>
> The accuracy and speed of response of *p***H** measurements stated in the literature assume that the thin, fragile gel layer of a glass electrode and the porous process junction of the reference electrode have had no penetration or adhesion (e.g., coating) of the process fluid and are in perfect condition at laboratory condition.

Fundamentals

Each *p***H** measurement and reference electrode has an internal galvanic half-cell. The measurement and reference electrode half-cells typically consist of a silver wire with a silver chloride coating at their end immersed in a solution with chloride ions, as shown in Figures 1-4a and 1-4c in Chapter 1, "The Essentials." The reference electrode fill is usually potassium chloride for aqueous processes, whereas the measurement electrode solution is usually a chloride buffer with a hydrogen activity equivalent to 7 *p***H**.

The actual sensing of *p***H** is accomplished by having a *p***H**-sensitive glass in contact with the internal fill, a 7 *p***H** buffer, and the external solution. The *p***H**-sensitive glass develops potentials per Equations 4-1a and 4-1d, which are Nernst equations, by the hydrogen ion (proton) exchange between hydronium ions in aqueous solutions and in the hydrated gel layer of the glass. The migration of hydronium ions through the boundary layer that surrounds the surface takes several seconds even under the best of conditions, but once the hydronium ions are in very close proximity to the surface, protons jump from the hydronium ions and become associated with water molecules on the surface. There is no migration of hydronium ions intact into the surface but rather a proton transfer that occurs for aqueous and nonaqueous solutions [1, 2, 3]. As protons move into one surface from a greater dissociation of hydrogen ions, they move out of the surface on the other side of the glass. Lithium and sodium ions in the dry interior of the glass provide electrical continuity between the internal and external hydrated gel layers. Thus, current flow through the glass consists of the conduction of single charged positive ions (cations). The potential developed is proportional to the difference in logarithms of the activity of hydronium ions in the solution and in the gel layer on each side of the glass membrane. If the gel layers have an equal number of sites for proton exchange, the K_{g1} and K_{g2} constants will be equal and Equations 4-1c and 4-1e can be combined to yield Equation 4-1f. If all the original lithium or sodium ions at these sites in the gel are also replaced by protons, activities a_{g1} and a_{g2} are equal, which yields Equation 4-1g [3]. Using the definition of *p***H**, the logarithms of hydrogen activity can be converted to *p***H**, which yields Equation 4-1h, where the difference in potentials is proportional to the difference in *p***H**. Also, because the internal fill has a hydrogen activity that corresponds to 7 *p***H**, the potential difference becomes Equation 4-1i, the equation typically shown in the literature. Examining this simplified equation for the ideal *p***H** measurement electrode yields the following conclusions:

- At 25°C, the output decreases from +480 millivolts (mV) at 0 *p***H** to –480 mV at 14 *p***H**.
- The output is 0 mV at 7 *p***H**.
  - There is no effect of temperature changes on the electrode output at 7 *p***H**.
- At 25°C, the output changes by 59 mV per *p***H** unit.

The general Nernst equation:

$$E_1 = E_0 + \frac{R \bullet (T + 273.16)}{F} \tag{4-1a}$$

For the exterior surface of the *p***H** glass electrode:

$$E_1 = K_{g1} + 0.1984 \bullet (T + 273.16) \bullet \log\left[\frac{a_1}{a_{g1}}\right] \tag{4-1b}$$

$$E_1 = K_{g1} + 0.1984 \bullet (T + 273.16) \bullet [\log(a_1) - \log(a_{g1})] \tag{4-1c}$$

For the interior surface of the *p***H** glass electrode:

$$E_2 = K_{g2} + 0.1984 \bullet (T + 273.16) \bullet \log\left[\frac{a_2}{a_{g2}}\right] \tag{4-1d}$$

$$E_2 = K_{g2} + 0.1984 \bullet (T + 273.16) \bullet [\log(a_2) - \log(a_{g2})] \tag{4-1e}$$

If $K_{g1} = K_{g2}$, then:

$$E_1 - E_2 = 0.1984 \bullet (T + 273.16) \bullet [\log(a_1) - \log(a_{g1}) + \log(a_{g2}) - \log(a_2)] \tag{4-1f}$$

If also $a_{g1} = a_{g2}$, then:

$$E_1 - E_2 = 0.1984 \bullet (T + 273.16) \bullet [\log(a_1) - \log(a_2)] \tag{4-1g}$$

Because $p\mathbf{H}_1 = -\log(a_1)$ for an external process and $p\mathbf{H}_2 = -\log(a_2)$ for an internal electrolyte:

$$E_1 - E_2 = 0.1984 \bullet (T + 273.16) \bullet (p\mathbf{H}_2 - p\mathbf{H}_1) \tag{4-1h}$$

Because the internal buffer of the glass electrode is 7 *p***H**:

$$E_1 - E_2 = 0.1984 \bullet (T + 273.16) \bullet (7 - p\mathbf{H}_1) \tag{4-1i}$$

where

a_1 = activity of hydrogen ions in external process fluid (normality)
a_2 = activity of hydrogen ions in internal fill fluid (normality)
a_{g1} = activity of hydrogen ions in the outer gel surface layer (normality)
a_{g2} = activity of hydrogen ions in the inner gel surface layer (normality)
E_o = potential developed at standard conditions (mV)
E_1 = potential developed at external glass surface (mV)
E_2 = potential developed at the internal glass surface (mV)
F = Faraday constant with units conversion (Joules per mV and moles)
K_{g1} = constant for the potential for outer gel surfaces layer (mV)
K_{g2} = constant for the potential for inner gel surfaces layer (mV)
$p\mathbf{H}_1$ = *p***H** of the external solution
$p\mathbf{H}_2$ = *p***H** of the internal solution (typically 7 pH)
R = universal gas constant (Joules per °K and moles)
T = solution temperature (°C)

Key CONCEPT The equation for calibrating the glass electrode assumes the point of 0 temperature effect and 0 millivolt input (isopotential point) is 7 *p***H**.

Water molecules in the glass surface are essential for the proton transfer to occur. Nonhygroscopic glasses such as Pyrex and quartz show no *p***H** response. Glasses that can absorb water but have lost their gel layer from contact with nonaqueous and highly acidic or basic solutions or from long-term exposure to air will lose their *p***H** response. It takes about 2 hours of immersion in water to replenish a completely dehydrated gel layer.

Key CONCEPT The time that glass electrodes are left dry or exposed to high and low *p***H** solutions must be minimized to maximize the life of the hydrated gel layer.

Instead of a sharp transition between the gel and dry layers, there is a continuous decrease in the number of hydronium ions and an increase in the number of lithium or sodium ions from the surface to the interior of the glass. As the electrode ages, the outside surface is stripped away, and the gel layer penetrates deeper into the glass. Eventually, there is a complete breakdown of the silicon oxygen network that forms the glass matrix [1]. For ideal conditions (25°C and zero ionic strength solutions between 6 and 8 *p***H**), a satisfactory glass response should last 2 to 3 years. Most of the variation in life expectancy for these conditions is associated with the manufacturing tolerances, with the greatest variability shown for manual glassblowing and assembly. Chemical attacks from strong alkaline and acidic solutions, high temperatures, and nonaqueous solutions greatly accelerate corroding and depleting the gel layer. The decrease in the electrode efficiency (slope of the millivolt versus *p***H** plot) during the aging process is usually quite moderate (<5%) until a complete loss of the response (zero efficiency and slope) occurs.

Accuracy and Precision

The accuracy and precision shown in Appendix A, Figure A-11a, attainable with *p***H** electrodes in a commercial laboratory environment under typical conditions, are not overly impressive. From an accuracy standpoint, the primary concern is drift or bias developed over time. When *p***H** electrodes are calibrated, the calibration line is characterized by a slope, noted as an efficiency, and an asymmetry potential, noted as an offset. The asymmetry potential is defined as the potential difference across the membrane when the external solution is at *p***H** 7 and 25°C. While the slope represents the *p***H** responsiveness of the glass, the asymmetry potential represents all the remaining deviations from ideality in both the measurement and reference electrodes. The drift in asymmetry potential for a standard set of electrodes sitting in a buffer solution under ideal conditions in an environmentally controlled laboratory is about 0.002 *p***H** units per day [4]. Tests for drift should be done with acidic buffers because the actual *p***H** of basic buffers exposed to air would continually drop from absorpting carbon dioxide. In practice, it is difficult to sort out the electrode drift from changes in the solution *p***H**.

In regard to precision, the short-term repeatability for the constant operating and equipment conditions of a standard set of electrodes in mild process solutions is about ± 0.01 *p***H**. Reproducibility (different conditions or equipment) is usually in the range of 0.03 to 0.04 *p***H** [4]. Repeatability is the spread in readings for repeated tests in the same solution when a *p***H** is approached from the same direction, whereas reproducibility is the spread in readings at any given *p***H** regardless of time, direction, and solution, and it includes drift.

Key CONCEPT Most accuracy statements and tests are for short-term exposure.

Installation effects, such as variable glass surface conditions, temperatures, ionic strengths, dissociation constants, streaming potentials, concentration gradients, and diffusion or liquid junction potentials, cause the *p*H measurement error in process applications to be an order of magnitude larger than the normally stated accuracy, which is really for electrodes sitting in buffer solutions at laboratory conditions [5]. Under the best of industrial conditions (e.g., dilute, low ionic strength solutions of weak acids and weak bases between 4 and 10 *p*H at constant temperature), the long-term accuracy, even with weekly calibration checks (average bias from a lab over a month of operation), is rarely better than ± 0.05 *p*H.

Key CONCEPT The long-term error in *p*H measurements installed in a process is an order of magnitude greater than the error normally stated in the literature.

Whereas the measured and controlled variable is *p*H, the actual process variable is hydrogen ion activity, which in turn is an indication of reagent concentration. Thus, the effect of mixing equipment and the reagent valve is first on the abscissa (x-axis) of the titration curve. The slope of the curve at operating conditions gives the sensitivity of the *p*H measurement as an inference of process concentrations. For *p*H system analysis, the abscissa is expressed in terms of a ratio of reagent to feed mass flows for continuous or fed-batch operation and reagent to vessel mass (reagent weight fraction) for traditional sequential batch operation. For the following discussion, we will simplify this to the percentage of reagent flow, similar to the simplification in Chapter 8, "Control System," for linear reagent demand control. To understand what is truly happening in a *p*H system, one must pay attention to the slope of the titration curve at operating conditions. This is particularly important in terms of limit cycles from stick-slip. The steeper slope, usually near the neutrality point (7 *p*H), amplifies the resolution limit, as shown in Figure 8-8a, for a system moderated by a weak acid and weak base. The amplification is extraordinary if there is no weak acid or weak base or conjugate salt to provide moderation. For example, the slope at 5 and 7 *p*H may well be 0.05 and 5 *p*H per percent of reagent valve capacity, respectively, for the final stage of a strong acid and strong base system. This means a 0.2% reagent valve stick-slip oscillation translates to a 0.01 and 1 *p*H amplitude oscillation at the set points of 5 and 7 *p*H, respectively. Conversely, a 0.1 *p*H electrode error corresponds to a reagent delivery error of 2% and 0.02% (concentration error) at the set points of 5 and 7 *p*H, respectively. Furthermore, for a mixing uniformity of 99.9% (standard deviation of 0.1%), the *p*H fluctuation would be 0.005 and 0.5 *p*H at the set points of 5 and 7 *p*H, respectively.

Measurements on the steep slopes of titration curves will show extreme sensitivity to concentration fluctuations. Vessels that are well mixed by conventional definitions can introduce 0.1 *p*H or larger amplitude noise into the measurement. Signal filtering is effective at filtering the high-frequency noise in pipelines but not the low-frequency nonuniformity in vessels. For strong acid and strong base systems, the steepness of the curve near the neutral point results in a reagent delivery error from the measurement error, even with noise that is generally less than the reagent delivery error from stick-slip.

Studies on mixing requirements are typically done on buffered systems or on the tails of the titration curve where the slope is relatively flat. Here, concentration gradients contribute much less to the variability in the measurement, so these measurements appear to be more accurate. Although the measurement error looks better on a flat curve because of less noise, the corresponding reagent delivery error is actually worse. The reagent delivery for any measurement error is larger, and the susceptibility of the glass gel layer and reference process junction to penetration from high concentrations is greater. The reagent delivery error from the measurement error is larger than the reagent delivery error from a stick-slip limit cycle that occurs for all proportional-integral or proportional-integral-derivative (PID) controllers in automatic. The fact that the measurement error is larger than the control error means that using Coriolis flowmeters and mass ratio control might be better than *p*H control in reactors because the concentration of reactants has tight specifications. For neutralization systems and feeds with variable concentrations, *p*H control remains the best solution.

Key CONCEPT The *p*H measurement error may look smaller on the flatter portion of a titration curve, but the associated reagent delivery error is larger.

Optimistic users and suppliers who provide one electrode per point in the process may believe they have achieved an accuracy of 0.02 *p*H or better. Results of installations with three electrodes per point in the process show that the error band from mixing noise and reference electrode drift over a month is, at best, 0.02 to 0.1 *p*H for a flat titration curve and 0.2 to 0.5 *p*H for a steep titration curve. Tests of 12 different electrodes in a recirculation line of a laboratory tank showed that right after buffer calibration, readings differed by more than 0.5 *p*H for water and were reduced to about 0.25 *p*H when buffer solutions were added to the tank. The author's plant experience has been that after a buffer calibration, three healthy electrodes will differ by 0.2 *p*H, where an electrode that is higher than the others may be lower within hours. This confirms some of the many advantages of middle signal selection (MSS) in terms of reducing the effect of these fluctuations and providing much more intelligence for calibration and diagnostics. If all three electrodes continually agree within 0.01 *p*H for more than a few hours in an industrial application, all three are coated, broken, or still have protective caps [5, 6].

The maintenance practice of removing and buffering electrodes is costly and often counterproductive because damage to the glass electrode's fragile gel surface and the upset to the reference electrode's thermal and ionic equilibrium can reduce measurement accuracy.It is possible to improve *p*H electrode performance and safety during *p*H maintenance, and reduce the cost of *p*H maintenance by an order of magnitude, through more realistic expectations and a better calibration policy that uses process standardization and relaxed tolerances (described in Section 4.7, "Dynamic Response of Electrodes, Holders, and Sample Systems").

> **Key CONCEPT** The cost of *p*H measurement maintenance can be reduced by a factor of 10 by having more realistic expectations and calibration policies.

Equation 4-1i shows that the potential difference will change at any *p*H other than 7 if the solution temperature changes. The magnitude of the *p*H error depends on the magnitude of the temperature error and the deviation of the *p*H reading from 7 *p*H. The error size is usually relatively small because the change in the process temperature is small compared to the 273.16 in Equation 4-1i used to convert to degrees Kelvin and because the set point of most *p*H systems is near the neutral point (7 *p*H).

The actual *p*H of the solution changes in response to the change in the ion dissociation constants with temperature changes. Until recently, compensation of only the change in mV generated per *p*H unit, or in other words, electrode temperature compensation, was offered. Microprocessor-based transmitters and receivers now offer solution temperature compensation. However, the relationship depends on the composition and operating conditions of the process stream. For strong base streams above 8 *p*H, the change is often about –0.3 *p*H per 10°C. Samples that cool down before they are measured in the lab will indicate a higher *p*H than the electrodes installed in the process.

The potential of interest in *p*H measurement is the difference between the potential developed at the outer and inner glass surfaces of the measurement electrode as defined by Equation 4-1g. All other potential represents an error. Figure 4-1a shows the physical location of the potential ($E_1 - E_2$) and many extraneous potentials for a combination double-junction electrode, which has a reference electrode as a concentric ring around the glass measurement electrode. The combination electrode is popular because it has fewer installation and spare parts requirements. The internal salt bridge

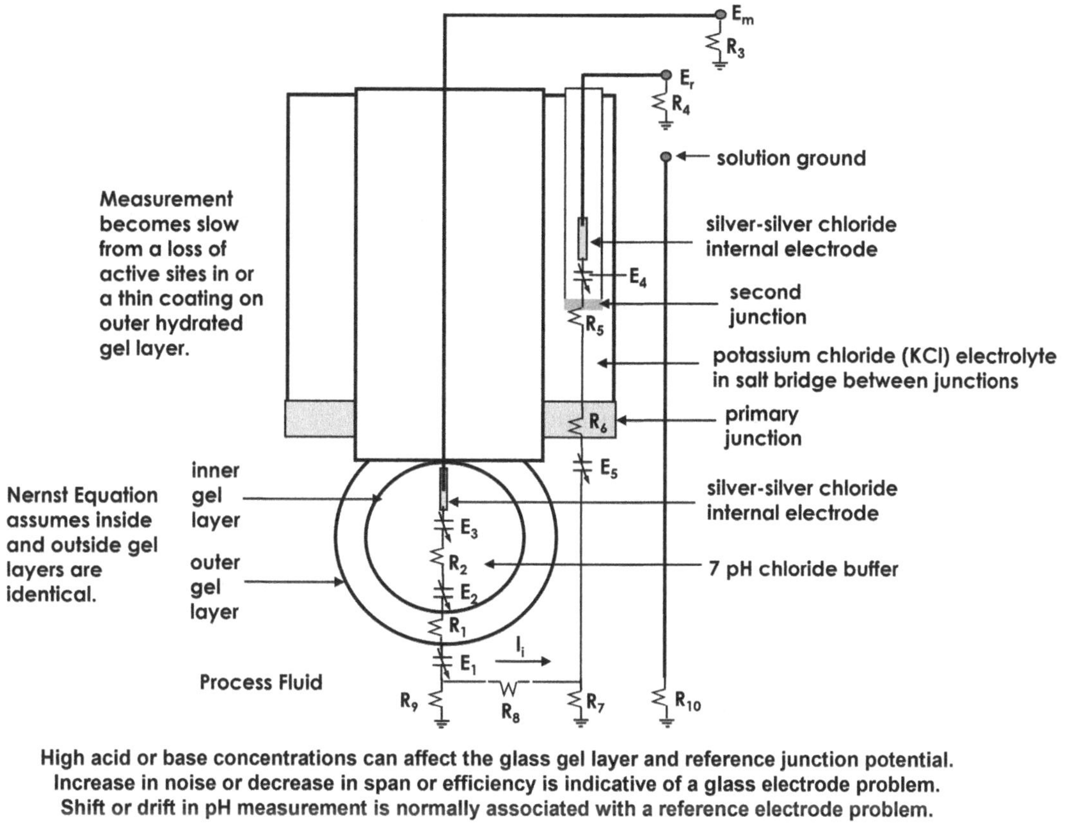

Figure 4-1a. Equivalent electrical schematic of a double-junction combination electrode.

between the outer and inner junction of the double junction increases the time it takes for process ions to migrate to the inner chamber where the internal silver–silver chloride electrode is located.

The following equation for the input potential (E_i) seen by $p\mathbf{H}$ transmitters reveals many of the more notable extraneous potentials that cause errors in the measurements:

$$E_i = E_1 - E_2 - E_3 + E_4 + E_5 - I_i \bullet (R_1 + R_2 + R_5 + R_6 + R_8) \tag{4-1j}$$

Normalizing Equation 4-1i for 25°C so that the slope is 59.1 mV per $p\mathbf{H}$ (0.1984 • 298):

$$E_1 - E_2 = 59 \bullet \frac{(T + 273)}{298} \bullet (7.0 - p\mathbf{H}_1) \tag{4-1k}$$

Lumping all the extraneous voltage offsets into one bias voltage:

$$E_{bias} = -E_3 + E_4 + E_5 - I_i \bullet (R_1 + R_2 + R_5 + R_6 + R_8) \tag{4-1l}$$

Substituting (4-1k) and (4-1l) into (4-1j), and solving for $p\mathbf{H}$:

$$p\mathbf{H}_1 = 7.0 - \frac{E_i - E_{bias}}{[59.1 \bullet (T + 273) / 298]} \tag{4-1m}$$

Using digital meter calibration factors with Nernst temperature compensation:

$$p\mathbf{H}_1 = p\mathbf{H}_{iso} - \frac{E_i - E_{std}}{[(S_{eff} / 100\%) \bullet S_{slope} \bullet (T + 273) / 298]} \tag{4-1n}$$

Adding solution temperature compensation:

$$p\mathbf{H}_{25^\circ C} = p\mathbf{H}_1 + K_1 \bullet (T - T_{ref}) + K_2 \bullet (T - T_{ref})^2 \tag{4-1o}$$

Adding digital transmitter calibration factors to transmit a controlled variable in percent:

$$CV_{\%} = \frac{p\mathbf{H}_{25^\circ C} - p\mathbf{H}_{0\%}}{p\mathbf{H}_{100\%} - p\mathbf{H}_{0\%}} \bullet 100\% \tag{4-1p}$$

Some digital transmitters have a $p\mathbf{H}_{span}$ adjustment defined as follows:

$$p\mathbf{H}_{span} = p\mathbf{H}_{100\%} - p\mathbf{H}_{0\%} \tag{4-1q}$$

where

$CV_{\%}$ = controlled variable for $p\mathbf{H}$ controller in percentage of scale (%)
E_i = potential that is the input seen by the $p\mathbf{H}$ transmitter (mV)
E_1 = potential developed at the external glass surface (mV)
E_2 = potential developed at the internal glass surface (mV)
E_3 = half-cell potential at the measurement internal electrode (mV)
E_4 = half-cell potential at the reference internal electrode (mV)
E_5 = liquid junction (diffusion) potential of the reference electrode (mV)
E_{bias} = asymmetry and offset voltages lumped as one bias voltage (mV)
E_{std} = electrode standardization potential (mV)
K_1 = solution temperature compensation first-order correction factor ($p\mathbf{H}$ per °C)

K_2 = solution temperature compensation second-order correction factor (p**H** per °C)

I_i = input leakage current of the preamplifier (milliamps)

$p\mathbf{H}_{iso}$ = isopotential point (point of zero temperature effect) (p**H**)

$p\mathbf{H}_1$ = p**H** at the electrode tip (p**H**)

$p\mathbf{H}_{25°C}$ = p**H** referenced back to p**H** at 25°C (p**H**)

$p\mathbf{H}_{span}$ = p**H** span for calibrating the digital transmitter's output (p**H**)

$p\mathbf{H}_{0\%}$ = p**H** at 0% of scale for calibrating the digital transmitter's output (p**H**)

$p\mathbf{H}_{100\%}$ = p**H** at 100% of scale for calibrating the digital transmitter's output (p**H**)

R_1 = measurement electrode glass resistance (Ω)

R_2 = measurement electrode internal fill resistance (Ω)

R_3 = measurement electrode cable insulation and connection resistance (Ω)

R_4 = reference electrode cable insulation and connection resistance (Ω)

R_5 = reference electrode internal fill resistance (Ω)

R_6 = reference electrode liquid junction resistance (Ω)

R_7 = resistance to ground at the measurement electrode tip (Ω)

R_8 = solution resistance between the measurement and reference electrodes (Ω)

R_9 = resistance to ground at the reference electrode tip (Figure 4-1a) (Ω)

R_{10} = solution ground resistance (Figure 4-1a) (Ω)

R_{11} = shunt resistance that causes a loss in electrode efficiency (Figure 4-1a) (Ω)

S_{eff} = electrode efficiency (%)

S_{slope} = electrode slope at 25°C (mV per p**H**) (e.g., 59.1 mV per pH)

T = temperature at the measurement electrode tip (°C)

T_{ref} = temperature reference point for the solution temperature compensation (°C)

Equation 4-1j shows that the effect of these potentials is additive. Whereas errors from changes in the parameters in Equations 4-1b and 4-1d result in horizontal shifts of the isopotential point, as shown in Figure 4-1b, the extraneous potentials in Equation 4-1j result in a vertical shift of the isopotential point, as shown in Figure 4-1c. Usually, there is no attempt to distinguish and compensate for a shift in the isopotential point that is horizontal from changes in the condition of the gel layer versus a shift that is vertical from changes in the path resistance and condition of the reference electrode.

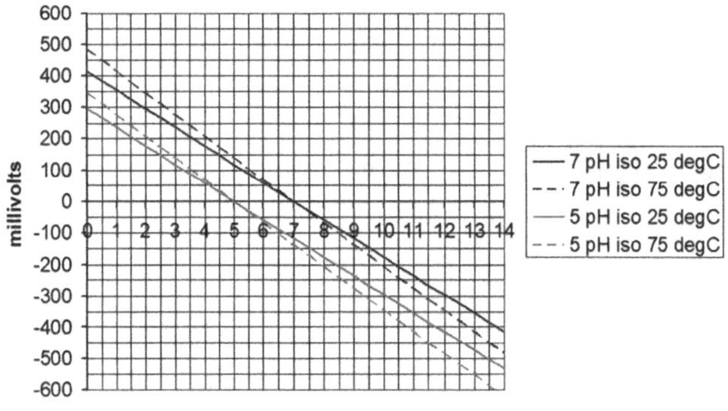

Figure 4-1b. Changes in the glass measurement electrode cause a horizontal shift of the isopotential point.

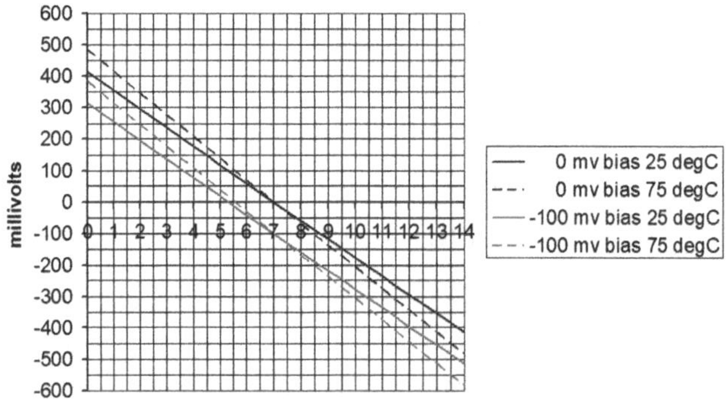

Figure 4-1c. Changes in the reference electrode and path resistance cause a vertical shift of the isopotential point.

Key CONCEPT **Horizontal shifts of the isopotential point from changes in the activity and composition of the gel layer are often unidentified or uncorrected.**

The measurement and reference electrode half-cell potentials in Equation 4-1j, which originate from an electrochemical reaction between the internal electrode and fill, are of opposite sign and ideally should be equal so their sum is zero. However, the half-cell potentials depend on the type of internal electrode, the internal fill concentration, and the electrode temperature. If the electrode type and fill are identical, the change in half-cell potential with temperature will cancel out unless a temperature gradient exists between the reference and measurement electrode locations. The measurement electrode normally has a silver–silver chloride internal electrode in a chloride buffer. The most common type of reference electrode also has a silver–silver chloride internal electrode and a potassium chloride electrolyte. The half-cell potential for a saturated silver chloride electrode varies with temperature from 193 mV at 40°C to 237 mV at 0°C.

Resistances

Some of the resistances in Equation 4-1j are large. Fortunately, the input leakage current (I_i) that flows through these resistances and the amplifier input is extremely small (about 1 picoamp or one trillionth of a milliamp). The current flows from the positive measurement electrode terminal to the negative reference electrode terminal, so the sign of the potential drop is negative compared to the measurement electrode's outer potential and causes the isopotential point to shift down and the *p***H** measurement to go upscale.

The measurement electrode glass bulb resistance is usually the largest resistance. It ranges from about 20 MΩ to 500 MΩ at 25°C, depending on the type and thickness of the glass. Rugged glasses, which are thicker than normal for greater resistance to breakage and abrasion and for a longer life in terms of time to failure of the silicon oxide matrix from penetration of the gel layer, often have a lower resistance per unit thickness so the increase in total resistance is moderated. The resistance dramatically increases at low temperatures and can exceed 1000 MΩ for the type of glass used that is usually reserved for concentrated acidic solutions at high temperatures. Because a greater electrical resistance per unit thickness typically translates to a greater resistance to etching and high temperatures, rugged electrodes generally are not suitable for hot alkaline solutions. However, advances are being made in glass formulations that increase the range of permissible operating conditions. Equation 4-1qq quantifies the exponential increase in resistance with temperature.

$$R_{m1} = R_{m0} \bullet 10^{-K_{exp} \bullet (T_1 - T_0)} \tag{4-1qq}$$

where

K_{exp} = exponential factor for glass type (e.g., 0.04)
R_{m1} = resistance at temperature at T_1 (Ω)
R_{m0} = resistance at temperature at T_0 (Ω)
T_1 = process temperature 1 (°C)
T_0 = process temperature 0 (°C)

The first-stage amplifier should have an input resistance 1000 times greater than the total resistance of the path from the measurement electrode through the reference electrode so the current resistance (IR) potential drop through the amplifier is insignificant. The connection and insulation resistance of the measurement electrode cable should be at least 100 times greater than the path resistance to ensure that the parallel resistance to ground (R_3) at the measurement input to the field-effect transistor amplifier has a negligible effect.

> **Key CONCEPT** The glass electrode requires exceptional cable insulation, amplifier input resistance, and connection integrity to reduce measurement noise.

Dehydration of the glass electrode can cause a significant increase in its resistance (R_1). However, the gradual penetration of the gel layer into the interior of the glass and thinning of the dry glass from the aging process or chemical attack should show up as a decrease in electrical resistance.

Measurement electrodes with a coated surface have a higher resistance (R_1), but the percentage increase is usually small and difficult to detect, particularly because the glass resistance is an exponential function of temperature per Equation 4-1q. Coatings are best detected as a dramatic increase in the time constant of the electrode because thin, almost imperceptible coatings greatly slow down the migration of hydronium ions. For example, a 1 mm coating on the glass can increase the T_{86} response time (time to reach 86% of the final response) of the glass electrode from 6 s to 5 min [7].

> **Key CONCEPT** The onset of the glass measurement electrode's coating shows up as a large increase in its time constant and response time.

Typically, the reference electrode junction resistance is only 1 to 10 kΩ. The onset of a coating or partial plugging of the porous junction can easily cause the resistance (R_6) to more than double. However, salts from the process precipitating in the junction will be highly conductive and will not show up as an increase in resistance.

> **Key CONCEPT** The onset of a nonconductive coating of the reference electrode shows up as a large increase in its electrical resistance.

The resistances of internal electrode fills and external process fluids usually are only a few thousand ohms, so the potential drop from leakage current flow is negligible. However, high-purity water (i.e., distilled water, deionized water, or steam condensate) and nonaqueous solutions have an extremely high resistance (R_8) that approaches the resistance (R_1) of the glass.

A high total electrical resistance from applications with high-resistance glass at low temperatures (<20°C), with dehydrated glass, or in nonaqueous and pure water streams causes the *p*H measurement to be sluggish and more susceptible to noise from ground paths, streaming potentials (static electricity), fluid velocity, and electrical interference.

Key CONCEPT Nonaqueous and pure water streams require extra attention to shielding, process path length, and velocity to minimize *p*H measurement noise.

The resistances in the path between the internal electrodes in the glass and measurement electrodes shown in Figure 4-1a are normally small compared to the glass measurement electrode resistance of 20 to 200 MΩ. The current flow is extremely small, making the potential drop across these resistances negligible. The process fluid resistance (R_8) is negligible except for nonaqueous solvents or ultrapure water. The resistances to earth ground should be large enough to suppress noise from electromagnetic interference and ground loops.

A decrease in the high-impedance shunt resistance (R_{11}) can cause a loss of the electrode slope (efficiency). The degree to which the measured potential is decreased depends on the relative magnitudes of the glass electrode impedance and the shunt impedance. In one case of bad sensors with low shunt resistances, the sensors had low slopes (short-spanned) at room temperature but had adequate slopes for a higher temperature. This was due to the decrease in glass impedance at the higher temperature, which decreased the effect of the high-impedance shunt [8].

How bad can it be? If you look at the resistance (R_{11}) of the high-impedance shunt as a multiple of the resistance (R_1) of the high-impedance glass electrode ($R_{11} = n \bullet R_1$), you can get an idea of how much the measured electrode efficiency would be lowered [8]:

n	Efficiency
100	99%
50	98%
20	95%
10	91%
5	83% (bad electrode)

The glass electrode impedance significantly changes with temperature, whereas the shunt resistance does not. As the temperature increases, the glass electrode impedance decreases, the effective *n* value due to the shunt resistance increases, and the effect of the shunt decreases. Conversely, as the temperature decreases, the effect of the shunt increases.

The decrease in the shunt resistance (R_{11}) can be accelerated by aging from the exposure of O-rings and seals to a harsh process, which could attack the construction materials and introduce some leakage, as well as from the effects of prolonged heating on the internal potting materials, which could break down over time. The decrease has little to do with the glass electrode itself but rather is related to the integrity of the sensor construction over prolonged periods. It is also likely that at least some high-impedance shunting may be built into the sensor [8].

Key CONCEPT Deterioration of O-rings, seals, and construction materials due to a harsh process leads to a decrease in the shunt resistance and, therefore, electrode efficiency that may be less noticeable at higher temperatures.

Reference Problems

The liquid junction potential is typically the largest source of error in the 4 to 10 *p***H** range if the measurement electrode is not extensively compromised as a result of dehydration, damage, age, or coating. If the measurement electrode is functional, the loss in efficiency does not translate to a large error in the neutral region.

Key CONCEPT The reference electrode is generally the source of the greatest *p***H** measurement error in the neutral region.

Ions migrate from areas of high to low concentration by diffusion. As ions move into the porous junction of the reference electrode, they establish a voltage known as the *liquid junction potential* or *diffusion potential* (E_5). The more mobile ions accumulate in the junction, the faster they build up an excess charge that slows down the further accumulation of these ions. The potential E_5 and, consequently, the *p***H** reading shifts until an equilibrium is reached. The potential E_5 for the standard KCl reference electrolyte is relatively small because the potassium (K^+) and chloride (Cl^-) ion electrolytes have about the same mobility, which means they accumulate in the junction at about the same rate. However, a KCl electrolyte is not usually used in process fluids with compounds such as sulfides and nitrates that cause the formation of precipitates that clog the junction.

A more common problem is the migration of process ions into the reference junction. Again, an excess charge of the more mobile ions accumulates until an equilibrium is reached that slows down the faster ions. The final potential E_5 established from the migration of ions into the junction can be estimated by the Henderson equation (Equation 4-1r), which shows that the potential increases with the difference in the mobility of the positive and negative ions and the concentration of ions in the electrolyte and process solution [2]. Thus, low and high ionic strength process solutions create a high junction potential.

$$E_5 = \frac{(I_{p+} - I_{p-}) - (I_{e+} - I_{e-})}{(E_{p+} + E_{p-}) - (E_{e+} + E_{e-})} \bullet \frac{R \bullet (T + 273.16)}{F} \bullet \text{In}\left[\frac{(E_{p+} + E_{p-})}{(E_{e+} + E_{e-})}\right] \tag{4-1r}$$

$$E_{e+} = \sum (c_{ei+} \bullet \mu_{ei+} \bullet z_{ei+}) \tag{4-1s}$$

$$E_{e-} = \sum (c_{ei-} \bullet \mu_{ei-} \bullet z_{ei-}) \tag{4-1t}$$

$$E_{p+} = \sum (c_{pi+} \bullet \mu_{pi+} \bullet z_{pi+}) \tag{4-1u}$$

$$E_{p-} = \sum (c_{pi-} \bullet \mu_{pi-} \bullet z_{pi-}) \tag{4-1v}$$

$$I_{e+} = \sum (c_{ei+} \bullet \mu_{ei+}) \tag{4-1w}$$

$$I_{e-} = \sum (c_{ei-} \bullet \mu_{ei-}) \tag{4-1x}$$

$$I_{p+} = \sum (c_{pi+} \bullet \mu_{pi+}) \tag{4-1y}$$

$$I_{p-} = \sum (c_{pi-} \bullet \mu_{pi-}) \tag{4-1z}$$

where

c_{ei+} = concentration of positive ion species i in the electrolyte
c_{ei-} = concentration of negative ion species i in the electrolyte
c_{pi+} = concentration of positive ion species i in the process
c_{pi-} = concentration of negative ion species i in the process
E_5 = liquid junction (diffusion) potential of the reference electrode
E_{e+} = accumulated charge of positive ions from the electrolyte in the junction
E_{e-} = accumulated charge of negative ions from the electrolyte in the junction
E_{p+} = accumulated charge of positive ions from the process in the junction
E_{p-} = accumulated charge of negative ions from the process in the junction

I_{e+} = accumulated concentration of positive ions from the electrolyte in the junction

I_{e-} = accumulated concentration of negative ions from the electrolyte in the junction

I_{p+} = accumulated concentration of positive ions from the process in the junction

I_{p-} = accumulated concentration of negative ions from the process in the junction

μ_{ei+} = mobility of positive ion species i in the electrolyte

μ_{ei-} = mobility of negative ion species i in the electrolyte

μ_{pi+} = mobility of positive ion species i in the process

μ_{pi-} = mobility of negative ion species i in the process

z_{ei+} = valence (charge) of positive ion species i in the electrolyte

z_{ei-} = valence (charge) of negative ion species i in the electrolyte

z_{pi+} = valence (charge) of positive ion species i in the process

z_{pi-} = valence (charge) of negative ion species i in the process

Key CONCEPT "Special electrolytes" can be selected that have virtually no difference in the mobility of their ions. One example is potassium nitrate (KNO_3).

Key CONCEPT Extremely diluted or concentrated process streams cause a larger junction potential because the process has a much lower or higher ionic strength, respectively, than the reference electrolyte.

The movement of process ions into the reference electrode will also plug its junction and eventually contaminate its internal electrolyte. To reduce the rate of contamination at the internal silver–silver chloride electrode and the shift of potential E_4, smaller reference junctions, second internal junctions, gel electrolytes, porous solid internals, and flowing junctions are used. Smaller junctions increase the rate of fractional plugging of the junction and therefore reduce the time to a totally plugged junction. Double junctions, gel electrolytes, and porous solid internals increase the time required for the reference junction potential to reach equilibrium after being inserted into a process stream or for a dramatic change in the stream, which might occur during start-up or batch operation. Thus, these designs work a compromise between the conflicting needs for a fast, uncontaminated, and unplugged electrode. Note that extremely slow references will ignore short-term fluctuations in stream concentrations from imperfect mixing and may actually appear to provide a more stable (lower noise) reading. Also, if the electrode is immersed in the process by an automatically retractable assembly only for a relatively short time (1 or 2 min), the reference may not have had time to change its potential.

> **Key CONCEPT** Reference junctions with large areas tend to decrease the junction's rate of plugging but increase the internals' rate of contamination.

> **Key CONCEPT** Double junctions, gel electrolytes, and porous solid internals decrease the rate of contamination but increase the time to equilibrium.

> **Key CONCEPT** Slow references may be more stable for short-term fluctuations from imperfect mixing and short exposure times from automated retraction.

Solid references with small pore sizes help prevent the movement of small process particles into the junction. If this design is combined with an internal helical reference pathway, it is possible to have a large reference junction area to reduce the rate of fractional coating without a detrimental increase in the rate of contamination. However, the precipitation reaction of the silver or potassium electrolyte can rapidly occlude the small pores, so these electrodes should not be exposed to cyanides, bromides, iodides, sulfides, or nitrates. Methods to immobilize the electrolyte are used to reduce the migration rate of the electrolyte ions into the junction, thereby extending the reference's life and decreasing the rate at which the reference potential changes.

The fastest and most accurate reference electrode has a flowing junction. A constant small flow of electrolyte purges the junction and prevents the process from moving into the junction. The result is a small, quickly established liquid junction potential that can meet the demands of high accuracy and batch applications. However, the reference flow must be kept small enough so as to not contaminate the process or excessively deplete the electrolyte by pressurizing the reservoir to maintain an appropriate pressure differential between it and the process. The external salt bridge provides the ultimate isolation of the reference's internal half-cell from the process. The additional installation and maintenance cost of this external pressurized reservoir makes it an unpopular choice with technicians unless it can be translated into reduced probe calibration and replacement or recognized process benefits from reduced variability.

> **Key CONCEPT** The fastest and most accurate reference has a flowing junction, but it requires regulated pressurization to maintain a small positive electrolyte flow.

Counterintuitive Solution

Because *p***H** measurement is challenging and the reality of what is happening is so discouraging, the first thought of maintenance is not to install more electrodes. This is also consistent with an increasing emphasis on minimizing project costs and the loss of technicians with practical, hands-on knowledge who recognize the increased intelligence offered by transmitter diagnostics and MSS.

The best way to deal with the dose of reality is to install three electrodes and use MSS to reduce the many adverse effects of the process application on the measurement and reference electrode. The revelation that the electrode readings will immediately differ by 0.1 *p***H** or more on installation is discouraging but must be viewed as an awakening. MSS provides much more intelligence on actual electrode performance, which decreases unnecessary calibrations and replacements and gives a faster, more accurate, and reliable *p***H**. The accepted practice in two large industrial chemical companies is to diligently use MSS for all *p***H** loops. Surprisingly, this is not a common practice in other companies. Even more confusing is installing two electrodes and then relying on inherently flawed logic and operator decisions as to which electrode is used for control.

For much more on how MSS inherently protects against a single failure of any type—even at the set point—and provides so many unrealized opportunities, see Section 4.4, "Great Expectations and Practical Limitations," and Section 4.6, "Failure Protection." We conclude with a critical insight that may be the most important takeaway on *p***H** measurement from this book.

> **Key CONCEPT** Electrodes used for MSS will pay for themselves within a year with more intelligent information and better reliability and responses that translate to lower maintenance costs and greater process efficiency and capacity.

4.2 Measurement Electrodes

A large surface area spherical glass bulb measurement and flowing junction reference electrode has the best accuracy because it has the fastest and lowest resistance glass and reference junction with the fewest extraneous potentials. A separate temperature probe also provides the fastest compensation for changes in process temperature. However, a flowing reference electrode requires pressurization and periodic filling of the reservoir. Furthermore, the accuracy and speed of response are only obtained with glass surfaces and reference junctions that are free from contamination, abrasion, and coatings.

Key CONCEPT

Spherical glass bulb and flowing reference electrodes offer the most accurate *p***H** measurement but have higher installation and maintenance costs. The lifecycle cost may be less if the value of a smaller *p***H** measurement error is significant and fouling, abrasion, and breakage of the electrodes are not an issue.

Manufacturers generally offer several different glass formulations that are tailored to meet a process requirement. Glasses designed to be more rugged and handle chemical attacks from strong bases and strong acids tend to have higher resistance and consequential accuracy limitations, particularly for wide *p***H** ranges. High measurement electrode resistance can also cause the *p***H** measurement to be sluggish and more susceptible to noise from ground paths, streaming potentials (static electricity), fluid velocity, and electrical interference. Progress has been made in reducing the impedance of glasses so they can be used at lower temperatures.

Glass formulation is a highly proprietary technology with many undisclosed effects, compromises, and consequences as to the ability to handle process conditions. Corrosion strips away the outer gel layer while aging leaches out earth ions, increasing the glass impedance. Fewer earth ions make *p***H** glass like regular glass. Both types of glass sense *p***H**, but regular glass impedance is too high. More doping with earth ions to decrease impedance per unit thickness makes glass softer and more easily dissolved. Glasses used for thicker glasses in rugged glass electrode design use doping to reduce the total glass impedance with the consequence of being more susceptible to etching. Even with more doping, the extra thickness can make the glass impedance still too high at low temperatures.

A high-temperature glass formulation (AccuGlass) developed by Rosemount Analytical since the last edition of this book, and can measure higher temperature streams with much greater life expectancy, less drift, and incredibly faster response, as shown in Figures 4-2a and 4-2b. The improvements are largely associated with greatly reduced glass aging. Figure 4-2a shows more than a 100% increase in life expectancy for process temperatures of 50°C, 75°C, and 100°C. Figure 4-2b shows a three orders of magnitude decrease in the 86% response time of new versus general-purpose glass electrodes after exposure to high temperatures. This relatively new high-temperature glass formulation performs as well as a traditional general-purpose *p***H** electrode for less difficult operating conditions [8].

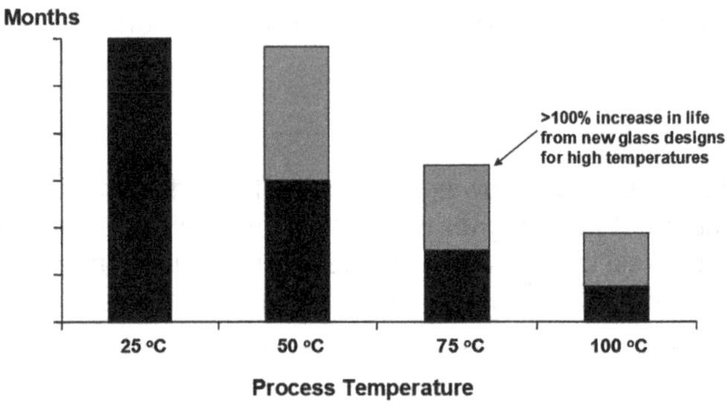

Figure 4-2a. Life expectancy of new high-temperature glass (AccuGlass) and general-purpose glass electrodes.

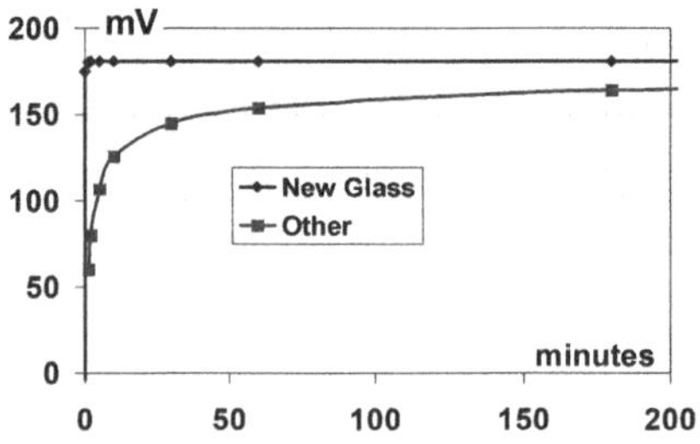

Figure 4-2b. Response times of new high-temperature glass (AccuGlass) and general-purpose glass electrodes.

Also available is a glass and electrode assembly designed to handle sterilization for applications in the food, beverage, and pharmaceutical industries. Figure 4-2c shows a factor of 5 decrease in electrode drift after sterilization by this design.

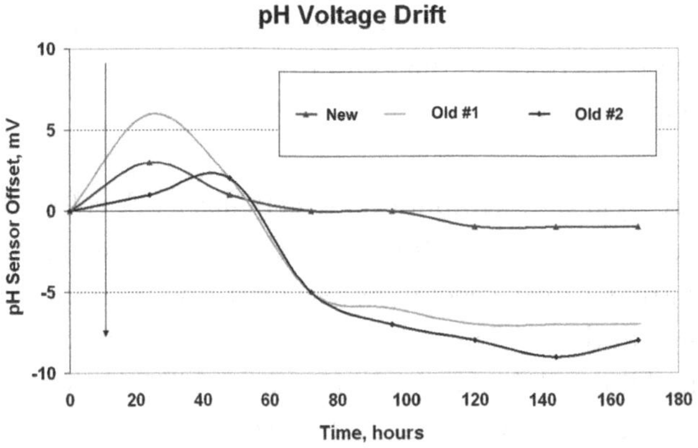

Figure 4-2c. Drift of new and old sterilizable *p***H** electrode technology.

Originally, measurement electrodes used a spherical glass bulb design with a 0.2 to 0.5 mm thickness for the *p***H** sensing membrane. Cracks tended to occur at the transition from the bulb to the straight side, although direct impingement of solids or debris caused cracks anywhere on the bulb. More durable semispherical and dome bulbs offer nearly the same performance. Most manufacturers now offer dome and flat glass sensors, depicted in Figure 4-2d, and ion-sensitive field-effect transistors (ISFETs).

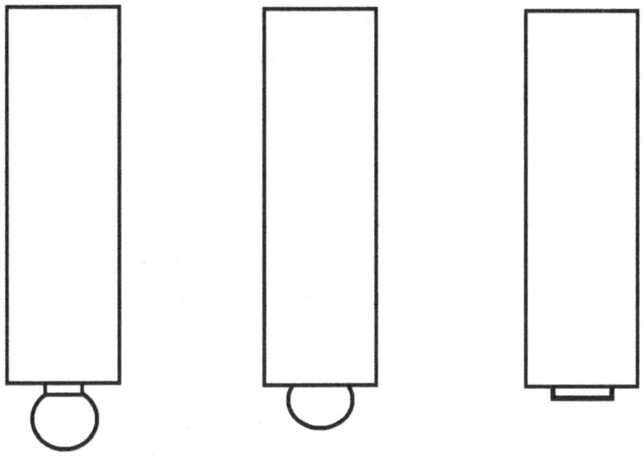

Figure 4-2d. Spherical bulb, dome, and flat glass sensors.

The prevalent reference electrode in the 1970s had a liquid electrolyte and a flowing junction and was separate from the measurement electrode. In the 1980s, combination electrodes incorporating double reference junctions, as shown in Figure 4-1a, created a throwaway concept by eliminating the need to refill reference solutions or troubleshoot separate electrodes. The electrolyte is a gel or is saturated in special wood or polymers to reduce the rate of contamination. The reference electrodes with porous solid materials impregnated with electrolyte are often called *solid-state* electrodes. In the 1990s, the pore size was reduced and the junction area increased.

In addition to the standard suite of measurement and reference electrodes just mentioned some special designs are intended to address application concerns that reduce the electrode's life or increase its installation and maintenance costs. Publications typically tout the advantages, and articles document the success stories. However, no electrode is perfect and a compromise is always at play. The following discussion will attempt to provide guidelines and explanations of the trade-offs based on limited testing, conversations with *p***H** specialists, and application experiences in the 1980s. For much greater detail on the construction of various electrodes, see Reference 4.7. All the electrodes presented here were developed in that time frame, although improvements have certainly been made. Before an electrode is used in any difficult or demanding application, it should be installed in the process on a test basis and monitored closely.

A glass dome or hemisphere is used to provide a stronger shape and a thicker glass membrane. These "rugged glass" electrodes are designed to provide longer life by reducing the chance of cracking during installation and maintenance and from exposure to aggressive fluids where abrasion, etching, or corrosion is a concern. If the dome has a higher total electrical resistance, it may not be suitable for low temperatures and may exhibit a slower response. If the electrical resistance per unit thickness is reduced to lower the total resistance, the glass may actually be more susceptible to chemical attack.

Flat glass electrodes have been developed to minimize direct impingement that reduces abrasion and to maximize a sweeping action that reduces fouling. For viscous fluids, a fat glass electrode also eliminates the separation of the flow field and stagnation associated with the downstream side of a spherical bulb. These electrodes generally have a stated range of 2 to 12 ***p*H**. Streams with high sodium ion concentrations and ionic strengths introduce a greater error for flat glasses. The change in glass formulation to make it flat is detrimental to response and reliability, particularly at higher ***p*H** and temperatures. The thin flat surface is vulnerable to cracks. Flat glass electrodes have been reported to exhibit a significant undocumented and uncompensated temperature effect [9, 10].

Pfaudler patented a measurement electrode made of Glasteel, a unique material formed by the fusion of glass and steel with an enamel coating. For operation below 5 ***p*H**, the probe has greater structural endurance, corrosion resistance, and a higher temperature rating of 140°C than conventional glass electrodes. It is designed to withstand the rigors of heat, pressure, and dynamic agitation in acidic reactors. This electrode must be paired with a transmitter that can be calibrated for a zero potential point between 1 and 3 ***p*H**. It provides a "relative" ***p*H** measurement in that the ***p*H** value sensed is not the same as that developed from a conventional ***p*H** electrode. However, its reproducibility and life expectancy may be better, and the lifecycle costs less because of its ability to handle higher temperatures and chemical attacks with reasonable accuracy in the 2 to 7 ***p*H** range despite a much higher initial cost for reactors. Nonetheless, the significant loss in performance at high ***p*H** and sodium ion concentrations is a concern.

Before the glass measurement electrode was developed for industrial measurement, the antimony electrode was used. Antimony is a hard, brittle material. Because of its toxicity, this electrode is no longer used. ***p*H** causes oxidation on the surface of other metal oxide electrodes currently in use. These electrodes require period cleaning because the oxidant accumulation deteriorates measurement accuracy and repeatability. Because oxidant accumulation on the electrode's surface deteriorates the measurement accuracy and repeatability, periodic cleaning is necessary. The electrode's response is nonlinear and requires narrow calibration spans or special polynomials for adequate accuracy. The metal oxide electrode is not recommended for operation below 2 *p***H**. It is more temperature sensitive than the glass electrode. The approximate 50 millivolt output per *p***H** unit increases from 1 to 3 mV per °C as the *p***H** increases. The isopotential point is about 2 *p***H**, so transmitters with an adjustable isopotential point are required for temperature compensation. Finally, these are actually oxidation-reduction potential (ORP) electrodes. They respond to oxidizing and reducing species. The reading can be affected by less than one part per million (ppm) chlorine. Metal oxide electrodes are used as a last resort in applications where severe abrasion from slurries or etching from hydrofluoric acid or other process operating conditions cause a rapid glass measurement electrode failure [10]. A more accurate possibility, even in these cases, is using an automated retractable electrode to limit process exposure time. Also, an electrode whose tip does not protrude into the pipeline when inserted can minimize abrasion if the cross-sectional profile of acid and base concentrations does not introduce an error from the electrode tip not being in the center of the pipeline.

Metal oxide electrodes have been developed that use a membrane designed to permit hydrogen ion migration but block the oxidizing and reducing species responsible for the ORP effect. Without a sufficiently thick membrane, reducing agents such as ascorbic acid decrease the millivolt potential developed and increase the *p***H** reading. Oxidizing agents such as hydrochloride have the opposite effect. There is some concern that high *p***H** exposure can dissolve the membrane and attack tantalum oxide. Ascorbic acid has been reported to affect the membrane. Increasing the membrane thickness decreases the speed of responses to *p***H**, which is already slow in the neutral region. These electrodes have no installed base and are presented here as an area for further research and development.

The Monsanto Company patented an iridium oxide *p***H** probe with a polymer coating that blocks negative ions. The ORP effect from positive ions is slowed down but not eliminated. The electrode requires a modified or microprocessor-based transmitter because the zero potential point is about 10 *p***H** and the isopotential point is near 4 *p***H**. It generates about 91 mV per *p***H** unit at 25°C. Iridium oxide has better linearity and repeatability than antimony and a potential temperature rating of 200°C. It also eliminates the problems of dehydration, etching sodium ions, and breakage associated with glass electrodes and can make measurements outside the 0 to 14 *p***H** range [10]. However, a reported drift of 0.1 *p***H** per day necessitates frequent calibration.

General Electric patented a high-temperature ***p*H** sensor whose yttria-stabilized zirconia tube enables the selective migration of hydrogen ions to an inner copper oxide electrode. The ***p*H** sensor agrees within 0.5 ***p*H** of calculated values at 285°C between 3 and 9 ***p*H** with uncertainty as to whether the discrepancies result from the calculation or the measurement. There is no alkalinity error. However, the membrane's response speed degrades with time [11, 12]. The sensor's expected life expectancy of a few months and a reported drift of 0.1 ***p*H** per day warrants investigation.

Leeds & Northrup patented the Durafet ***p*H** electrode. It uses an ion-sensitive field-effect transistor ISFET instead of glass to provide a low-impedance, breakage-resistant measurement electrode. Most electrode manufacturers now offer an ISFET ***p*H** probe when glass is not permitted for applications less than 10 ***p*H** and temperatures less than 50°C. The characteristic drift of laboratory versions of ISFET ***p*H** sensors has been reduced. It is much faster than glass because the response is by an electrical field rather than moving hydrogen ions into a membrane. The sensor has integrated automatic temperature compensation and no sodium ion error or known ORP interference. It needs no hydration and uses a conventional reference electrode. The slope of about 59 mV per ***p*H** unit is close to that for glass, but the isopotential point shift necessitates a special calibration. The thin silicon nitride crystal covering is susceptible to degradation if faced into the flow. Unlike glass sensors, the ISFET requires no hydration and generally has a longer life if left dry. There has been some discussion but no documentation of the life expectancy in process solutions or the interference from stray electrical fields and ground potentials. Also, a partial process coating of the sensor due to fouling can cause a partial voltage response, unlike glass electrodes. Using an ISFET is not advisable for clean-in-place and in-line sterilization. A retractable ISFET is an option for these applications.

Most ***p*H** measurement electrodes use glass bulbs with various shroud designs to provide bulb protection (from breakage and abrasion) and bulb exposure to flow (to ensure a representative reading and prevent areas of stagnation and material buildup), as shown in Figure 4-2e. The standard design has slots or holes to enable the process to flow around the bulb. Shrouds where the bulb is recessed without a slot or hole are used for pulp and very abrasive applications. However, the recessed area can become clogged or caked up, covering the bulb and reference junction. When material breaks loose, it can take the bulb with it. A bubble can also get trapped in the shroud. A tapered shroud is favored for coating applications because it creates eddies to help keep the electrode surfaces clean [13].

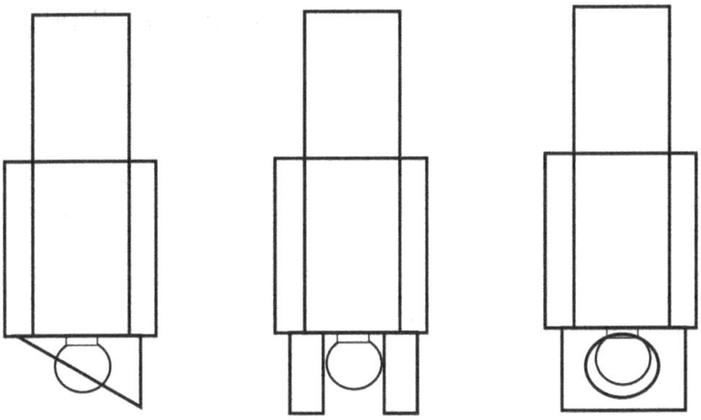

Figure 4-2e. Shroud designs work a compromise between the needs for bulb protection and bulb exposure to flow.

4.3 Reference Electrodes

The construction features of a particular reference design in a reference electrode present a trade-off in the effort to quickly establish and minimize a constant junction potential while resisting contamination and coating. The user must decide which is most important for a given application. A decrease in the area of the reference junction reduces the rate of contamination but increases the fractional rate of coating and plugging. A thickening or solidification of the internal fill, the addition of double and triple junctions, and a decrease in the junction porosity slow down contamination but delay equilibrium and often increase the magnitude of the junction potential.

The most popular reference electrode has a silver–silver chloride inner electrode and a potassium chloride electrolyte. The potassium and chloride ions have about the same mobility, which minimizes the junction potential from a difference in diffusion rates. However, the potassium chloride solution is saturated and tends to crystallize, especially at low temperatures, which reduces the diffusion rate and causes a drift in the associated potential at the junction. Also, silver from the silver–silver chloride internal element gets into the potassium chloride fill, reacts with sulfides and nitrates, and clogs the junction.

Rosemount Analytical patented a TUpH sensor (shown in Figure 4-3a) that has a solid reference with micron-sized helical pathways, a triple-seal barrier, and a reference junction that covers the whole lower sleeve area of the combination electrode to resist plugging and contamination. It also uses the new high-temperature glass formulation with a low sodium ion error that resists breakage and aging, especially at high temperatures, the formulation performance. are illustrated in Figures 4-2a and 4-2b. The design also protects the electrode from vibration and shock by transferring energy away from the glass-metal seal, which should translate to significantly lower maintenance costs in harsh, hot, caustic, dirty, and abrasive streams. It is not used where precipitation of the internal electrolyte could plug the small pores or where a rapid reference equilibrium is required.

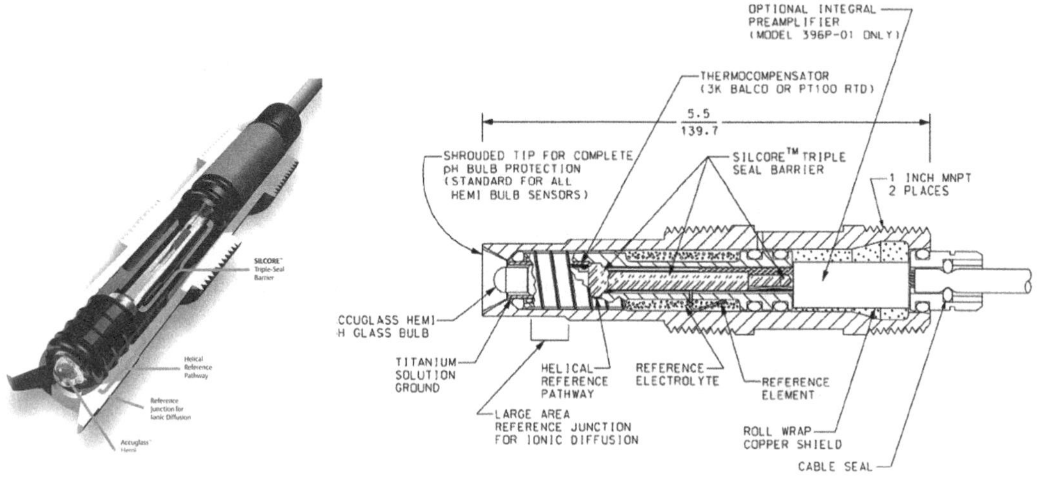

Figure 4-3a. High-temperature glass (AccuGlass) electrode with a solid-state reference electrode.

Rosemount has also developed a reference electrode that has a replaceable outer reference electrode chamber that can be easily removed and refilled (see Figure 4-3b). The outer chamber electrolyte can be chosen to be more compatible with a process to minimize the liquid junction potential. Problems with coatings and contamination are readily dealt with by quickly replacing the outer electrode chamber. The electrode also uses the new high-temperature glass formulation which has resulted in the electrode's performance advantages illustrated in Figures 4-2a and 4-2b.

Figure 4-3b. High-temperature glass (AccuGlass) electrode with removable reference junction.

In 1976, TBI patented a solid-state reference electrode that is nearly impervious to process contamination. The reference electrode consists of four wooden dowels saturated with potassium chloride solution. Process material must work its way up into the innermost wooden dowel, where the internal silver–silver chloride element resides, to create a drastic offset in the *p*H reading. Teflon is substituted for the dowel in contact with the process for applications requiring a sanitary surface or with chemicals that attack wood, such as hydrogen peroxide and hot, concentrated sodium hydroxide. Teflon is more difficult to keep saturated and slows down the time to equilibrium. This slowness is an advantage for continuous *p*H control, where it is desirable to ignore fluctuations in ionic strength from the steady state. However, it is difficult to get the electrode to match buffers within 0.1 of a *p*H for a two-point calibration. Because installation effects of more than 0.2 *p*H are normal, tight agreement with buffer solutions is not indicative of the electrode's installed performance. For high ionic strength process streams, high ionic buffers are recommended to precondition the electrode. A process calibration is essential for solid-state electrodes. For some batch applications, the reference does not have time to reach equilibrium, and *p*H errors of 0.5 *p*H and larger occur. In batch operations, it is important to use a wooden dowel, if possible, for the reference junction to make it faster.

Ingold patented a Xerolyte stiff polymer reference electrode. It uses an aperture diaphragm to provide direct contact between the process and the reference electrolyte. It has performance advantages in streams with high salt concentrations, emulsions, suspensions, proteins, sulfides, hot alkali, and pressure fluctuations. However, it is not suitable for stream sterilization, large temperature changes, or extremely acidic media (below 2 *p*H).

Great Lakes Instruments patented a differential reference electrode using a second glass measurement electrode (behind a double junction) in a buffer solution for the reference electrode. The glass bulb isolates the silver–silver chloride element of the reference from sulfides and nitrates in the process. The buffer solution is chosen to be compatible with process fluids to ensure a constant buffer *p*H and a constant reference potential. A ground rod eliminates errors and noise from ground potentials. The smaller positive ions in the buffer solution tend to move through the reference junction out into organic and pure water streams. This creates an offset in the *p*H reading. The differential electrode has an excellent performance record in streams with inorganic acids and bases, and sulfides such as wastewater. The reference buffer solution has much smaller positive ions than negative ions, whereas most reference electrode fills are designed to have the same size ions. The positive ions can move out of the junction more quickly, causing a large *p*H offset from a large junction potential, particularly for organic and pure water streams.

Amagruss patented the Refex nonporous reference electrode that uses an ionic electrochemical conductive salt loaded into a polymeric matrix. There is no contact between the process liquid and the internal KCl electrolyte or Ag/AgCl half-cell. Because there is no porous junction, there is no migration of electrolyte ions or process ions into the junction. Consequently, there is no threat of contamination and no diffusion potential or time required to reach equilibrium. Thus, it is a candidate for batch, high and low ionic strength, and nonaqueous processes and streams with sulfides or cyanides that might cause precipitation. Also, coatings have no effect on accuracy other than via an increase in circuit resistance because there is no penetration or clogging of a junction. However, coating the glass bulb will dramatically slow down the response as it will for any measurement electrode. The reproducibility for different process conditions and replacement electrodes needs further documentation because a different attempt at this design concept suffered from a variability of matrix loading during manufacturing.

4.4 Great Expectations and Practical Limitations

The *p***H** electrode provides an incredible concentration measurement of hydrogen ion concentration, potentially covering 14 orders of magnitude with a precision of 9 decimal places at 7 *p***H** corresponding to a *p***H** change of just 0.02 *p***H**. Although a glass electrode in a well-organized lab setting is capable of this precision, more realistic goals are important for *p***H** control in industrial applications. The goals and consequential *p***H** measurement selection, implementation, and support must be set based on an understanding of the effects of the electrode design, age, operating conditions, installation, calibration, and maintenance practices. These goals require a more intensive understanding of measurement response and reliability than is typically detailed in the literature. Suppliers may not offer any alert to the limitations. Here we provide a concise view of what affects the *p***H** measurement slope, zero offset, drift, range, temperature compensation, measurement response time, reference equilibration time, repeatability, and life expectancy. Glass formulations and electrode designs are offered with the promise of addressing particular concerns. Unfortunately these improvements often compromise measurement response and reliability in an undisclosed way. A reality check is essential. The realization of the need for knowledge of limitations and the consequential value of MSS is critical.

The type of electrode used depends on physical and chemical conditions. Therefore, in the same plant and even in the same process, many different types are used and are not necessarily interchangeable. When replacing an electrode (which could occur frequently in harsh conditions), one should verify the type of replacement electrode and its ability to handle process operating conditions.

One must consider the particularly challenging effects of low and high temperature, low and high conductivity, pure water, low water content, nonaqueous solvents, low and high *p***H**, poisoning, bubbles, solids, tars, slime, and non-ideal mixing. Situations and practices such as allowing electrodes to sit dry, frequently removing or roughly handling electrodes, chasing calibrations, using nonrepresentative samples, employing wrong or contaminated buffers, and a nonequilibrium and incomplete response during a batch and calibration can cause even the best design to perform poorly.

Using three electrodes with MSS offers many advantages and the ability to improve the response and reliability and to address the many challenging effects. MSS can even help identify the best electrode design and the best time to calibrate or replace electrodes and understand what is going wrong. MSS allows an electrode to be removed from service for buffer calibration without having to put the *p***H** control loop in manual if the remaining two electrodes are in close agreement. MSS inherently protects against a single failure of any type, including at a 7 *p***H** set point, and avoids the least accurate, slowest, and noisiest electrode. MSS reduces drift, the effect of deterioration in measurement electrode efficiency and offset, and reference electrode junction potential.

> Key CONCEPT
>
> In addition to providing better performance and reliability, three electrodes and MSS reveal what is truly happening with more practical expectations and the ability to address limitations, which leads to better design, installation, and maintenance.

For applications that demand an exceptional installed reproducibility of better than 0.05 *p***H** on the surface, the best technical choice is a spherical glass measurement electrode and a flowing junction reference electrode. The glass must be selected to match the *p***H** and temperature range of the process. The reference electrode aperture area and reservoir pressure must be chosen to maintain a small constant flow of electrolyte into the process. However, these types of electrodes have special installation and maintenance requirements that make them an extremely unpopular choice compared to throwaway solid-state combination electrodes. Spherical bulbs are more prone to breakage and must be carefully handled and installed with a properly designed protective shroud. The separate reference electrode is also generally made of glass and requires a separate reservoir regulated to a constant differential pressure above the process pressure. The reservoir electrolyte level must be carefully monitored and maintained.

> Key CONCEPT
>
> A glass measurement electrode with a spherical bulb and a reference electrode with a flowing junction offer the best accuracy, but semispherical or dome bulbs with the best glass formulation and special reference designs with MSS can provide sufficient accuracy with less attentive maintenance.

A technician may not understand active gel layers and diffusion junction potentials but will recognize broken glass bulbs and bodies, empty electrolyte reservoirs, additional electrodes, and a mess of tubing and accessories. If the accuracy is not justified and properly explained to the people who deal with the installation daily, the electrodes will soon end up in the junk pile along with your reputation. Remember that any electrode forced on maintenance and operations will not only fail but fail miserably no matter how well the selection matches up with the application. I have seen electrodes that were technically not the best choice succeed because the technicians and operators believed in them. The best bet is to get these people involved in the selection process and, if possible, make them think they made the choice.

Fortunately, few applications require a reproducibility of better than 0.05 *p***H**. Most chemists and process engineers set accuracy requirements that are unrealistic in terms of both practical value and attainment. Because they generally think in terms of what they can do in a laboratory environment and often lack first-hand experience with errors in field installations and documentation of the actual effects on the process operation, the requirements are often unrealistic. The result is often overcalibration and excessive maintenance costs. One solution is to make this chapter required reading before a general discussion of the real accuracy requirements.

> **Key CONCEPT** The accuracy requirement stated by process engineers and chemists is often unrealistic and should be corrected to a more reasonable expectation.

In any unproven process application, it is critical to test a selected electrode at actual process conditions before committing to a permanent installation. Of course, this is best done by a field trial in which the electrode is inserted into the process at the proposed point of *p***H** control and compared to lab samples for at least a month. Manufacturers are willing to provide the electrodes and assistance on the basis that you order electrodes and spares if the installation is successful. If there is great uncertainty as to whether an electrode will work at all in the harsh conditions of the process, a preliminary test in the lab at the operating temperatures and velocities is warranted unless the chemicals are too hazardous or difficult to handle or maintain at process conditions. It is not advisable to ship chemicals to the electrode manufacturer for testing because of safety and liability concerns. For the more common chemicals, the manufacturer may be able to develop its own samples or test data for a given range of compositions and operating conditions.

> **Key CONCEPT** In any unproven application, it is critical to test an electrode at actual process conditions before committing to a permanent installation.

Batch Processes

For *batch* processes, there is usually insufficient time for a solid-state reference electrode with a porous junction to reach ionic equilibrium. However, if the electrolyte's ions are immobilized and the junction has no pores, a solid-state reference is a candidate because there is no diffusion potential to be established. However, the reproducibility required of *p***H** measurements in batch processes is usually higher than in continuous processes, and the flowing reference junction and spherical glass bulb electrode are generally the best choices. The flowing reference reduces the possibility of penetration and entrapment of microorganisms or products in the junction and possible contamination of batches. For less demanding batch applications in the food and water treatment industry, a nonporous or differential reference electrode and ISFET may be an option. For sanitary applications, the electrode must be able to withstand repeated sterilizations, and the installation assembly and connection must not have any crevices.

> Key CONCEPT Batch applications require greater measurement reproducibility that necessitates the reference's diffusion potential to be small and rapidly established.

Continuous Processes

For *continuous* processes, the slowness of a solid-state reference electrode to reach equilibrium can be an advantage by the reference electrode ignoring short-term changes in process ion concentration and type. However, in high ionic strength streams, it may take hours for the reference to reach ionic equilibrium after installation. In these applications, it is beneficial to calibrate the electrodes and soak them in buffers loaded with salt ions of a concentration and type similar to what exists in the process stream and to minimize the number of times the electrodes are removed for cleaning or calibration. In general, electrode selection for continuous processes depends more on what type of electrode will last the longest and require the least service. For chemical processes, the operating conditions can be particularly harsh.

> Key CONCEPT For continuous chemical processes, there is a greater need to increase life expectancy and decrease maintenance costs despite high temperatures and pressures and high concentrations of solids, ions, and hazardous materials.

Before we address how to deal with specific challenging applications, we will discuss the metrics not commonly detailed in literature to get an overview of which process conditions affect the metrics. MSS can improve the ability of a *p***H** installation to meet all these metrics by automatically selecting the best electrode.

Isopotential Point

The *p*H where the temperature effect on mV generated is zero is known as the *isopotential point*. For glass electrodes where the voltages associated with the inner chamber of the measurement glass electrode and the reference electrode are the same, the isopotential point is usually 7 *p*H. There can be horizontal shifts and vertical shifts for problems with the glass and reference electrode, respectively, of the isopotential point. Contamination of the reference electrode's inner chamber can shift the isopotential point, which is one of the many reasons flowing junctions, multiple chambers, and solid-state reference electrodes are used. These shifts are depicted on plots of mV versus *p*H for different temperatures, as depicted in Figures 4-1b and 2-1c, but the error introduced is usually less compared to the errors originating from changes in slope and offset and from changes in solution *p*H with temperature changes. When glass electrodes are not used, it is critical to adjust the isopotential point. The isopotential point of the Pfaudler Glasteel material typically lies between 1 and 3 *p*H. The isopotential point of ISFET and metal oxide electrodes depends on the electrode design and, consequently, the supplier. Tests may need to be done by varying the *p*H and temperature, taking into account changes in the solution *p*H with temperature changes to determine the *p*H where the changes in mV with temperature changes is nearly zero. This is a challenging task, so it is hoped the supplier of non-glass electrodes will provide guidance. Also, some *p*H transmitters designed specifically for glass electrodes may have the isopotential point fixed at 7 *p*H.

An equation was developed, tested, and patented for the isopotential point of an experimental ISFET from measurements in the lab. By changing the drain current of the ISFET, the isopotential point of the sensor can be changed. An ISFET sensor can be "tuned" to have a specific desired isopotential point.

Offset (Zero)

The offset adjustment often called standardization and done to eliminate any nonzero potential like what is done for the isopotential adjustment, the effect on the millivolt response is quite different. The change in mV from an offset adjustment (standardization) is affected by the slope adjustment and actual efficiency, as shown in Equation 4-1m. For this reason, the offset adjustment is done first and then checked after the slope adjustment.

The offset at 7 *p*H and 25°C is 0 mV only if the activity of the hydrogen ions on the inner and outer glass surface hydrated gel layers are equal and the reference electrode junction potential is 0. This is quite a challenge because the glass outer surface and reference junction are exposed to non-ideal fluids and difficult operating conditions. If the glass has been cleaned, hydrated, and rejuvenated, the offset is typically associated with a nonzero liquid junction potential. Offset is the most frequent calibration adjustment needed because reference junction potentials often change significantly because of process ions migrating into the porous reference junction.

Slope (Efficiency)

A loss in efficiency shows up as a decrease in the slope and span of the *p***H** measurement. Abrasion from solids, etching from hydrofluoric acid attack, scratching, loss of Li^+ ions, and aging from high-temperature exposure and all other conditions and designs that result in an increase in electrical resistance of the glass also decreases efficiency. Lithium and rare earth ions leave the outer layer when exposed to low and high *p***H** and low water content. These are replenished by lithium and rare earth ions from the interior glass. The gradual loss in efficiency ends up as a rapid drop to zero efficiency when the supply of these ions from the interior glass is depleted.

Some low sodium ion error glasses may show hysteresis or memory, causing a bowing of the final *p***H** response curve and consequential change in the slope and span when exposed to acids, such as a 4 *p***H** buffer or hydrochloric acid wash.

A significant decrease in shunt resistance R_{11} from deterioration in an electrode holder not designed for specific harsh process conditions can cause a decrease in efficiency.

Although the effect of a shortened span may not be noticeable for operation near 7 *p***H** because the decrease in *p***H** change is small, possibly within normal fluctuations, an increase in the response time often associated with a loss in efficiency is problematic.

Drift

A gradual change in the *p***H** reading that is not a result of a change in process *p***H** can have many causes. Most notable is the reference's equilibration time, often due to solid-state design, coatings, and contamination, and the long response time of the measurement electrode, often due to coatings, dehydration, and aging. The response and equilibration time shows up as drift after the electrode has been exposed to samples or buffers for calibration and to different fluids because of upsets, transitions, startups, and batch operations. The drift may be continual for continual reference electrode contamination and measurement electrode glass surface deterioration.

Temperature changes can also cause drift resulting from temperature compensation response time or failure to include the effect of temperature on water, acid, and base ion activities and dissociation constants. When the temperature sensor for compensation is in the electrode, there is a significant heat transfer lag, particularly for plastic electrode holder bodies. It may take 10 to 20 min for the sensor to read the process temperature. If the process temperature is changing, the temperature compensator may never catch up.

Temperature Compensation Error

Temperature compensation offered in transmitters is typically for the change in mV generated for a *p***H** per the Nernst equation for a glass electrode. Smart transmitters have offered temperature compensation based on the actual *p***H** of the process solution changing with temperature. The literature gives examples of how the change in water dissociation constant affects the actual process *p***H** for solutions without any weak acids, bases, or other solvents that can be potentially significant. Consequently, the required temperature compensation can be a function of process composition that may not be readily known or approximated by a simple first-order equation. Where temperature compensation errors must be minimized, lab tests must be conducted to provide data points on how the actual *p***H** changes with operating conditions. A signal characterizer offered in most controllers can be used to provide a piecewise fit to the curve that is a plot of sample *p***H** versus sample temperature for a given composition and after correction for the Nernst equation. Based on intelligence as to the expected composition for a given operating point, the most representative curve is used to provide a *p***H** measurement corrected for actual *p***H** change with temperature. It is important for maintenance and operations that the *p***H** before and after the correction is displayed.

The temperature sensor measurement lag must be considered in the lab tests. Using a separate bare element resistance temperature detector (RTD) in the lab and the process can make the temperature measurement lag negligible.

Measurement Electrode Response Time

It is critical that the percentage of final responses be noted in any response time statements. Unfortunately, this is rarely the case, and the term *final response time* is used. The most representative response time is 86% of the final response (T_{86}) because the plot of *p***H** versus time can flatten out dramatically for the last 10% of its response. For feedback control, the controller sees enough in T_{86} response time to correct sufficiently for the problem, especially considering that the accuracy of the measurement is probably less than the last 10% of the response. Because the dead time in the electrode response is usually negligible, the T_{86} response time corresponds to two time constants. The T_{86} response time is used in ISA-75.25.02, *Technical Report for Control Valve Response Measurement from Step Inputs*, for many of the same reasons.

The general-purpose glass spherical bulb has the fastest T_{86} response time, about 3 s for velocities 1 ft/s and faster in buffer solution. The spherical and dome bulb and the new high-temperature AccuGlass are almost as fast. If the solution is not stirred or is not buffered, the response time can increase by a factor of 2 or more. Table 4-7a shows how velocity and the direction and magnitude of the *p***H** change affect the electrode time constant in a buffer solution. Flat glass and rugged glass electrodes can have a response time of 5 to 50 times larger. Aging and fouling can cause a response time to be 100 to 1000 times larger. Section 4.7 discusses electrode and installation effects on the dynamic response of electrodes.

The detrimental effect of slower measurement electrode response on control-loop performance is unfortunately not generally recognized. Nearly one-half of the T_{86} response time shows up as an equivalent dead time in the *p***H** control loops on well-mixed vessels recommended for tight *p***H** control. The large process time constant in these vessels is desirable for slowing down disturbances and enabling larger PID gains and tighter control. The total effective dead time is the nemesis because it delays recognizing and correcting a control error. As discussed in Section 8.10, "Dynamics and Performance," the minimum peak and integrated errors are proportional to the dead time and dead time squared, respectively. The total effective dead time is the summation of all the pure dead times and all the effective time constants from the small time constants, as shown in the Figure 8-1a block diagram illustrating the sources of dynamics in a *p***H** control loop. For *p***H** systems, it is extremely important to minimize the excursions on the titration curve that subject the loop to the nonlinearities of the changing process gain by tighter control from a smaller total loop dead time.

RULE of Thumb *For tight control in well-mixed vessels, the* T_{86} *response time should be less than 0.2 min.*

The T_{86} response time should be identified before the electrode selection is finalized. A test electrode can be inserted into buffers, and controller tuning software can be used to identify the electrode time constant and dead time by connecting the *p***H** measurement to a controller and manually stepping the controller output when the electrode is inserted into the buffer.

Key CONCEPT The electrode T_{86} response time can be identified by connecting the electrode to a controller and using tuning software to identify the dynamic response when it is inserted in the buffer.

Some transmitters offer a stabilization feature that will not accept a *p***H** value entered until the *p***H** measurement has stopped changing. A computation of the rate of change can also be configured in the *p*H controller using a dead time block with its dead time set to be longer than the suspected T_{86} response time. The input to the block is subtracted from the output of the block and divided to get the measurement rate of change. This provides an immediate update to the rate of change, whereas conventional methods that wait, delay updates by the wait time. This rate of change computation is part of the future value block described in Chapter 9, "Digital Twin," that can be used to predict *p*H values at optimum times into the future, which is particularly useful for optimizing set point response and batch control. The rate of change result is useful during commissioning and normal process operation to compare electrode responses for MSS.

> **Key CONCEPT**
>
> A simple rate of change calculation that is part of a future value block can be configured in the *p*H controller to monitor the effect of the measurement electrode response time and reference electrode equilibration time during calibration and during commissioning, and to provide greater intelligence in the comparative analysis of electrode health offered in MSS.

During process operations, increases in the number of process oscillations and the period of these oscillations can be construed as a possible sign of an increased T_{86} response time. For installations in in-line systems or small vessels with small residence times, the increase in electrode T_{86} response time may show up as smaller amplitude oscillations because of the filtering action of the measurement time constant.

> **Key CONCEPT**
>
> A greater appearance and persistence of oscillations with a longer oscillation period but possibly smaller amplitude is a symptom of a slow electrode.

Measurement Electrode Life Expectancy

The average life expectancy of an electrode is 3 years in a lab, 1 to 2 years in a nice process and installation, and anywhere from days to months in challenging applications. This short life span puts greater emphasis on the need for MSS to recognize and protect against electrodes nearing the end of their life expectancy. Ideally, the electrodes used in MSS should have different ages to reduce the likelihood that more than one electrode will have a problem.

> **Key CONCEPT**
>
> Measurement electrode life expectancy is much less than commonly expected, especially in challenging applications.

RULE of Thumb *Use MSS to help make decisions on electrode health and life. The greatest protection is offered by using electrodes of different ages.*

Aging is caused by the leaching of earth ions accelerated by high temperatures. Formulations with more earth ions tend to make glass softer and more easily dissolved, which can reduce life expectancy. Alkali-resistant glass tends to be more fragile.

Corrosion and erosion from exposure to chemicals and repeated dehydration from exposure to nonaqueous solutions or from being left dry, most notably between batch cycles and during equipment downtime, reduce life expectancy.

Reference Electrode Plugging and Poisoning [8]

To properly function, a reference electrode should maintain a standard potential, which is a function of the silver ion dissolved in its fill solution, and maintain electrolytic contact with the process by diffusing potassium chloride ions through its reference junction. The potassium and chloride ions in a reference fill solution diffuse through a reference at an equal rate, so there is a net charge balance across the liquid junction. Potassium chloride is said to be equitransferent. The junction pores or aperture enables process ions to migrate into the reference electrode.

- Ions from the process diffusing into the reference junction are likely not equitransferent, so a junction potential develops in opposition to the charge imbalance created by unequal diffusion rates of positive and negative ions.
- Certain ions in the process can precipitate or complex the silver ions when diffusing into the reference electrode. The most common are sulfide, bromide, iodide, and cyanide, and they can affect the functioning of the reference electrode in two ways:
    - Poisoning – Reducing the silver ion concentration in the innermost portion of the reference electrode causes silver ions on the silver–silver chloride wire to dissolve. After all the silver chloride has dissolved, a large offset in the potential (and the *p*H measurement) is produced and the reference electrode must be replaced.
    - Plugging – A second effect can be the precipitation of silver within the liquid junction, blocking the diffusion of ions into the process and thus opening the *p*H measuring circuit and causing the *p*H measurement to drift.
- Reference electrodes have been designed to retard the migration of process ions into the reference electrode by using multiple liquid junctions, an outer reference

fill of potassium nitrate to lower silver ion concentration, gelled fill solutions to prevent ion transport by thermal convection, and solid-state designs.

- A reference with solid internals, an embedded electrolyte, and a helical tortuous pathway, as shown in Figure 4-3a, can drastically slow down the contamination process. These reference electrodes typically have a large reference junction area that helps prevent coatings from greatly affecting the reference junction resistance and potential. However, they can have a large equilibration time.
- In extreme cases where there are high concentrations of harmful ions in the process, a flowing reference is required, which uses a flow of fill solution into the process to make electrolytic contact to prevent the incursion of process components into the reference. Flowing reference electrodes are maintenance intensive because of the need to keep an adequate reservoir of fill solution and to maintain pressure on the fill that is sufficient to overcome process pressure.

> **Key CONCEPT** Reference electrodes develop a junction potential because of the migration of process ions into the junction pores or aperture that creates an offset and can lead to plugging and poisoning.

Reference Equilibration Time

The time for the reference electrode to reach a constant junction potential, known as *equilibration time,* is usually not stated in the literature or considered in the selection, installation, and calibration of electrodes. It can range from minutes to hours to days. The future value block discussed in Chapter 9 can provide the knowledge that is sorely needed. Better decisions can be made as to the type of reference electrode and the preparation of buffer solutions to help the reference junction potential reach equilibrium. Multiple blocks can be employed with different dead time settings to monitor how the equilibration time is changing. This knowledge can be used to decide how long new or removed electrodes must be conditioned by process conditions to achieve reference electrode equilibration before being used for batch or closed-loop control.

> **Key CONCEPT** Knowing the reference electrode equilibration time in the process application is essential for proper electrode selection, preparation, and maintenance.

Usually, the reference electrode equilibration time is negligible only for water solutions that are close to 7 *p***H** and 25°C. Because this is rarely the case, it is critical that the equilibration time and its impact be recognized and dealt with. The reference electrode potential must be constant before the *p***H** electrode is used for process control.

This means reference electrodes should be conditioned with solutions matching the challenging effects of the process application before use and stay exposed to process fluids after insertion. To ensure a constant electrode potential for batch operations and for start-ups and transitions, the electrodes must be continually immersed in a representative process fluid. If there is any risk of fouling, the process fluid velocity must be about 5 ft/s, which typically requires recirculating process fluid past the electrodes. For MSS, the equilibration of a single electrode can inherently occur because it will be effectively ignored in use for control by the other electrodes at equilibrium.

Key CONCEPT MSS will enable a reference electrode to be put in service that has not reached equilibrium if the other two electrodes have reached equilibrium.

RULE of Thumb *Reference electrodes that develop a significant junction potential in the process application must be conditioned before being inserted and continually exposed after insertion to stirred or flowing fluid representative of the process.*

Challenging Applications

There are many challenging applications. Here we provide a concise view of the problems and some possible solutions posed by high and low temperatures and *p*H, salts, solids, nonaqueous solutions, low-conductivity solutions, and reference poisoning agents. We present the consequences and how electrode design and installation can to minimize them. If necessary, they can also be minimized by using automated retractable assemblies to provide scheduled electrode exposure time, cleaning, regeneration, and calibration.

High and Low Temperature

For *high-temperature* process streams, the glass type and reference internals must be rated for the time duration of the highest temperature, acid, and base concentrations. The alkalinity error is larger and the chemical attack is dramatically greater at high temperatures. The life expectancy of the electrode falls off exponentially based on the temperature, acidity, and alkalinity of the process streams. Figure 4-4a shows a rough approximation of the operating conditions where using general-purpose glass electrodes requires caution or is not advisable. The relatively new glass formulation (AccuGlass) improves the performance in the "fair life expectancy" ratings and extends those ratings into conditions causing "poor life expectancy" ratings. The table assumes the correct type of glass is selected for the *p*H and temperature range. Notice that the ratings deteriorate more quickly for high *p*H. For streams above 12 *p*H and 45°C, the life expectancy of glasses is unacceptable for continuous exposure. For streams below

130°C, the upper limit of the glass electrode, an automatically retractable assembly can reduce the process exposure. For temperatures less than 120°C within an operating range of 1 to 6 *p*H, the Glasteel electrode is an option to increase life expectancy. For higher temperature streams, a sample cooler is required.

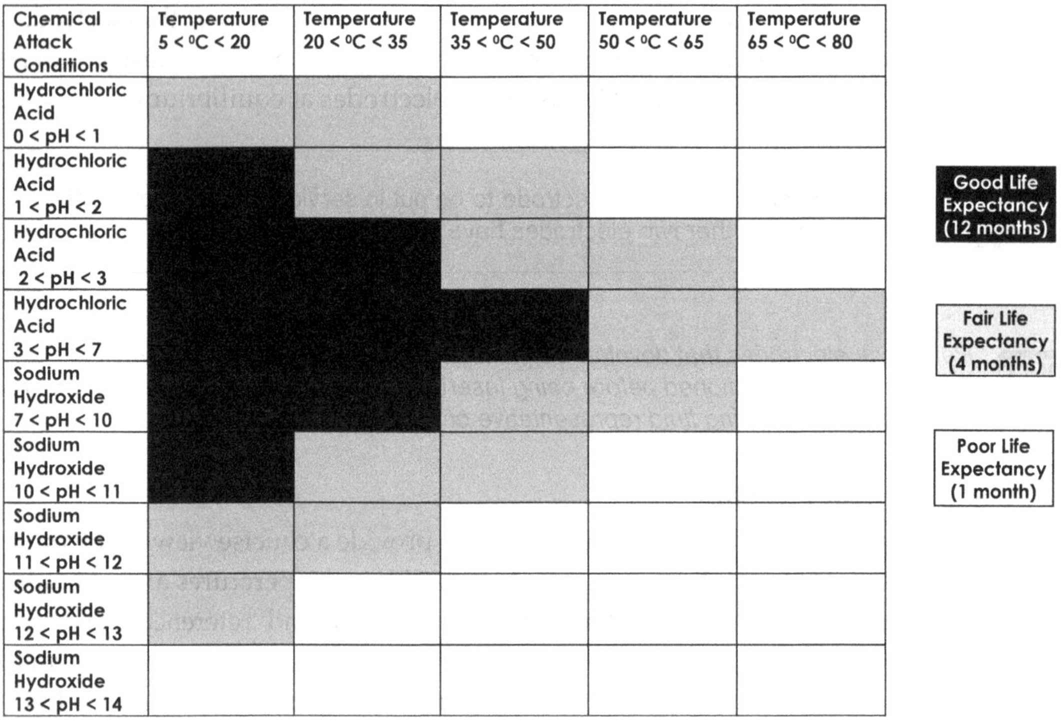

Figure 4-4a. Temperature ratings of general-purpose glass electrodes in terms of life expectancies.

Key CONCEPT: Life expectancy exponentially declines with high temperature, and the combination of a high temperature and high caustic concentration is disastrous.

The most viable alternative for prolonging the life expectancy of electrodes in a hot stream is an automatic retractable assembly. As with any electrode assembly, the construction materials must be checked for their resilience to chemicals and temperature. Corrosion increases at higher temperatures, so more attention must be paid to all wetted parts. Solutions can be switched to condition, diagnose, and calibrate the sensor, and the insertion versus the retraction time can be optimized. To avoid thermal shock, the probe should be kept in a relatively hot but neutral solution for the condition. For the most severe service ratings, the electrode will end up being retracted for 2 min or more because the T_{86} response time (time to 86% of the final value) is at least 0.2 min for even a clean electrode. This sample-hold methodology adds an average dead time of about half the retraction time and can only attenuate the peak error of upsets whose time constant is larger than the dead time. Thus, a large back-mixed volume at or upstream of the control point is essential.

RULE of Thumb *For moderately high process temperatures and large back-mixed volumes, use automatic retractable assemblies to reduce process exposure.*

If a heat exchanger is installed to cool the sample, the process temperature must be accurately measured at the electrode and the *p***H** measurement corrected for the effect of the cooled temperature on the dissociation constants. Lab samples must be run at the same cooled temperature. Also, the additional dead time in the loop from the transportation delay through sample lines and the sample cooler must be minimized.

RULE of Thumb *For extremely high process temperatures and large back-mixed volumes, use tightly temperature-controlled coolers and accurate electrode and process solution temperature compensation.*

The additional dead time from the cycle time of an automatically retractable sensor or the transportation delay of a sample cooler make the *p***H** loop a candidate for an enhanced PID (discussed in Appendix E) and dead time compensation (described in Section 8.12, "External Reset Feedback"). If an enhanced PID is not used, the PID execution should be synchronized with the cycle of the retractable assembly, and the dead time block should be updated to the current cycle time to minimize the variability of the dead time from the cycle time. Alternately, using MSS and scheduling so that only one electrode is retracted can prevent the retraction cycle time from increasing the *p***H** control system dead time.

Key CONCEPT An enhanced PID and dead time compensation can help deal with the consequences of the dead time created by retractable assemblies and sample cooling.

Key CONCEPT MSS and smart retraction scheduling can eliminate the large and variable dead time from retractable assemblies.

Some manufacturers, such as AMEL Instruments, have started to commercialize zirconia *p***H** electrodes for temperatures higher than 160°C. These electrodes can potentially handle much faster disturbances, but the long-term drift in actual process applications needs further investigation. Even though this electrode's response speed may degrade with time, it has eliminated the large transportation delay associated with sample systems.

At low temperatures, the dramatic decrease in glass resistance results in signal noise. Glass electrode resistance doubles for every 7°C decrease, going from about 100 MΩ at 28°C to about 1600 MΩ at 0°C [14]. The reference junction potential may also be affected. Lower process temperatures lower the activity and the dissociation constants of solvents, acids, and bases, lowering the actual solution *p*H. Buffers and samples should be at process operating temperatures, which can be challenging for very cold solutions.

Key CONCEPT At low temperatures, the dramatic decrease in glass resistance results in signal noise.

Solids

For *solids*, the goal is to keep the glass surface and reference junction clean and in good condition. The best inherent method to prevent the start of many coatings is to increase the fluid velocity to 7 ft/s. However, high velocities are not effective at removing existing coatings. Thus, the electrodes must never be exposed to low velocities. The velocities at the electrode tips of submersion assemblies and flow-through cell holders rarely exceed 1 ft/s. High velocities are usually achieved by using retractable or insertion assemblies in recirculation lines. Abrasion increases with velocity along with the risk of damage from clumps of material and foreign substances. Therefore, it is important that electrodes be installed downstream of pumps that have suction strainers. Also, velocities higher than 10 ft/s create more noise and reduce the life of the glass.

RULE of Thumb *For nonabrasive solids, installation in a recirculation line with a velocity of 5 to 9 ft/s downstream of a strainer and pump may prevent coatings.*

Flat surface glass electrodes can provide an effective sweeping action at a much lower velocity and eliminate the separation of the flow stream and stagnation that occurs on the back side of a glass bulb. Thus, for abrasive, sticky, and viscous streams, flat surface glass may decrease the fouling and stagnation and enable a more representative measurement of changes in the 4 to 10 *p*H range. The 1 ft/s velocity in a flow-through holder may be sufficient to keep the glass clean. However, the flat glass formulation has a high resistance requiring minimizing thickness Tests show significant vulnerability to cracking, slower response, greater vulnerability to chemical attack, and loss of accuracy outside the neutral region. For these reasons, semispherical or dome glass bulbs with a protective shroud and a higher fluid velocity are preferred and, if necessary, jet washing.

RULE of Thumb *For abrasive solids and viscous fluids, a flat glass electrode can minimize coatings at low velocities and stagnant areas in the neutral region, but response and reliability are seriously compromised.*

If a spherical or dome glass bulb measurement electrode is used, integral jet washers can be used to provide in-service cleaning. High-velocity nozzles pointed at the bulb are an option for submersion and flow-through assemblies, as shown in Figure 4-4b. The *p*H controller should automatically switch to manual during the jet washing because the signal will be erratic. If the cleaning solution contaminates the process, automated retractable assemblies can be used to withdraw the electrodes from the process into a chamber where it is flushed, cleaned, calibrated, and conditioned by switching in appropriate streams. The controller, in this case, should also automatically switch to manual when the electrodes are retracted. Note that it is not sufficient to just freeze the last *p*H measurement because it is never exactly equal to the set point, and reset action in the *p*H controller will continue to drive the output to eliminate the offset. The last thing you want in terms of maintenance cost, electrode life, and safety is for technicians to manually clean electrodes. Rubbing and handling electrodes reduces their life expectancy and reproducibility. Brush cleaners attack the outer gel layer and smear sticky stuff onto the bulb, not really cleaning the reference junction.

A. Cleaning solution
B. Solenoid valve or manual valve (supplied by others)
C. Corrosion-resistant tubing (supplied by others)
D. Polypropylene ¼-in. (6.4 mm) compression fitting
E. ¼-in. (6.4 mm) 316 stainless steel
F. ¼-in. (6.4 mm) polypropylene
G. Stainless set screw for adjustable spray nozzle height
H. 2-in. (50.8 mm) NPT threads
I. Sensor
J. 1-in. (25.4 mm) PVC coupling for submersible applications (supplied by others)
K. 1-in. (25.4 mm) PVC or stainless conduit (supplied by others)
L. Cable
M. Timer supplied by others

Figure 4-4b. Electrode with an integral jet washer pointed at the glass bulb.

RULE of Thumb *When the fluid velocity is insufficient to sweep electrodes clean, use an integral jet washer or a cleaning cycle in a retractable assembly.*

For nonflowing reference electrodes, it is important that the reference junction has small pores or no pores and an active surface area that is as large as possible. When particles get stuck in the pores, they are extremely difficult to remove. Plugging the reference junction isolates the electrolyte from the process and increases the electrical resistance and time to equilibrium. A larger junction surface area increases the chance that part of the junction is still free but increases the contamination rate from process ions moving into the reference. Gel-filled reference combination electrodes slow down the contamination rate but not as much as solid-state references, which have a large junction area that is either the sleeve or annular ring around the glass measurement electrode. Cone and sleeve annular flowing junctions not only prevent plugging and contamination but also offer a faster and more reproducible reference potential.

RULE of Thumb *To prevent particles from plugging a nonflowing reference junction, use a solid-state reference electrode with small or no pores.rodes clean, use an integral jet washer or a cleaning cycle in a retractable assembly.*

High p*H and Alkalinity Error*

For *high* *p*H streams, the concerns are alkalinity error and chemical attack of the glass. While alkalinity error is often plotted versus *p*H, it does not have to be associated with high *p*H because it could be caused by the presence of a high concentration of a salt, such as sodium chloride, instead of sodium hydroxide (caustic).

Alkalinity or *sodium ion error* is caused by alkali ions such as sodium ions penetrating the measurement electrode silicon oxygen network and displacing the hydrogen ions in the gel layer, which causes a decrease in the hydrogen ion activity of the gel layer relative to the process and a decrease in the measured *p*H. The high *p*H end of the millivolt versus *p*H line bends upward, which results in an indicated *p*H that is lower than the actual *p*H. The error is greatest for glasses with alkali ions with an equal or larger radius than that of the alkali ions in the solution. A glass with small lithium ions is called a *tight matrix*. Alkalinity error is called *sodium ion error* because sodium ions are more prevalent and more of a problem because of the popularity of sodium hydroxide as a reagent and the smaller radius of the sodium ion and, therefore, the greater capability to penetrate the matrix compared to the next most common alkali ion, the potassium ion. The alkalinity error of Corning 015 glass for process streams with lithium and potassium ions is about twice and one-fifth, respectively, of that for sodium ions because of their relative size. The alkalinity error also increases with temperature. For example, the alkalinity error of Corning 015 glass starts at 9.5 *p*H at 30°C, 8.0 *p*H at 50°C, and between 5 and 6 *p*H at 60°C. The heat of neutralization from sodium hydroxide can cause a significant unsuspected alkalinity error for a 7 *p*H set point. Glass formulations have been made that reduce the alkalinity error but with some sacrifice as to nonlinearity over the 0 to 14 *p*H scale. The correction *p*H increases exponentially with *p*H above some threshold level. Note that the solution *p*H is equal to the measured *p*H plus the correction *p*H and that the *p*H error is equal to the negative of the correction *p*H.

Low p***H***

For *low* *p***H** streams, the concern is the acid error and dehydration of the glass surface from an acidic environment. Although high temperatures do not greatly increase the acid error, extremely hot acidic streams dramatically decrease the glass's life expectancy.

The cause of the *acid* error at the low end of the *p***H** scale is not as well understood. Tests show that the gel layer absorbs acid molecules and increases its hydrogen activity relative to the process, which increases the *p***H** reading [15]. A reduction in the thickness of the outer hydrated gel layer of the measurement electrode marks the onset of the acid error. The low *p***H** end of the millivolt versus *p***H** line bends downward, as shown in Figure 4-4c, which results in an indicated *p***H** that is lower than the actual *p***H**. The sign of the *p***H** error is the opposite of that for the alkalinity error. The error is independent of temperature but increases with time. Thus, significant upscale *p***H** measurement drift at the low end of the *p***H** scale is symptomatic of the acid error. The initial acid error of Corning 015 glass electrodes is 0.2, 1.7, and 8.9 *p***H** for 3, 12, and 18 molar sulfuric acids, respectively. Acid error is not a problem for a feedback loop because it does not start above 1 *p***H** assuming the loop *p***H** set point is higher than 1. However, some *p***H** feedforward loops of strong acid influent have developed significant errors and failed. Figure 4-4c shows that the initial *p***H** correction magnitude increases exponentially as the *p***H** decreases below 0.5 *p***H**.

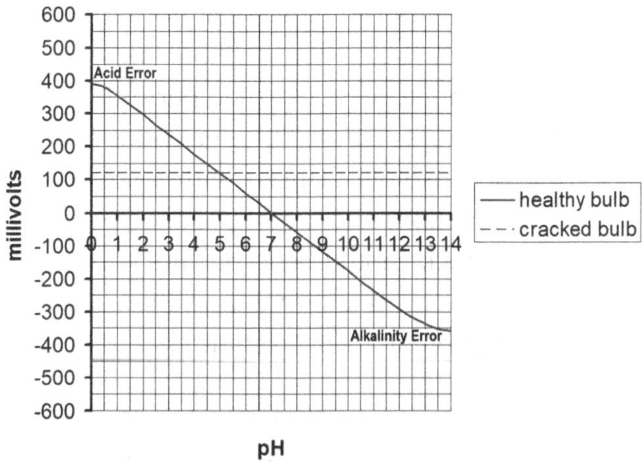

Figure 4-4c. Acid and alkalinity errors bend the millivolt versus *p*H line.

RULE of Thumb *For high and low* p**H** *streams, use special glass formulations and retractable electrodes or sample coolers to prolong the glass's life expectancy.*

Hydrofluoric acid will etch glass below 3 *p***H**. Because most *p***H** controller set points are higher, the solution is to ensure the *p***H** control is tight enough to reduce the number and duration of *p***H** excursions below 3 *p***H**.

Metal oxide measurement electrodes are a last resort when even with retractable sensors, there is significant chemical attack of caustic, etching of hydrofluoric acid from extended operation below 3 *p***H**, wear and cracking from abrasive particles, or the degradation from high temperatures. However, all metal electrodes develop an oxide coating that must be periodically scrubbed off and act as an ORP probe. A membrane can reduce the ORP effect but will significantly slow down the measurement electrode response time.

RULE of Thumb

*For hydrofluoric acid, the best bet is to use a periodically hydrated glass by the soaking of the electrode in water that is possible with an automatically retractable assembly and an operating point above 3 p**H**.*

Nonaqueous Solutions

For *nonaqueous* solutions where there is a different solvent than water, the issues are bulb dehydration, electrical resistance, electrolyte precipitation, and the relative nature and expanded range of the *p***H** scale.

To minimize the effect of bulb dehydration and the increase in solution electrical resistance, use a large area bulb with a low-resistance glass. Because this type of bulb is more susceptible to breakage, it is important to shield it from direct impingement and reduce the fluid velocity as much as possible without increasing the propensity for coatings. The hydrated gel layer must be replenished periodically, which is best accomplished by an automated retractable sensor. The electrodes should be cleaned with a solvent that will dissolve process solutes; rinsed with a volatile solvent, such as acetone, to remove the cleaning solution; and, finally, soaked in plant water to replenish the gel layer. Any calibration check should be in buffers with the same solvent as the process stream. As with other low-conductivity fluids, there is a heightened sensitivity to fluid velocity and streaming potentials from static charges.

The normal potassium chloride (KCl) saturated in water electrolyte should not be used because KCl will precipitate in a nonaqueous solvent, the water will form a separate phase, and the ions migrating into the junction will have unequal mobility, setting up a large diffusion potential. It is easiest to change the electrolyte in the outer chamber of an internal salt bridge in a double-junction reference electrode. An external reservoir and salt bridge can be used to minimize the frequency of refilling the electrolyte and provide a great level of protection. Common choices for electrolyte solutions for nonaqueous processes are lithium chloride (LiCl) in acetic acid, ethanol, or methanol, and potassium iodide (KI) in acetic acid, acetone, or methanol. Candidate solutions should be tested by soaking the electrodes overnight in the process and measuring the drift.

Large low-resistance bulbs and nonaqueous electrolytes are used in nonaqueous solutions that require cleaning, soaking, and buffer solutions.

The *p***H** of a nonaqueous solution can be below 0 *p***H** for an acidic solvent, such as acetic acid, and above 14 *p***H** for a basic solvent, such as an amine. The *p***H** measurements can only be compared to measurements in the same solvent because the dissociation constants of acids, bases, and solvent; the solvent's dielectric constant; and the ionic mobility and consequential activity coefficients will all be different. The neutral point is not necessarily 7 *p***H**. For example, at 25°C on the water-based scale, the neutral point of methanol is 8.42 *p***H**, and it is 9.55 *p***H** [16] for ethanol. The buffer solvent should be as similar as possible to the process solvent. If a batch process moves from an aqueous to a nonaqueous composition, the readings cannot be compared between phases, but an automatic retractable assembly can enable cleaning, buffer, and soaking solutions that match the sensor calibration and conditioning with the batch phase. If the water content drops below 5%, the solution can be considered to be nonaqueous [2].

High Ionic Strength

High ionic strength streams increase the reference junction potential and decrease the activity of the hydrogen ions. Unless the ions can penetrate the glass, the biggest adverse effect is the drifting potential of the reference. It may take hours to days for a nonflowing porous reference junction to reach ionic equilibrium. In general, a gel-fill reference electrode is slower than liquid-fill reference electrode, a double-junction reference electrode is slower than a single-junction reference electrode, and solid state is the slowest of all the reference electrodes. To improve the accuracy of the electrodes and hasten the approach to equilibrium, high ionic strength buffer solutions should be used for calibration and soaking (preconditioning). To create these buffer solutions, salts similar to the ones in the process that are compatible with the buffer are added so that the ionic strength (defined in Chapter 2, "The Chemistry") of the buffer is close to that of the process solution.

For high ionic strength solutions, use a fast reference electrode that is calibrated and preconditioned in high ionic strength buffer solutions.

Low Ionic Strength

In *low ionic strength* streams, such as pure water, the electrical resistance of the solution is high, making the measurement more susceptible to electrical noise and streaming potentials. Low-resistance glass, large spherical bulbs, and an upstream liquid reference electrode should be used. The electrode assembly should be shielded from electrical fields and isolated from ground potentials [17]. The solution temperature should be measured. The sample flow rate must be regulated to keep the velocity low and constant. Figure 4-4d shows an integrated assembly with these features for pure water applications, such as boiler feed water, demineralized water, and deionized water.

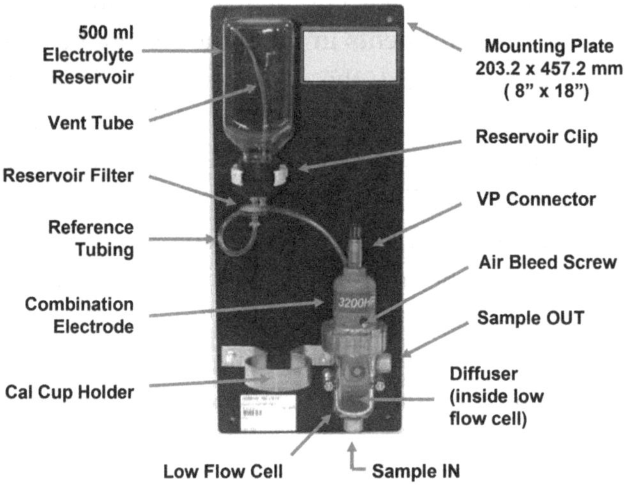

Figure 4-4d. Pure water assembly to reduce noise from streaming potentials, ground potentials, and electrical interference.

RULE of Thumb *For low ionic strength solutions, use a large, spherical, low-resistance bulb and minimize the noise from potentials and electrical interference.*

Cyanides, Bromides, Iodides, Sulfides, or Nitrates

For *cyanides, bromides, iodides, sulfides, or nitrates,* the concern is the reactive precipitation of the silver and potassium chloride in the electrolyte, which is observable as a blackening of the junction. The solution is to use an electrolyte in an internal replaceable or an external salt bridge that is compatible with the process, a differential electrode, or a solid-state reference electrode with an immobilized electrolyte or a nonporous junction.

RULE of Thumb *For sulfides and nitrates, use a compatible electrolyte in an internal replaceable salt bridge, an external salt bridge, a differential electrode, or a solid-state electrode.*

For *fluctuating and high pressures,* the principal concern is the separation and contamination of the reference electrode. Gel-filled electrodes can develop voids. Liquid-filled electrodes will suck in process fluids when the pressure in the process is greater than in the reference. Solid-state or pressurized reference electrodes with an internal or external salt bridge should be used. Ideally, the differential pressure between the process and the salt bridge should be constant to maintain a constant flow of electrolyte.

RULE of Thumb *For fluctuating and high pressures, use a solid-state or pressurized reference electrode with an internal or external salt bridge.*

For *concentrated surfactant solutions,* the reference junction potential rapidly shifts, and the surfactants tend to "salt out" in the reference junction when they contact the concentrated electrolyte solutions. Because the surfactants are conductive, neither of these effects correlates with reference junction impedance. The only way to detect these problems is to remove the sensor from the surfactant, rinse it with deionized water, and place it in a buffer solution. Depending on the concentration of the surfactant and the length of process exposure, you may observe an offset in the range of 0.2 to 0.8 *p***H** [4].

4.5 Smart Transmitters

Smart *p***H** transmitters not only do the standard central processing unit (CPU), continuity, and input/output (I/O) checking but they also measure the circuit resistances between the measurement and reference electrode inputs and the solution ground. The diagnostics must include changes in glass resistance with temperature, and low-conductivity process streams should also include the change in fluid resistance with temperature. A large decrease or increase in the resistance of the glass indicates a broken or dehydrated electrode, respectively. A large increase in the resistance of the reference electrode indicates that it is coated or plugged with a nonconducting material.

The best way to detect a coated measurement electrode is not from electrical resistance but from the speed of response to *p***H** changes. The additional electrical resistance from the process coating is typically small compared to the resistance of the glass. A thin, almost imperceptible film can dramatically slow down the migration of hydronium ions and increase the time constant of the electrode from a few seconds to several minutes. For loops in automatic, a slow measurement electrode shows up as either an increase in the loop's oscillation period or the settling time, depending on the controller tuning and coating thickness. If there is more than one *p***H** electrode, the coated electrode always will lag behind the clean electrode. For a ramping disturbance, the lag time is approximately the additional time constant of the coated electrode. If an automated retractable assembly is used, the time constant in a buffer solution can be measured. However, the flushing and cleaning cycle may remove the coating. To diagnose the coating that existed while in service and the effectiveness of the cleaning cycle, the time constant of the electrode should be checked before and after flushing and cleaning. As previously mentioned in the section on electrode selection for solids, it is not sufficient to just freeze the reading. The *p***H** controller must be put in manual.

There must be extensive communication between the control system, performance monitoring system, smart transmitter, cleaning system, and maintenance system. Ideally, a smart transmitter or performance monitoring system signals an impending need to clean or replace a probe. The control system decides if this is a good time based on production and process requirements and the severity of the problem, puts the *p***H** controller in manual, and signals the cleaning system to start its cycle. For batch operations, there are certain steps where the *p***H** electrode is not used, and for continuous operations, there may be times when the unit can run at reduced rates. The cleaning system diagnoses the "before" and "after" time constant and accuracy of the probe and tells the *p***H** control system when it is done and if the electrode is good enough to go back in service. The control system then decides when to put the *p***H** controller back in automatic. All the diagnostics from operation and cleaning are communicated to the automated maintenance system. A suggested flow of information is shown in Figure 4-5a.

Key CONCEPT Performance monitoring, cleaning, control, and automated maintenance systems and smart transmitters must not make unilateral decisions.

RULE of Thumb *The control system should schedule automated maintenance based on the severity of the problem and production and process requirements.*

A *p***H** measurement is a prime candidate for smart transmitters and automated maintenance systems because the life expectancy decreases and the performance (e.g., efficiency, offset, repeatability, 86% response time, etc.) degrades more with process conditions than any other measurement commonly used in industry.

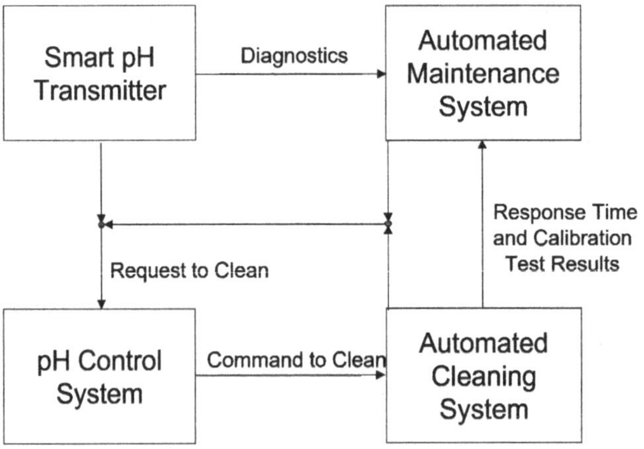

Figure 4-5a. The flow of information for automated maintenance of *p***H** electrodes.

4.6 Failure Protection

It is difficult to predict the failure mode of *p*H electrodes because they can fail anywhere on or off the *p*H scale. However, the middle selection of three *p*H measurements will inherently ride out the first measurement failure of any type. It will also ride out a second measurement failure if it is in the opposite direction of the first. The implementation is straightforward in modern control systems because there is a Fieldbus function block that offers MSS that can be used in the field or in a field-based DCS. It is important to use the middle measurement and not a median or average measurement. In misguided attempts to do something better, configurations have been developed to reject one signal based on expert system–type rules and then average the remaining two measurements, but these rules cannot possibly cover all the bases and will eventually end up doing more harm than good. MSS is simple and foolproof. Equipment and piping should have connections for three probes. The user must realize that the multiple measurements will never agree and that this realization is the key to stopping chasing short-term transient deviations between electrodes. In the only case where multiple measurements were consistently observed to have the same readings, the electrode tips still had their protective caps filled with a buffer solution [18]. If there are Coriolis mass flowmeters and the titration curve can be accurately identified or modeled, an online *p*H estimator can possibly be developed to replace one of the electrodes in a system with constant feed and reagent concentrations.

Key CONCEPT *p***H** measurements can fail anywhere on or off the *p***H** scale, but MSS will inherently ride out a single electrode failure of any type.

Key CONCEPT If multiple *p***H** measurements all have the same value, the electrodes are broken or disconnected, or the protective caps are probably still on the electrode tips.

Using three electrodes and MSS also has numerous other advantages, as cited in the many Key Insights in this book. It will reduce noise and eliminate spikes much more effectively than a transmitter damping adjustment or a process variable filter without adding a lag time to the loop. It will enable predictive maintenance based on the long-term trend of an electrode's deviation from the middle value. A single transmitter can be taken out of service without switching the controller to manual. A single electrode can also be manually or automatically retracted for cleaning, rejuvenation, and calibration with the controller in automatic unless there is a big spread between the remaining measurements. More intelligent maintenance reduces lifecycle cost, and better measurement performance offers benefits from better process efficiency and capacity.

Using two *p***H** electrodes raises more questions than answers because they rarely agree, and there is no referee other than the lab, which has its own set of problems. However, installations of two electrodes are much more prevalent than three. Usually, a manual switch enables operations to select the current favorite, but the selection is based more on "war stories" than logic. High- and low-signal selectors are used when a downscale or upscale failure, respectively, clearly poses a greater safety hazard, but such predictions are not reliable and do not address the day-to-day deterioration in response time and accuracy. Some control system engineers attempt to develop more customized logic, but this is exceptionally difficult because of many challenges, and it is prone to do more harm than good. MSS and the collaboration of smart transmitters with automated maintenance, cleaning, control, and performance monitoring systems, shown in Figure 4-5, will lead to much better predictive maintenance and failure protection.

4.7 Dynamic Response of Electrodes, Holders, and Sample Systems

The dynamic response of a clean *p***H** measurement electrode without any problems directly immersed in an aqueous solution can be approximated by the time required to reach 63% of the final response (T_{63}). Because the electrode dead time is negligible compared to other dead times in the system, T_{63} can be viewed as approximately the same as an open-loop time constant (τ_o) whose value varies between 2 and 30 s. Using a time constant is more effective than using a response time (T_{98}) (time to 98% of the final value) because T_{98} is difficult in practice to measure in industrial applications because of noise, *p***H** electrodes exhibit an unusually long tail, and the first part of the response is most important for process control. Using T_{86} (86% response time) is more effective. One should get the manufacturer's definition of response time (because it may not be T_{63} or T_{86}) and then convert it to a time constant. The percentage of the final value definition for response time can make a huge difference because T_{63}, T_{86}, T_{95}, and T_{98} theoretically correspond to 1, 2, 3, and 4 time constants, respectively, for a response with negligible dead time.

RULE of Thumb *Get the manufacturer's definition of response time in terms of the percentage of the final response and convert it to a time constant for control system analysis.*

The major contributors to an electrode's time constant are a boundary layer of fluid around the electrode bulb that slows down the diffusion and migration of ions to the *p***H**-sensitive glass and a double-layer effect at the surface of the glass that poses an energy barrier to the proton [19]. The hydronium ion moves from regions of high to low concentration (diffusion) and from regions of high to low positive charge (migration). The boundary layer, which slows down this movement, increases in thickness as the velocity of the stream past the electrode decreases. The time constant for velocities below 1 ft/s is large enough to deteriorate loop performance. Because the velocity in pumped pipelines is between 3 and 7 ft/s, whereas the velocity in sample lines and agitated vessels is usually less than 1 ft/s, the pipeline injector or insertion type of electrode described in the next section should be considered whenever a location in a pumped pipeline close to the mixing equipment outlet is available. The double layer has the greatest effect in nonaqueous solutions where the measurement electrode time constant may be 100 s or more. It is probably responsible for the time constant for a decrease in *p***H** being about twice as large as the time constant for an increase in *p***H** and for the time constant for buffered solutions being much smaller [20]. The time it takes for the proton to move to its site in the glass once it jumps the double-layer energy barrier has been measured to be about 30 ms by using a jet stream to eliminate the boundary layer and double-layer effect [21]. Table 4-7a summarizes some lab test results for electrodes with different stream velocities and *p***H** direction changes for a solution with a high buffer index (flat portion of a titration curve). Not shown is the fact that the time constants can be several times larger for steep curves. For example, the time constant of solutions with strong acids and strong bases is about two and four times larger for positive and negative changes, respectively, than what is shown in Table 4-7a. The table reveals that the best case 86% response time (T_{86}) is about 3 s in weak acid and weak base applications with a fluid velocity of about 4 ft/s.

Table 4-7a. Electrode time constants for a 3 *p***H** change in a buffered solution.

Direction	Velocity (ft/s)	Time Constant (s)
Positive	4	0.75
Negative	4	1.5
Positive	2	1.5
Negative	2	3
Positive	1	2
Negative	1	4
Positive	0.5	3
Negative	0.5	12

Key CONCEPT

The low velocity in vessels, flow-through cells, and sample line tees and solutions with strong acids and strong bases increase the glass electrode time constant.

A slime coating of 1 mm thickness slows down the migration of hydronium ions to the glass surface enough to increase the measurement electrode's response time (T_{86}) from 6 s to 5 min [7]. Chapter 8 shows how a slow electrode response greatly increases the loop's oscillation period and the tendency to oscillate. The amplitude of the measurement oscillations is an attenuated version of the real-world *p***H** fluctuations. For thick coatings, a hydronium ion concentration is trapped and the *p***H** reading freezes.

A change in salt concentration will change the electric field and alter the migration of hydrogen ions enough to change the *p***H** until a concentration gradient develops and the diffusion of the hydrogen ions reestablishes the original hydrogen ion activity at the measurement electrode surface. For example, adding 0.1 mol/L of sodium chloride to a 3 *p***H** solution of hydrochloric acid will cause the *p***H** to temporarily drop to 2.9 *p***H** and return to 3 *p***H** about 6 s later [19]. The measured *p***H** goes through a transient even though the true *p***H** has not changed. The control loop must ignore rather than chase such transients, which are essentially noise.

Near the neutral point, the titration curve is steeper and fluctuations that might be considered insignificant concentration fluctuations are amplified by the high process gain. This can be visualized by plotting a fluctuation on the x-axis and translating it to the y-axis of the titration curve. An electrode, transmitter damping, or signal filter time constant will attenuate this noise and make a trend in manual look better, but it also slows down the *p***H** controller's ability to see upsets and the results of its own actions. The result is an increase in the loop's oscillation period and a tendency to oscillate that increases the loop's settling time. A single back-mixed volume and MSS are the preferred methods for smoothing out noise and reducing loop variability in automatic.

> **Key CONCEPT** The trend for a large electrode time constant looks less noisy for a loop in manual but shows slow, less-damped oscillations for a loop in automatic.

The time constant and dead time (time delay) from a flow-through cell or chamber and the dead time from a sample line are frequently larger than the electrode time constant. The chamber is partially back mixed from the turbulence of the incoming flow, so the residence time (volume divided by flow) is split between a time constant and an equivalent dead time. If the turbulence creates an agitation flow equivalent to the sample flow, the split is equal so the time constant and dead time are equal to one-half the residence time per Equations 4-7a through 4-7d. If these values are small compared to the process time constant, as would be the case for a vessel, the distinction of the split is of no consequence and the whole residence time can be considered a dead time because small time constants become equivalent to dead time as shown in Chapter 8.

The dead time can be assumed to be equal to the residence time of the sample line per Equation 4-7b because the time constant from the velocity profile is usually small compared to the total transportation delay so that the sample line is essentially plug-flow (no back mixing). Because the volume of the sample line is equal to the length multiplied by the cross-sectional area, the dead time can be decreased by decreasing the length or the diameter of the line or increasing the sample flow. The same relationship holds for the transportation delay from a pumped pipeline. A common mistake is to install a large agitator to reduce equipment dead time and then install the electrodes in a flow-through holder or chamber with a long length of intervening sample line or pumped pipeline so the resulting measurement dead time is greater than the equipment dead time. A heat exchanger in the sample line to lower the sample temperature or in the pipeline to remove the heat of neutralization can add significantly to the flow path volume and, therefore, the transportation delay. Chapter 8 discusses how the total loop dead time deteriorates loop performance.

The process open-loop time constant is the residence time of a back-mixed volume:

$$\tau_p = \frac{V_m}{F_s} \tag{4-7a}$$

The process time delay (dead time) is the residence time of a plug-flow volume:

$$\theta_p = \frac{V_p}{F_s} \tag{4-7b}$$

For a 50% back-mixed equipment volume (flow-through cells):

$$V_m = 0.5 \bullet V_e \tag{4-7c}$$

$$V_p = 0.5 \bullet V_e \tag{4-7d}$$

For plug-flow (sample lines, pipelines, static mixers, and heat exchangers):

$$V_m = 0 \tag{4-7e}$$

$$V_p = A \bullet L \tag{4-7f}$$

where

A = cross-sectional area of a sample line, pipeline, static mixer, or exchanger (ft^2)
F_s = volumetric flow rate (ft^3/min)
L = length of a sample line, pipeline, static mixer, or exchanger (ft)
τ_p = process open-loop time constant (min)
θ_p = process time delay or dead time (min)
V_e = equipment volume (ft^3)
V_m = back-mixed volume (ft^3)
V_p = plug-flow volume (ft^3)

4.8 Installation Practices

The sensor location should provide the most representative, most reliable, and fastest measurement. The electrodes should measure the *p*H of a perfectly mixed combination of feed or influent streams and reagents with minimum maintenance, time delay, and lag.

In general, *p*H sensor location should be chosen to meet the following objectives [8]:

- Safe and convenient access for calibration and maintenance
- Permissible process conditions (e.g., temperature and pressure)
- Representative mixture (no short-circuiting of reagent, stagnation, or layering)
- Optimal fluid velocity for minimum coatings and abrasion
- Minimum transportation delay

> Key CONCEPT — The sensor location should provide the most representative, reliable, and fastest measurement.

Electrodes should be installed in homogeneous, well-mixed zones without bubbles. In rare cases, electrodes are intentionally inserted in a water layer of a vessel used for phase separation, such as a decanter, because there is no liquid mixture, and oil or nonaqueous solvents cause maintenance, speed of response, and accuracy problems.

If an electrode is located too close to the reagent's point of entry into a vessel, short-circuiting occurs, where the electrodes see the effect of the reagent before it is mixed with the contents of the vessel (Mistake 6 in Figure 1-1e in Chapter 1). If it is located in a stagnant area of a poorly mixed vessel, then it is seeing old and inconsistent mixtures and is more susceptible to coatings (Mistake 10 in Figure 1-1e).

The additional dead time from a delayed and slow measurement increases the loop's oscillation period, control error, and sensitivity to nonlinearity. The installation must minimize the number of times the electrodes must be removed for maintenance (e.g., calibration and cleaning). Removing and manually handling the electrode increases errors and reduces the life of the electrode. The fragile gel layer is altered, and the equilibrium achieved by the reference junction is upset.

Key CONCEPT The more frequently an electrode is manually removed from service, the more often it will need to be removed.

It might seem that the submersion assembly is best for control. However, velocities below 1 ft/s dramatically slow down the electrode measurement response because of the increased boundary layer near the glass surface and promote the formation of deposits that can further slow down the measurement. The bulk velocity in even the most highly agitated vessels rarely exceeds 1 ft/s and is often much lower. The result is coating problems and a slow response. Removing a submersion assembly is cumbersome and time-consuming because of the assembly's long length. Jet nozzles help reduce the number of times the submersion assembly must be removed. Long assemblies in highly agitated vessels require a protective tube with side supports.

Before the 1990s, the most prevalent electrode installation assembly was a flow-through cell chamber or a simple pipe tee into which a probe was threaded. A small sample stream was diverted from the tank or pipeline, entered the bottom of the flow chamber, and exited the side. Strainers, filters, jet washers, and ultrasonic cleaners are added to reduce coating problems. Unfortunately, plugged strainers and filters can become as much of a maintenance headache as coating on the electrodes. Figure 4-8a shows many options of how an electrode is installed in sample lines. Winterizing all the fittings and items is difficult, and the temperature control is typically lacking to keep the sample temperature equal to the process temperature. Today sample systems are viewed as undesirable because of the additional transportation delay and increased installation and maintenance costs.

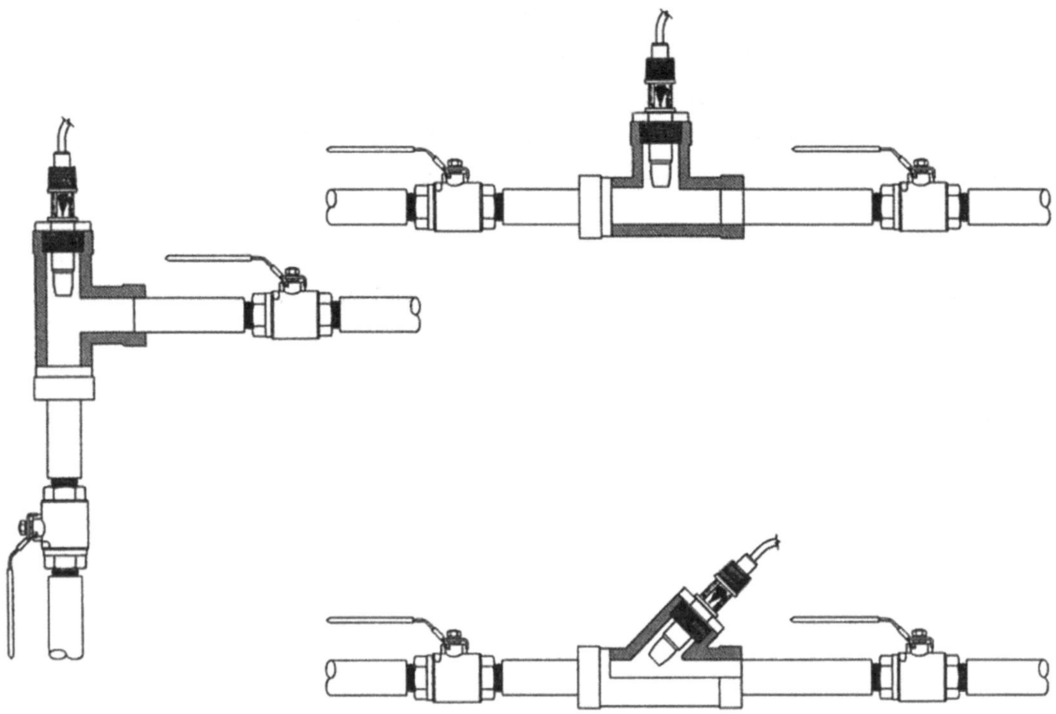

Figure 4-8a. Inserting *p*H probes in sample lines.

Key CONCEPT Electrodes in submersion assembly flow-through cells and sample lines tend to have more coating problems and a slower response.

Key CONCEPT Changes in the sample temperature from winterization and ambient conditions change the dissociation constants and the solution *p*H.

The development of the retractable injector probe, shown in Figure 4-8b, enabled inserting electrodes into and removing electrodes from pressurized vessels and pipelines. The high velocity in pumped pipelines decreases the coating rate and the time constant of the electrodes. The electrodes can be removed, even though the pipeline is pressurized, by slightly loosening the fitting that grips the electrode's metal sheath, slowly withdrawing the sheath until it clears the ball valve, closing the ball valve, and withdrawing the sheath the remaining distance. In some respects, this type of assembly is safer than a flow chamber with an integral terminal enclosure because the sealing method is easily checked and any leaks are visible. The restraining strap must never be removed to prevent the probe from becoming a projectile because of worn or missing fittings. A tee and an additional ball valve can be provided to flush the assembly before removing the sheath. The retractable assembly enables installing electrodes in pipelines for vessel recirculation and downstream of static mixers, and it eliminates the need for isolation and bypass valves.

The best location, except for very abrasive service, is a recirculation line close to the outlet. Installing an electrode downstream of the pump is preferred because the strainer will block and the pump will break up clumps of material that might break the electrodes (Mistake 11 in Figure 1-1e in Chapter 1). The retractable probe provides the most straightforward and economical solution. The electrode tips must be pointed down to prevent the air bubble inside the electrode fill from residing in the tip and drying out the inside surface. An installation angle of 20 to 90 degrees from horizontal is sufficient to keep the bubble out of the tip. Some electrode designs eliminate the bubble via a flexible diaphragm for fill contraction and expansion.

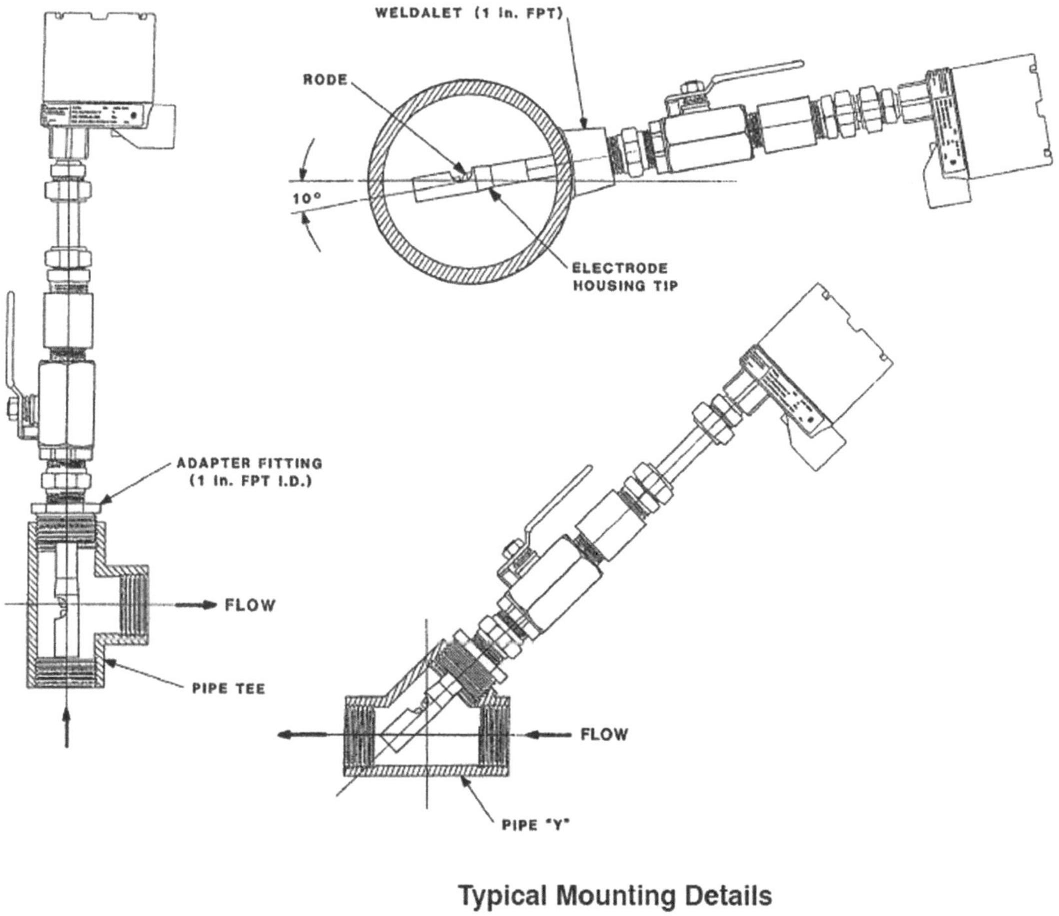

Figure 4-8b. Retractable *p*H probe for pipelines.

Key CONCEPT Retractable probes in pumped pipelines have cleaner and faster electrodes and open the opportunity to just leave the probes alone.

Retractable injector–type probes are usually installed in a horizontal pipeline unless vertical lines are needed to help drain solids. Parallel lines enable the probes to be isolated, but the velocity and distribution of solids are not equal. Consequently, there tends to be an offset and a difference in life expectancy and coating rate between probes. Installing probes in series ensures that each probe sees the same mixture and velocity, which leads to more consistent readings and maintenance requirements. For coating and abrasive service, the slot or hole in the shroud is facing into and away from the flow stream, respectively. For all lines, the velocity is adjusted, and in vertical lines, the angle of insertion is varied based on whether the goal is to prevent coatings or abrasion. Figures 4-8c and 4-8d show the arrangements.

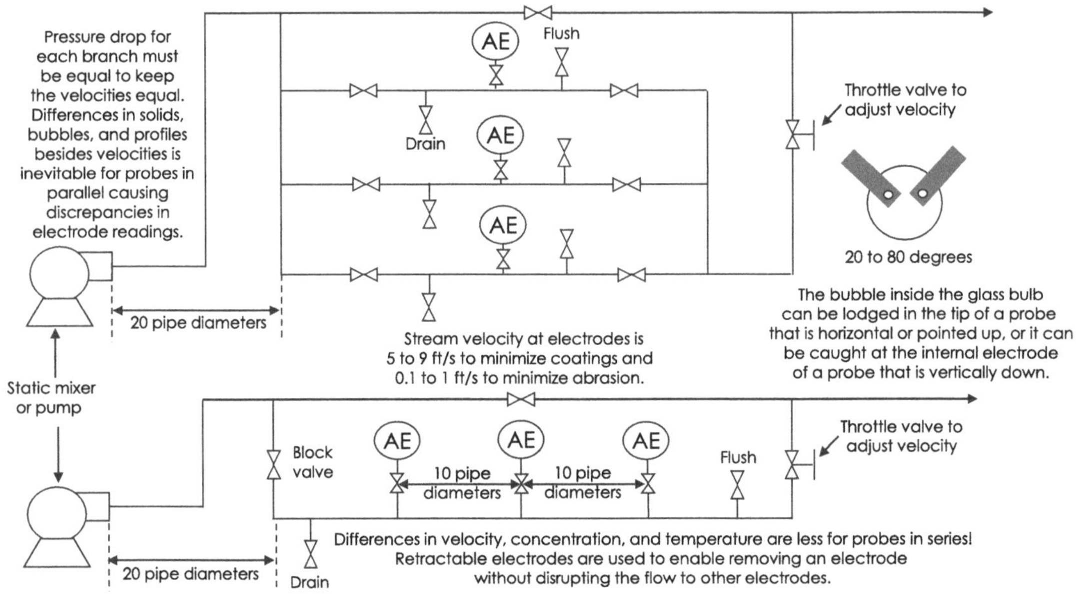

Figure 4-8c. Horizontal piping arrangements for injector probes.

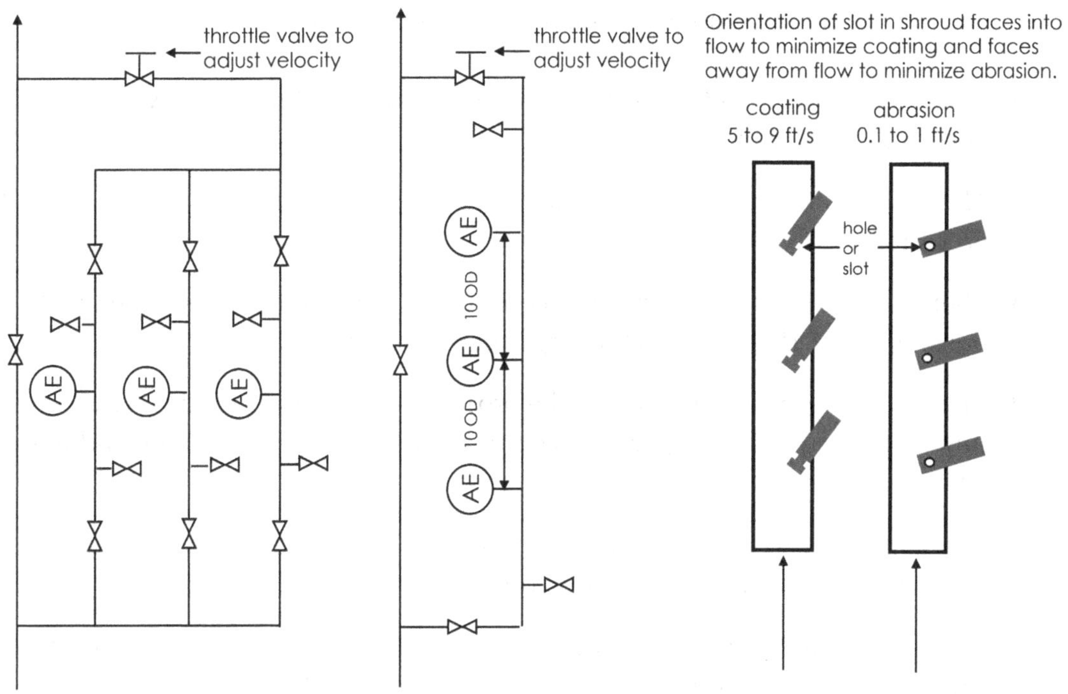

Figure 4-8d. Vertical piping arrangements for injector probes.

Key CONCEPT Installing multiple probes in series ensures each electrode will see the same velocity and mixture that is important for consistent performance.

Electrodes do best if they are left in service, exposed to and in equilibrium with the process. In severe services where the rate of fouling or degradation is excessive, the electrodes must be removed and serviced periodically. Automated procedures provide much more consistent results, especially when it comes to *p*H electrodes. Automated retractable assemblies have double-acting pneumatic piston actuators that can withdraw and insert an electrode on command by energizing or de-energizing solenoid valves on the supply and exhaust air lines. In the retracted position, the electrode tip sits in a chamber. Solenoid valves on various solutions to the inlet of the chamber are energized to flush, clean, calibrate, and soak the electrode. A solenoid valve on the outlet drains the chamber.

RULE of Thumb *If velocity cannot keep the electrodes clean, the process exposure time must be reduced, or the gel layer rejuvenated, using an automated retractable probe system.*

The following is a compilation of common installation pitfalls.

- The measurement electrode cable is frayed or its connector is wet. (A low impedance path to ground is created.)
- The shield to the preamp is grounded. (A ground loop to the solution ground is created.)
- There is no provision for continuous circulation and self-draining the pipeline or sample line. (Electrodes sit in stagnant process fluid.)
- The temperature measurement of the process fluid at the electrodes is not transmitted. (The solution temperature compensation in the control room and in the lab is inaccurate.)
- The sample line is not winterized. (The sample line freezes during water batching or when the sample flow is shut off.)
- There is no throttle valve and no meter for adjusting or measuring velocity at the electrodes. (Partially plugged lines or improper velocities are not detected.)
- The electrode tips are in the top of a line or near the surface of a vessel. (Air bubbles collect near the electrode tips, drying out the internal gel layer and causing a noisy signal.)
- The slot in the sheath of an injector probe is pointed into the flow for a slurry stream. (Electrode abrasion develops.)
- The slot in the sheath of an injector probe is not pointed into the flow for a fouling stream. (Electrode coating develops.)
- The electrodes are not kept wetted for batch or intermittent operations. (Dehydration of the electrode occurs.)
- The submersion electrode assembly is near the bottom of the vessel and is not supported. (The assembly breaks due to the whipping action from agitation.)
- The submersion electrode assembly is installed in a stilling well. (The response is sluggish, and a coating develops due to insufficient velocity past the electrodes.)

4.9 Calibration Procedures

In general, electrodes should have a "two-point buffer calibration" before they are commissioned and after they are cleaned or the gel layer is rejuvenated. When an electrode has been in service long enough for the reference electrode to reach equilibrium, a "grab sample" process standardization should be done. Thereafter, process standardizations should only be performed when three or more samples—preferably spread over several days—show the same relative offset to prevent the common problem of chasing calibration adjustments. Samples that come back with alternating error signs are a sure sign of mixing or measurement noise. Chasing noise will cause additional errors and create additional work. Data analysis and selecting the middle value of three electrodes or two electrodes and one estimator can be used to improve the understanding of what is noise and what is real to reduce the frequency of process samples. In the "grab sample" method, the measured *p*H must be time synchronized with the time the sample was taken from the process (not when it was measured in the lab). Also, the sample must be at the same point in the process as the *p*H measurement electrode, and the lab measurement must be made at the same temperature or corrected to the same temperature in the process as the *p*H measurement electrode. All the concerns about the lab sample mixture changing because of absorption, evaporation, dissolution, and reaction discussed in Chapter 3, "Titration Curves," should be addressed.

Key CONCEPT Buffer calibrations upset the ionic equilibrium of the reference.

RULE of Thumb *After an electrode is in service, a careful process standardization based on long-term offsets is preferred over a buffer calibration unless the electrode must be removed to be cleaned or rejuvenated.*

The transmitter calibration for the electrodes and the output signal to the control room should be checked before they are installed to verify the integrity as received and again before start-up to verify the integrity as installed. The electrode and output calibration may not have all the calibration adjustments shown in Equations 4-1m and 4-1o or may use different names. The compensation by the isopotential *p*H adjustment for the change in mV generated per *p*H due to a change in electrode temperature has a vastly different effect than an offset adjustment that may be known as a *standardization potential*. The electrode slope setting is sometimes noted as an efficiency adjustment. The transmitter output span setting is entered either directly or via *p*H values for 100% and 0% scale. The 0% scale value is sometimes called an *elevation adjustment*.

Buffer Calibration [8]

The standardization and slope adjustments interact for any *p*H range that does not start at the isopotential point. Thus, these adjustments are repeated until the desired accuracy is achieved. This interaction can be eliminated by adjusting the isopotential point.

The basic two-point buffer calibration procedure is as follows:

1. Clean the sensor.
2. Rinse the electrode with distilled water.
3. Enter the isopotential *p*H.
4. Insert the electrodes into a buffer solution that matches the isopotential *p*H value and adjust the offset (standardization potential) to make it match the buffer. Make sure the buffers are fresh. Use high ionic strength or nonaqueous buffers as necessary to match the process composition and allow enough time for the reference to reach ionic equilibrium. Measure and note the time constant of the measurement electrode response.
5. Rinse the electrode with distilled water.
6. Insert the electrodes in a 4 *p*H buffer for a set point below 7 *p*H or in a 10 *p*H buffer for a set point above 7 *p*H and adjust the span, slope, or efficiency setting. Make sure the buffers are fresh and a 10 *p*H buffer has not been sitting exposed to air for more than a few hours. Use high ionic strength or nonaqueous buffers as necessary to match the process composition and allow enough time for the reference to reach ionic equilibrium. Measure and note the time constant of the measurement electrode response.
7. If there is a severe interaction between the slope and standardization adjustment, go back to step 4 and adjust the isopotential or offset setting.
8. Store the *p*H electrode in either a 4 *p*H solution to condition the gel layer of the glass or a 7 *p*H buffer to reduce the loss of ions from the reference junction. Electrodes should never be stored in distilled, deionized, or pure water because these will cause a large loss of ions from the glass and the reference junction.

Additional Guidance [8]

*p*H transmitters are factory calibrated for an accurate millivolt potential measurement, but due to variations in *p*H sensors and aging effects, initial and routine *p*H sensor calibration is necessary. The transmitters can be set up and calibrated using handheld communicators or an asset management system (AMS). The AMS windows can provide past calibration results and guidance, as shown in Figure 4-9a buffer calibration tips.

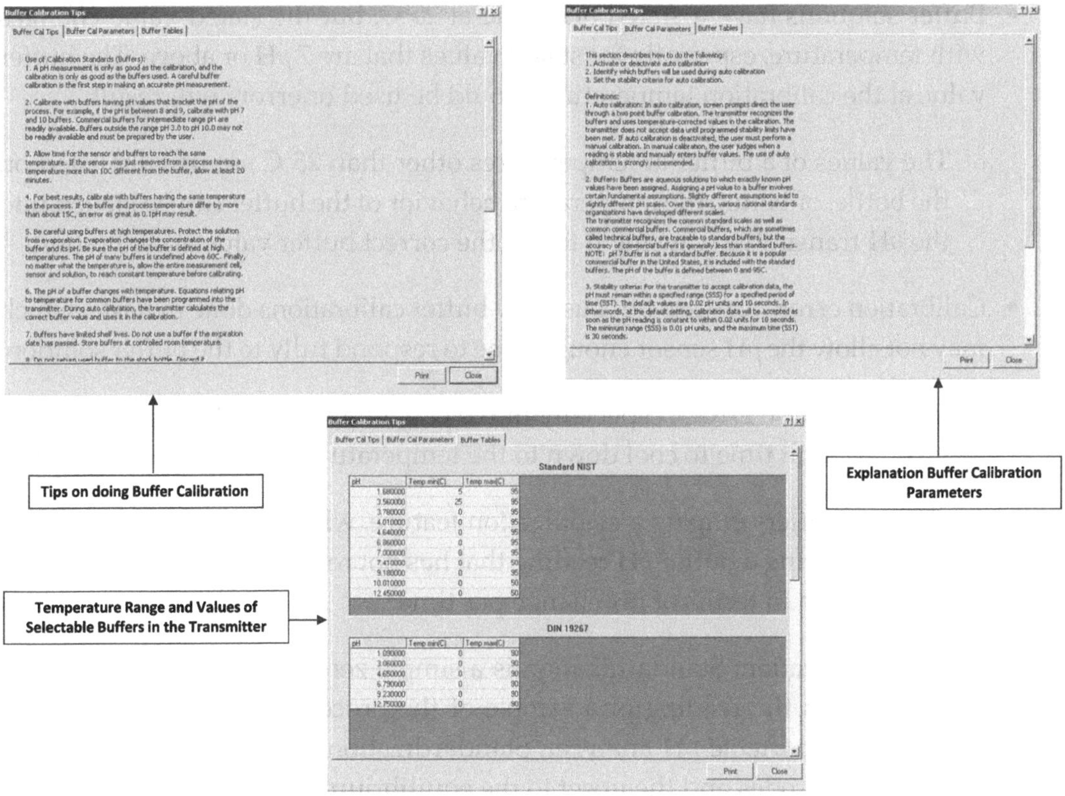

Figure 4-9a. AMS window for buffer calibration tips.

- Buffer calibrations use two buffer solutions, usually at least 3 *p***H** units apart, that cause the *p***H** analyzer to calculate a new slope and zero value, corresponding to the particular characteristics of the sensor, to more accurately derive *p***H** from the millivolt and temperature signals.
    - The slope and zero value derived from a buffer calibration provide an indication of the condition of the glass electrode from the magnitude of its slope, while the zero value gives an indication of reference poisoning or asymmetry potential, which is an offset within the *p***H** electrode itself.
        - The slope of the *p***H** electrode tends to decrease from an initial value relatively close to the theoretical value of 59.16 mV/ *p***H**, largely due, in many cases, to the development of a high-impedance short within the sensor, which forms a shunt of the electrode potential.
        - Zero offset values generally lie within + 15 mV due to the liquid junction potential. Larger deviations are indications of poisoning.

- Buffer solutions have a stated *p***H** value at 25°C, but the stated value changes with temperature, especially for stated values that are 7 *p***H** or above. The buffer value at the calibration temperature should be used or errors will result.
  - The values of a buffer at temperatures other than 25°C are usually listed on the bottle, or better, the temperature behavior of the buffer can be loaded into the *p***H** transmitter, causing it to use the correct buffer value at calibration.
- Calibration errors can also be caused by buffer calibrations done in haste, which may not allow the *p***H** sensor enough time to respond fully to the buffer solution.
  - This will cause errors, especially in the case of a warm *p***H** sensor not being given enough time to cool down to the temperature of the buffer solution.
  - *p***H** transmitters employ a stabilization feature, which prevents the analyzer from accepting a buffer *p***H** reading that has not reached a prescribed level of stabilization in terms of *p***H** change per time.
- *p***H** Standardization: Standardization is a simple zero adjustment of a *p***H** analyzer to match the reading of a sample of the process solution made using a laboratory or portable *p***H** analyzer. Standardization eliminates removing and handling electrodes and the upset to the equilibrium of the reference electrode junction. Standardization also takes into account the liquid junction potential from high ionic strength solutions and nonaqueous solvents in chemical reactions that would not be seen in buffer solutions. For the greatest accuracy, samples should be measured immediately at the sample point with a portable *p***H** meter. If a lab sample measurement value is used, it must be time-stamped and the lab value compared to a historical online value for a calibration adjustment. The MSS value from three electrodes of different ages can be used instead of a sample *p***H**. The periodic standardization based on data analytics can be automated, greatly reducing the need for samples and freeing up operations, lab, and maintenance technicians.
  - Standardization is most useful for zeroing out a liquid junction potential, but some caution is warranted when using the zero adjustment.
    - A simple standardization does not demonstrate that the *p***H** sensor is responding to *p***H**, as does a buffer calibration, and in some cases, a broken *p***H** electrode can result in a believable *p***H** reading, which may be standardized to a grab sample value.

- A sample can be prone to contamination from the sample container or even exposure to air; high-purity water is a prime example. A referee measurement must be exposed to a flowing sample using a flowing reference electrode.
- A reaction occurring in the sample may not have reached completion when the sample was taken, but it will have completed by the time it reaches the lab.
- Discrepancies between the laboratory measurement and an online measurement at an elevated temperature may result from the solution *p***H** being temperature dependent. Adjusting the analyzer's solution temperature compensation (not a simple zero adjustment) is the proper course of action.
- Remember that the laboratory or portable analyzer used to adjust the online measurement is not a primary *p***H** standard, as is a buffer solution, and although it is almost always assumed that the laboratory is right, this is not always the case.

4.10 Troubleshooting Logic

Even with combination electrodes, it is helpful to diagnose whether the main problem is with the measurement or reference electrode so the design can be improved. For example, it might be best to change the reference rather than the glass type or shape. If the *p***H** indication has high-frequency noise, a slow response, or a shortened span, there is most likely a measurement electrode problem. If the *p***H** indication drifts or develops an offset, it is probably a reference electrode problem. If the offset is very large, the reference electrode is probably contaminated. If you have multiple electrodes, you can swap wires to try different pairs of measurement and reference electrodes to identify which electrodes are the culprit and which combination most closely matches the process sample or buffer *p***H**.

RULE of Thumb *Spikes, fast noise, slow response, and a shortened span are symptomatic of glass electrode problems, whereas drift and an offset are more symptomatic of reference electrode problems.*

Spikes or rapid changes in *p***H** (see Figure 4-10a) are often seen in *p***H** recordings. The user must first determine whether the spike represents a real process change. If there are three electrodes and a middle value is selected, the control system

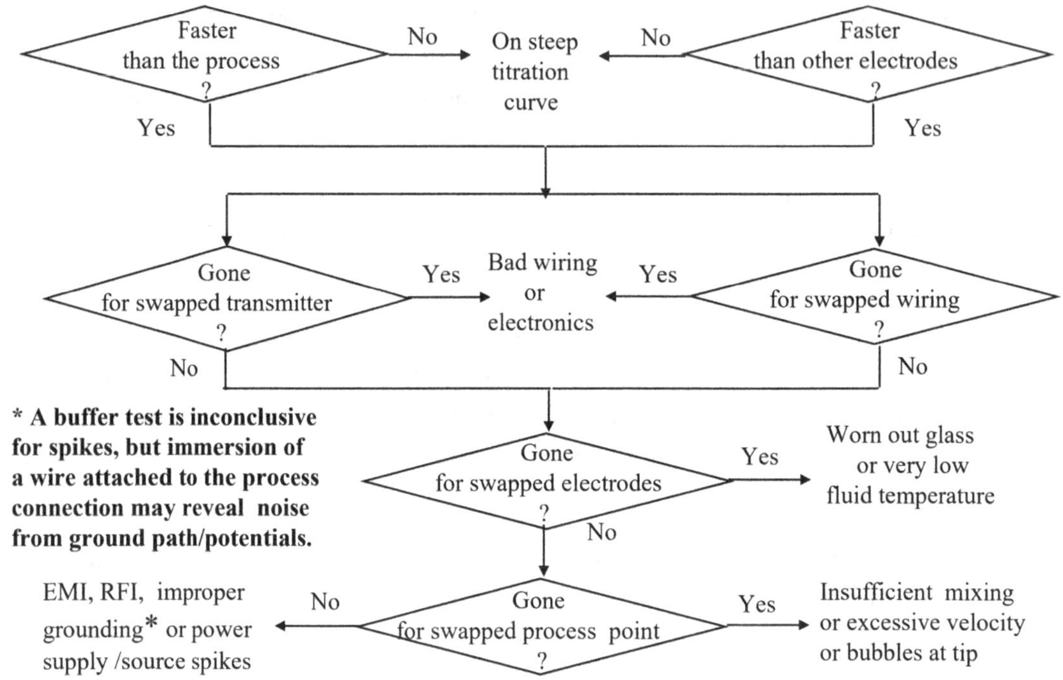

Figure 4-10a. Troubleshooting logic for spikes and rapid changes.

automatically ignores unreal spikes. Also, the user can readily infer a spurious change (fast or slow). For a single electrode, the user must figure out how fast a concentration change can occur and then translate from the abscissa to the ordinate of the titration curve to estimate the possible rate of *p*H change. This is not easy, and most people do not anticipate how fast *p*H can change from movement on the steep part of the titration curve.

A dehydrated or worn-out gel layer, or a temperature lower than the glass rating, will cause high glass impedance and a noisy response. One indication of this condition is a fluctuation in the *p*H reading caused by moving the lead wires by wind or by hand.

Swapping transmitters, preamps, and circuit wiring can isolate problems in the measurement circuit but they do not rule out grounding and power problems or electrical interference. If a grounded wire inserted into a buffer solution shifts the reading, there is a ground path. This is best done in the field near the actual installation so that the effect of local ground potentials is included.

RULE of Thumb *To test for ground paths, insert a wire that is connected to a local ground into the buffer beaker to mimic the solution ground shifts and determine if the* p**H** *reading shifts.*

Swapping process connections can help track down the source of the spike to concentration gradients from poor mixing, streaming potential electrostatic charges from high fluid velocity, or upscale noise from bubbles at the tip of the measurement electrode.

A sluggish *p*H measurement (see Figure 4-10b), as indicated by an increase in the loop period that is not caused by a change in controller tuning, is easily tracked down to a coated or broken glass measurement electrode tip if there are three electrodes and using the middle select prevents the loop period deterioration. Tough coatings can be removed by briefly soaking or spraying with a dilute hydrochloric acid solution. Glass bulbs should never be manually scrubbed. A buffer test for response time will show the slowness if the coating was not washed off in preparing for or conducting the buffer test.

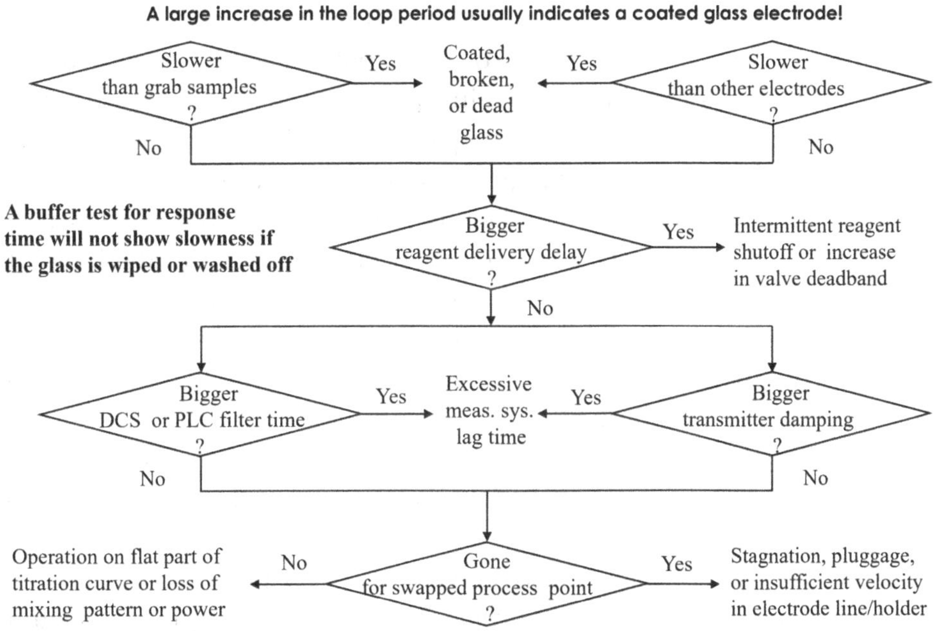

Figure 4-10b. Troubleshooting logic for a sluggish response.

If the electrodes are diagnosed to be clean and in good condition, the most common causes are an increase in reagent delivery delay from intermittent shutoff and thus backfilling the reagent line per Chapter 6, "Control Valves," and an increase in dead time from valve deadband (backlash) and stick-slip (friction) per Chapter 7, "Reagent."

An increase in a filter time in a programmable logic controller or a DCS or a damping setting in a transmitter will add a measurement time constant, which contributes to loop dead time and, therefore, loop period.

The equations developed in the first section of this chapter enable a more sophisticated type of troubleshooting in terms of changes in the *p*H calibration curve. Performance monitoring systems will eventually be designed to diagnose these systems based on information from smart transmitters and automated cleaning and calibration systems.

Chunks of solids impinging on the electrode, rough handling, chemical attacks, or old age can lead to *cracks in the glass*. If the measurement electrode bulb breaks, process fluid will be in contact with both the inside and outside *p*H-sensitive glass layers. The hydrogen activity, and therefore, the potential, will be about equal so the difference in potentials is nearly zero, which corresponds to a constant 7 *p*H reading. However, the measurement electrode glass resistance is bypassed (paralleled) by a much smaller process fluid resistance so that R_1 decreases and the measured potential increases per Equation 4-1j. The increase in measured potential is seen as a decrease in *p*H. The result is a shift of the isopotential point vertically upward and a counterclockwise rotation of the millivolt versus *p*H line to a horizontal position, as shown in Figure 4-8a. To determine the *p*H reading for this line position, the original line position used in the calibration must be known. For example, if the calibration is represented by the 45-degree line, the horizontal line in Figure 4-8a at 120 mV would give an indication of 5 *p*H. Field experience shows that a broken measurement electrode causes a constant *p*H reading between 4 and 6 *p*H. If the reference electrode breaks, the internal fill is contaminated with the process fluid. The reference electrode half-cell potential will change and shift the isopotential point vertically upward or downward, and the *p*H span will accordingly be biased in the opposite direction. Unlike the measurement electrode, the reference electrode fill can be contaminated without breakage due to its liquid junction. Pulsating pressures or high ion concentrations can force ions from the process fluid through a porous junction into the internals of the reference electrode.

Slurries can cause bulb *abrasion*, nonaqueous fluids or air pockets can cause *dehydration*, and hydrofluoric acid can cause *etching*. The response symptoms for bulb abrasion, dehydration, and etching are the same. In each case, the outer glass layer is less sensitive to hydrogen activity. In other words, a given activity will be measured as a lower activity that corresponds to a higher *p*H (*p*H is the inverse logarithm of hydrogen ion activity). The response span is shortened with an upscale offset. The outer glass voltage for a given activity and the change in voltage for a change in activity will be less. The isopotential point is shifted left, and the millivolt versus *p*H line slope is decreased by rotating the line counterclockwise toward the horizontal position. The increase in the glass bulb resistance due to dehydration will cause the isopotential point also to shift downward vertically, which aggravates the upscale shift of the *p*H span. The response may become noticeably slower and erratic because of the reduced and irregular area for proton exchange between the fluid and the outer gel layer.

Oils, tars, gums, polymers, and particles can cause the coating of measurement and reference electrodes. A *partial coating* on the measurement electrode forms a layer that slows down the diffusion of the hydrogen ions from the process fluid to the glass surface. The result is a very slow measurement response. A *complete coating* on the measurement electrode forms a barrier that stops all diffusion of the hydrogen ions from the process fluid to the glass surface. The result is a constant *p*H reading value that depends on the hydrogen activity of the coating. Because the hydrogen activity of most coatings is low, a constant upscale *p*H reading is common. This corresponds to a horizontal millivolt versus *p*H line at a negative millivolt level. A partial coating of the reference electrode will change the liquid junction potential and the resistance at the electrode tip. Usually, the junction potential decreases and the resistance increases so that the isopotential point is shifted down and the *p*H span is shifted upscale. A complete nonconductive coating can cause an open circuit between the reference and measurement electrodes. The transmitter output will go off scale up or down, depending on the *p*H meter manufacturer and model number. The more typical failure mode is upscale.

The presence of a nonaqueous solvent is indicated by a *p*H beyond the normal 0 to 15 *p*H range. An *acidic solvent* acts as a proton donor causing the proton activity to increase and the *p*H scale to shift down. For example, the *p*H scale for an acetic acid solvent is –6 to –1 *p*H, and for formic acid, the solvent is –9 to –2 *p*H. A *base solvent* acts as a proton acceptor, decreasing proton activity and shifting the *p*H scale up. For example, the *p*H scale for an ammonia solvent is 16 to 49 *p*H. An *alcohol solvent* acts as both a proton donor and an acceptor, like water, moderating changes in proton activity. The *p*H scale occupies about the same region as water but may extend further upscale and downscale. For example, the *p*H scale for an ethanol solvent is –4 to + 16 *p*H. A *hydrocarbon solvent* acts neither as a proton donor nor an acceptor; the solvent is passive to changes in proton activity. Consequently, the *p*H scale upper and lower limits are usually spread further apart than those for water. For example, the *p*H scale for an acetone solvent is –5 to 20 *p*H. Often, nonaqueous solvents have a higher resistance than water. So, the isopotential point is shifted downscale, and the *p*H span is shifted upscale. This shift is counteracted by the more positive liquid junction potential for acid solvents but is accentuated by the more negative liquid junction potential for base solvents. It is important to realize that dehydration, with all its attendant symptoms and errors, will also occur for all nonaqueous solutions unless the measurement electrode is periodically rinsed with water to replenish the hydronium ions in the gel layer. When dehydration starts, the response slows down (an increase in the loop oscillation period) and the *p*H measurement drifts upscale. In a control loop, the measurement drift would not be noticeable until a base reagent valve had drifted all the way open or an acid reagent valve had drifted all the way closed. Thus, the controller output should be monitored for early detection of *p*H measurement drift.

Pure *water solutions* have a high solution resistance like nonaqueous solutions. The addition of a small concentration of ions changes both the conductivity and *p*H tremendously, resulting in an erratic measurement besides the upscale shift of the *p*H span. Most pure water streams will absorb enough carbon dioxide on exposure to air to exhibit a buffering effect below 7 *p*H.

Changes in *ionic strength* will cause changes in the liquid junction potential of the reference electrode. If the concentration of positive ions increases, the liquid junction potential increases and shifts the isopotential point upscale and the *p*H downscale. If the concentration of negative ions increases, the liquid junction potential decreases or becomes more negative and shifts the isopotential point downscale and the *p*H upscale.

Gas *bubbles* can result from gas entrainment from excessive agitation, gas evolution from a reaction, or a gas reagent. The gas bubbles cause a high resistance and reduced hydrogen activity at the measurement electrode on impact. The result is an intermittent shift of the isopotential point to the left and down, which creates upscale *p*H noise.

A *short* from the measurement to the reference electrode terminals will cause a zero mV input and, thus, a 7 *p*H reading. In fact, the standby position on some *p*H meters connects a shorting strap between these terminals to give a 7 *p*H reading and to prevent polarization of the electrodes while immersing the electrodes or applying power to the transmitter.

Moisture on the measurement electrode terminal, which decreases resistance R_3 in Figure 4-1a, is more of a problem than moisture on the reference electrode terminal, which decreases resistance R_4 in Figure 4-1a. The equations in Appendix G estimate the error that results from the voltage divider effect of moisture on these terminals. For resistances R_3 and R_4 reduced to 10 MΩ due to moisture, the error for a 100 MΩ measurement electrode is 10%, whereas the error for a 10 kΩ reference electrode is only 0.1%.

Returning electrodes to the manufacturer for troubleshooting is ineffective and raises safety concerns from contamination of the wetted surfaces. It may be impossible to eliminate hazardous chemicals trapped in a porous junction. Also, washing the electrodes gets rid of coatings and rejuvenates the glass surface, which severely reduces the scope of the troubleshooting possible by the electrode manufacturer. Most electrode troubleshooting is best done on-site.

Smart Transmitter Diagnostics [8]

- The source of a questionable *p*H measurement can be determined to be the temperature measurement or the millivolt measurement by examining their values, as in the case of a failed temperature sensor, which indicates a process temperature that is fantastically low or high.

- *p*H transmitters also provide internal diagnostic routines that can give the user real-time fault and warning information. These routines can be categorized as follows:
  - Transmitter diagnostics alert the user to electronic and CPU problems with the transmitter itself, which can indicate either that the validity of the measurement has been compromised or, in extreme situations, that the transmitter electronics should be replaced.
  - Sensor diagnostics can indicate problems with the *p*H sensor or the process, as indicated in Figure 4-10c, AMS *p*H variables dashboard.
    - These include millivolt measurements out of range, which indicate that a sensor has been wired incorrectly or there is a major sensor problem.
    - Temperature measurements will trigger alerts if they are too low or high and outside the operating range of the measurement. Failures in the temperature element such as shorts, opens, or incorrect wiring will also trigger alerts.
    - Calibration errors are triggered by sensor calibration constants (slope and zero) being out of the normal range, which indicates a sensor failure or procedural errors by the user.
    - Advisory messages do not indicate a problem with the transmitter or sensor function but do let the user know that calibrations, simulations, and other events are taking or have taken place.
  - Glass *p*H electrode impedance diagnostics need to consider the following:
    - Glass *p*H electrode impedance is in the range of tens to hundreds of megohms and is strongly temperature dependent.
    - A crack or break in the glass electrode creates a short, which results in an impedance of 1 megohm or less.
    - Detecting broken or cracked *p*H electrodes requires impedance measurement circuitry with temperature compensation and setting a low-impedance alarm.
  - Calibration history data can provide more intelligent maintenance:
    - Having factory calibration and calibration data sets along with the associated time-stamp is useful for establishing the service trend and scheduling preventive maintenance and/or sensor replacement.

- Figure 4-10d shows the calibration history data that can be used for preventive diagnostics. In this example of glass electrode aging, the glass impedance increases and the sensitivity of the sensor (slope) gradually decreases with time. This trend indicates that sensor replacement must be scheduled within a couple of months.

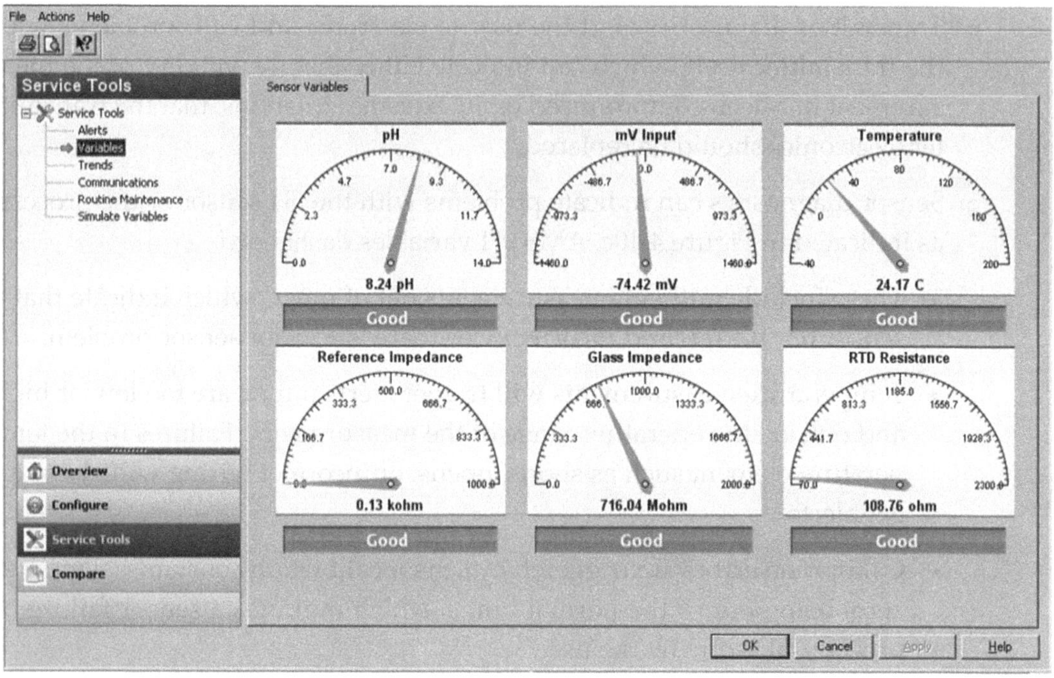

Figure 4-10c. AMS *p*H variables dashboard.

History	Time Stamp	Calibration Method	Slope (mV/pH)	Offset (mV)	Glass Impedance, (MΩ)	Reference Impedance, (kΩ)	Temperature (°C)
1-Latest	120 days	Auto-Cal	56.0	11	820	13	25
2	90 days	Auto-Cal	57.5	10	760	10	28
3	60 days	Auto-Cal	57.1	9	670	15	26
4	30 days	Manual	58.4	10	550	10	27
5-Oldest	2 days	Manual	59.0	3	480	3	30
Factory Calibration	0.0	Auto-Cal	59.3	2	500	2	25

Figure 4-10d. Advanced *p*H diagnostics calibration record.

4.11 Best Practices

One must be attentive to all that is best regarding electrode design, installation, calibration, and maintenance to achieve the true potential of *p*H electrodes [22].

1. Choose electrode holder materials of a construction and design that meet the requirements of worst-case corrosive and temperature operating conditions. An often-serious overlooked problem is the incompatibility of the standard O-rings and gaskets for process conditions.

2. Select the best glass that will prevent an appreciable increase in drift, error, failure rate, and the 86% response time for the worst-case temperature, *p*H, and chemicals that can attack glass (e.g., select general-purpose, high *p*H, low *p*H, high-temperature, sterilizable, or hydrogen fluoride -resistant glass).

3. Use a semispherical bulb with best glass formulation to improve accuracy and response time.

4. For *p*H <1 or >12, use conductivity or Coriolis meter density to get a better concentration measurement of a single component in water for operating points outside of 4 to 10 *p*H (e.g., concentrations greater than 1% sulfuric acid or caustic in water are best measured by conductivity because the *p*H corresponding to these concentrations is at the extremes of the scale where the *p*H slope is flat).

5. Use the best reference design and fill that will prevent an appreciable increase in drift, error, failure rate, and the 86% response time for the worst-case temperature and composition. Examples of worst-case composition include low water or pure water solutions with low conductivity and salt and chemical concentrations that can change junction potential, plug junctions, and poison reference internals. Certain ions in the process can precipitate or complex the silver ions when diffusing into the reference electrode, the most common being sulfide, bromide, iodide, and cyanide. These ions cause plugging and poisoning (attacking the silver–sliver chloride internal electrode).

6. If the reference electrode accuracy and equilibration speed deterioration causes poor continuous or batch performance or excessive maintenance, use a pressurized flowing junction to provide the most constant reference potential and the fastest junction equilibration rate, eliminating a variable liquid junction potential, plugging, and poisoning. When inserting these electrodes in samples, ensure the junction flow rate is small and the sample size is large and stirred so the sample *p*H does not change because of the reference electrolyte entering the sample.

7. Use double- and triple-junction and solid-state references to slow down the contamination rate if the slower equilibration and greater drift of the reference potential compared to the flowing junction are acceptable. This is of particular concern for high ionic strength streams.
8. Use special electrolytes to prevent precipitation of silver salt from contact with a process component (e.g., silver cyanide precipitate from cyanide contact with silver).
9. Use a removable outer reference junction chamber to enable an electrode to be rejuvenated (reference junction and fill quickly replaced) before plugging and poisoning becomes a problem.
10. Use a solid-state reference electrode to reduce the effect of plugging, contamination, and poisoning if a slower equilibration of reference potential is not a problem.
11. If a chemical attack causing premature aging from high temperature or dehydration (nonaqueous solvents or low water concentrations) is severe, reducing electrode life to a matter of days, use an automated, retractable insertion assembly to limit sensor exposure to the process just long enough to get a periodic *p*H measurement then flush to clean and rejuvenate the electrode while retracted. A buffer solution can be pushed into the sensor assembly after flushing to provide an automatic calibration before reinsertion.
12. If the solution conductivity is extremely low (e.g., condensate, boiler feedwater, deionized water), use a flowing junction reference electrode in a special assembly with a low sample flow, diffuser, and electrolyte reservoir.
13. Use a Variopol (VP) or equivalent connector to easily disconnect the electrode cable, eliminating the need to disconnect the transmitter and retract the cable through conduit.
14. Use a smart transmitter with a stored calibration record.
15. Use a solution ground for impedance diagnostics and ground potential elimination.
16. Use a smart transmitter to detect glass and reference problems.
17. Use solution *p*H temperature compensation in addition to standard Nernst compensation developed in laboratory testing at process conditions.
18. Test several different process connections and electrodes to identify the best electrode type and location (shortest dead time and least noise/bubbles).

19. Ensure the electrode is installed with the tip pointing down above a 20-degree angle to horizontal to prevent bubbles from residing in the tip or being caught on the internal electrode.

20. Ensure electrodes are always wetted (a common problem for batching and shutdown).

21. Use MSS to eliminate a response to a single failure and reduce the effect of noise, coatings, aging, abrasion, and contamination (assuming different lengths of times each electrode is in service) to provide more intelligent diagnostics of the deviant electrode and to enable removing a single electrode for cleaning, rejuvenation, and calibration or replacement without having to put the *p*H control loop in manual. Only if the same electrode consistently shows poor performance is it removed. After inserting a calibrated new electrode or a cleaned and rejuvenated existing electrode, the speed and efficiency of other electrodes can be compared to quantify improvement and schedule maintenance for the other electrodes. The more intelligent maintenance and the faster, smoother, and more accurate response greatly reduce lifecycle costs and increases process performance benefits.

22. Ensure the stream velocity and tip protective shroud design are the best for the process conditions. Use a low velocity (0.1 to 1 ft/s) and protect the tip from flow impingement to reduce abrasion or contamination, and use a high velocity (5 to 10 ft/s) and sweeping action at the tip to reduce coatings.

23. Ensure the electrode tip extends into the pipe centerline and past vessel baffles.

24. Choose an electrode location free of bubbles (e.g., not near the sparger ring).

25. Locate the electrode sufficiently downstream from a pump or static mixer to reduce concentration fluctuations, but not so far as to increase the dead time by more than 6 s.

26. Put insertion electrodes in series in the same line (at least 10 pipe diameters apart) rather than in parallel lines so each electrode sees the same velocity and same composition.

27. Choose an electrode location free from flashing (e.g., not in pump suction or valve discharge).

28. Ensure the insertion electrode assembly has a ball valve and restraining strap to safely withdraw (retract) the electrode from a pipe or vessel, preventing blowout.

29. Ensure the electrode and transmitter locations are safely accessible for maintenance.

30. Ensure the electrode and transmitter signal connections are always dry in terms of location and type of connection seals and, if necessary, by using a dry air purge. Avoid submersion assemblies and transmitter enclosures or electrode holder cable entries that are not sealed and top entry of cables into holders.

31. Determine the solution *p***H** compensation needed by making changes to the sample temperature and noting the change in solution *p***H** caused by the changes in dissociation constants and ion activity for samples that cover the expected *p***H** range.

32. Use an accurate RTD temperature measurement and a built-in smart transmitter process temperature compensation or input signal characterizer block to give a piecewise linear fit to the plot of process *p***H** correction versus process temperature.

33. New, cleaned, and rejuvenated existing electrodes must be calibrated in fresh buffer solutions of the same ionic strength and solvent as the process.

34. The reference junction potential must be relatively constant (junction reached equilibrium) before new, cleaned, and rejuvenated existing electrodes are used for closed-loop control (automatically achieved by MSS).

35. Because a 10 *p***H** buffer solution *p***H** is more readily changed by temperature and carbon dioxide absorption, a 4 *p***H** buffer solution is generally preferred to set slope after a 7 *p***H** buffer solution is used to set offset. However, for electrodes designed for high *p***H** operation, a fresh 10 *p***H** buffer solution may be best.

36. During calibration, the full measurement response must be achieved, and the changes in the reference junction must be negligible before an offset or slope setting is made.

37. Once an electrode is in service, the electrode is removed or the offset is adjusted only after consistent deviation of an electrode response compared to others in MSS or a representative process sample.

38. Samples used for standardization must be at the same temperature and composition as the process. The electrode used must be verified to be fast and accurate with the sample *p***H** result compared to the field *p***H** measurement at the time the sample was taken.

References

[1] Roger G. Bates, *Determination of pH; Theory and Practice*, 2nd ed. (New York: Wiley, 1973).

[2] Peter Meier, "Practice and Theory of pH Measurement," Metler-Toledo publication CH-8902 (1989).

[3] Douglas A. Skoog and Donald M. West, *Principles of Instrumental Analysis*, 2nd ed. (Saunders College, 1980).

[4] Dave Mills, personal correspondence (2004).

[5] Gregory K. McMillan, "Understand Some Basic Truths of pH Measurement," *Chemical Engineering Progress* (October 1991).

[6] Gregory K. McMillan, "pH Measurement: The State of the Art," *InTech* (February 1993).

[7] Ralph L. Moore, "Good pH Measurements in Bad Process Streams," *Instruments and Control Systems* (December 1990).

[8] Gregory K. McMillan, *Essentials of Modern Measurements and Final Elements in the Process Industry* (Research Triangle Park, NC: ISA [International Society of Automation], 2010).

[9] Gregory K. McMillan, "pH Electrode Performance – An Investigative Study," Monsanto Integrated Manufacturing Committee, IMC Report 92-1 (1992).

[10] Gregory K. McMillan, et al., "pH Measurement," *Analytical Instrumentation*, Chilton, (1994).

[11] Leonard W. Niedrach, "A New Membrane-Type pH Sensor for Use in High Temperature-High Pressure Water," *Journal Electrochemical Society* 127, no. 10 (October 1980).

[12] Leonard W. Niedrach, "Electrodes for Potential Measurements in Aqeous Systems at High Temperatures and Pressures," *Angewandte Chemie International Edition in English* 28, no. 3 (March 1987).

[13] Helmuth Galster, *pH Measurement – Fundamentals, Methods, Applications, Instrumentation* (New York: Wiley, 1991).

[14] Clark C. Wescott, *pH Measurements* (New York: Academic Press, 1978).

[15] J. C. Vogelzang, "pH Theory and Practice," Yokogawa publication TE 12B6A0-E-H (1987).

[16] Martin S. Frant, "How to Measure pH in Mixed and Nonaqueous Solutions," *Today's Chemist at Work* (June 1986).

[17] C. Rushton and A. Bottom, "Measuring pH in Low Conductivity Waters," *Kent Technical Review,* Great Britain, no. 24, Coden: KTTRAH (March 1979).

[18] Gregory K. McMillan, *Process Control Case Histories* (New York: Momentum Press, 2010).

[19] Hugo Z. Hershkovitch and Thomas J. McAvoy, "Dynamic Modeling of pH Electrodes," *Canadian Journal of Chemical Engineering* 56 (June 1978).

[20] A. L Guisti and J. O. Hougen, "Dynamics of pH Electrodes," *Control Engineering* 8, no. 4 (April 1961).

[21] A. Disteche and M. Dubuisson, "Transient Response of the Glass Electrode to pH Step Variations," *Review of Scientific Instruments* 25, no. 9 (September 1954).

[22] Gregory K. McMillan and P. Hunter Vegas, *Industrial/Process Instruments and Controls Handbook,* 6th ed. (New York: McGraw-Hill Education, 2019).

5

Mixing Equipment

5.1 What Was Good Might Be Bad

Neutralization systems are generally extremely dilute aqueous streams with set points on a relatively steep portion of the titration curve. Consequently, small deviations in the concentration that might be negligible for most other systems are significant for *p***H** control. For example, a 0.0002% deviation in acid concentration that is 50 times smaller than the 0.01% deviation that corresponds to a uniformity of 99.99% associated with good mixing would cause the *p***H** to swing between 4 and 10 *p***H** in a strong acid and strong base system per Figure 1-1b in Chapter 1, "The Essentials." Most studies in mixing are based on a process gain of 1 because the controlled variable is considered to be the concentration of interest. The logarithmic relationship of *p***H** can create process gains that are orders of magnitude larger. Studies of mixing requirements in *p***H** control systems have typically been conducted for buffered systems, *p***H** set points on a flat portion of the titration curve, and diluted weak reagents. Consequently, they do not reveal the sensitivity and dispersion problem of mixing a few drops of a strong, concentrated reagent in a large volume.

> **Key CONCEPT** *p***H** control requires a degree of mixing that goes well beyond the norm.

> **Key CONCEPT** A system considered to be well-mixed for concentration control may be poorly mixed for *p***H** control.

The percent concentration deviation of the reagent from its average concentration in a volume should be converted to an equivalent deviation in the reagent-to-influent feed ratio, which is the x-axis of the titration curve. This deviation in ratio should be multiplied by the slope of the titration curve at this point, which is the process gain, to get the deviation in *p***H**. Equation 5-1a shows the definition of the ratio of reagent-to-influent mass flow. Equation 5-1b shows how this ratio can be computed from the ratios of weight fractions per Equation 2-1i in Chapter 2, "The Chemistry." The average ratio can be computed for the average concentrations, which gives the operating point on the titration curve. The change in ratio can be computed from this same equation if the deviation is substituted for the average in reagent concentration in the effluent, as shown in Equation 5-1c. The deviation in *p***H** is then this deviation in ratio multiplied by the process gain, which is the slope of the curve at the operating point as denoted in Equation 5-1d. This deviation about the operating point must be well within the control band (minimum and maximum allowable *p***H**) for the system to be considered "well-mixed" for *p***H** control.

$$R_{ri} = \frac{F_r}{F_i} \tag{5-1a}$$

$$R_{ri} = \frac{x_{re} \bullet x_{ii}}{x_{rr} \bullet x_{ie}} \tag{5-1b}$$

$$\Delta R_{ri} = \frac{\Delta x_{re} \bullet x_{ii}}{x_{rr} \bullet x_{ie}} \tag{5-1c}$$

$$\Delta p\mathbf{H} = \Delta R_{ri} \bullet K_{cv} \tag{5-1d}$$

where

F_i = influent mass flow (kg/h)
F_r = reagent mass flow (kg/h)
K_{cv} = process gain for the controlled variable ($\Delta p\mathbf{H}/\Delta R_{ri}$)
$p\mathbf{H}$ = average *p***H** operating point (*p***H**)
$\Delta p\mathbf{H}$ = deviation in *p***H** from the average *p***H** operating point (*p***H**)
R_{ri} = average mass flow ratio of reagent to influent
ΔR_{ri} = deviation from the average mass flow ratio of reagent to influent
x_{ie} = average weight fraction of influent acid or base in the effluent stream
x_{ii} = average weight fraction of influent acid or base in the influent stream
x_{re} = average weight fraction or reagent acid or base in the effluent stream
Δx_{re} = deviation in average weight fraction or reagent acid or base in the effluent stream
x_{rr} = average weight fraction of reagent acid or base in reagent stream in the plant

RULE of Thumb *For a system to be rated "well-mixed" for pH control, the deviation in the reagent-to-influent flow from nonideal mixing multiplied by the process gain must be well within the control band.*

5.2 Mixing Dynamics

As flows enter and proceed through a back-mixed volume from inlet to outlet, molecules on the leading edge recycle back and mix with molecules upstream. In other words, the molecules recycled back take longer to exit the volume. The standard deviation of the distribution of residence times of molecules is larger. The folding back smooths out the change in concentration, which translates to a process time constant. In plug flow, eddies and vortices that recycle back molecules are negligible. The result is a sharp transition in concentration and *p*H, which translates to a process dead time. The standard deviation of the distribution of molecule residence times is nearly zero, which means all the molecules at the leading edge exit at the same time. Plug flow requires radial mixing (mixing over the cross section) to ensure the concentration along the leading edge is uniform. Back mixing requires axial mixing (mixing along the axis of flow).

Key CONCEPT Back mixing creates a beneficial process time constant, and plug flow creates a detrimental process dead time for *p*H control.

When the time for a reaction to go to completion approaches or exceeds the residence time, plug flow is desirable to reduce the residence time distribution, to decrease the size of the volume needed for the reaction to go to completion, and to improve the yield by reducing by-products from secondary reactions. In *p*H systems, the neutralization rate is so fast that it is complete in less than 1 s, and there are no secondary reactions. Residence time distribution is of no significance for *p*H control unless there are solids or bubbles that must dissolve.

A controller has more time to catch up to a disturbance if the change is slow and gradual, which corresponds to a process time constant that is larger than the process dead time. Conversely, a controller has difficulty dealing with abrupt, rapid, and delayed changes from plug flow, which corresponds to a process dead time that is much larger than the process time constant. The section on loop performance in Chapter 8, "Control System," explains how the peak error from a load upset is proportional to the loop's dead time to time constant ratio. The integrated error is the product of this ratio and the dead time, which makes it proportional to the dead time squared. Thus, the magnitude of the dead time can be more significant than the ratio. A small plug flow volume, such as a static mixer, with a poor dead time to time constant ratio but a small dead time, can have a smaller integrated error than a large back-mixed volume, such as an agitated vessel.

For *p*H systems, the dead time to time constant ratio determines how far the operating point moves on the titration curve and how much of the nonlinearity is seen by the *p*H controller. In other words, a small ratio will enable a *p*H controller to stay close to the set point and only be exposed to the change in process gain with curvature in a small operating region. The implication is that *p*H control systems will be much more nonlinear for plug flow than for back-mixed volumes, and the linearization techniques presented in Chapter 8 will become more important.

Key CONCEPT A small plug flow volume may have a smaller integrated error than a large back-mixed volume because the dead time magnitude is smaller.

RULE of Thumb *For large volumes, maximize back mixing and minimize plug flow to minimize the error from load upsets and the effects of the pH nonlinearity.*

The residence time is the volume divided by the volumetric throughput flow. The residence time becomes a process time constant for a completely back-mixed volume. For a plug flow volume, the residence time becomes a dead time. Real systems for *p*H control can be approximated as a combination of plug flow and back-mixed volumes. Thus, the residence time is split between a process time constant and process dead time.

A static mixer provides good radial mixing and little back mixing and has a residence time distribution that is close to plug flow. In turbulent flow through a pipe, eddies propel molecules forward and backward, so there is some smoothing in the axial direction, but radial mixing is not as good as the static mixer. In laminar flow through a pipe, there is poor radial mixing and a nonuniform velocity profile. In a vessel with axial agitation, geometry, and feed entry and exit points that prevent stagnation, channeling, or short-circuiting, back mixing is promoted, and the *p*H changes are smooth and gradual.

For plug flow, there tends to be more short-term fluctuations (noise) from a lack of smoothing. Consequently, signal filtering and middle-value selection are generally more important for *p*H control in a static mixer than in an agitated vessel.

RULE of Thumb *For plug flow volumes, use linearization techniques, filtering, and middle signal selection to help the pH controller handle noise and upsets.*

5.3 Agitated Vessels

In vessels, there are many different internal flow patterns from agitation and many different parameters to quantify the amount of agitation. For *p***H** control, the flow pattern should be axial, as shown in Figure 5-3a, where the fluid is pulled down from the top near the shaft, circulated along the bottom to the sidewalls, and pulled back up to the top near the sidewalls. If the agitation breaks the surface without froth, its intensity level is in the ballpark for good *p***H** control. The pattern is called *axial* because of the vertical up-and-down flow pattern parallel to the axis of the shaft. Baffles that are 90 degrees apart, extend vertically along the entire length of the sidewall, and are one-sixtiethof the diameter in width are recommended because they help establish the vertical flow currents to reduce vortexing, swirling, and air induction from the surface, plus they increase the uniformity of the flow pattern and assure a more consistent power draw. Tests made with a pitched blade turbine showed that adding baffles reduced the time for complete dispersion (mixing time) from 28 to 20 s and the standard deviation from 4.8 to 2.4 s [1]. Propeller and pitched blade turbines provide an axial flow pattern. A double-spiral blade and a tangential jet nozzle cause an undesirable corkscrew axial pattern because the concentration change has a long and slow corkscrew flow path to follow before dispersion is complete.

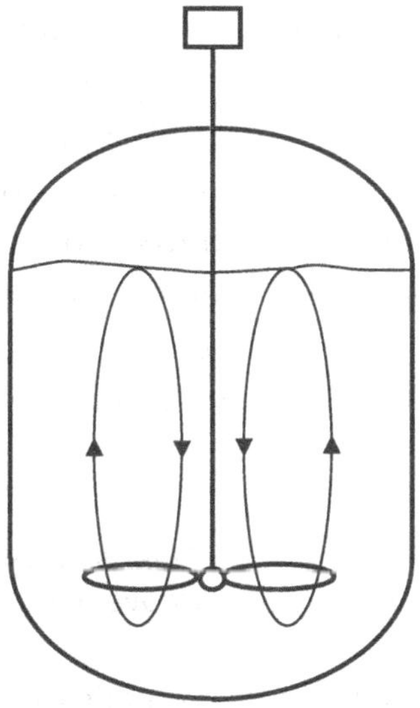

Figure 5-3a. Axial agitation flow pattern recommended for *p***H** control.

A flat blade or bar turbine causes a radial flow pattern (as shown in Figure 5-3b) in which the fluid is pushed and pulled radially out from the axis of the agitator shaft. The radial pattern is undesirable because it results in an area of fluid near the surface that is not mixed as vigorously as required for *p***H** control. A side-entry axial

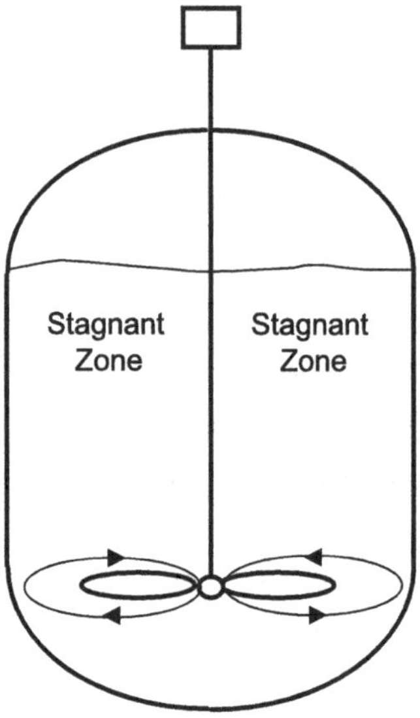

Figure 5-3b. A radial agitation flow pattern is detrimental to good *p*H control.

agitator will produce an undesirable radial flow pattern at the entry point and should be avoided from the standpoint of maintainability as well.

For a vessel with axial agitation, baffles, and a liquid height that is about the same as the diameter, the equipment dead time is approximately the turnover time that can be estimated as the liquid mass divided by the summation of the influent, reagent, and recirculation flows plus the agitator pumping rate per Equation 5-3a [2]. While dead time is theoretically (on average) half of the turnover time, in practice, that time delay is appreciably larger than the turnover time. Consequently, in Equations 5-3b and 5-3c, no credit has been taken for induced flow from secondary vortices. Also, Equation 5-3a assumes the reagent has been injected below the surface and does not include a reagent delivery delay that can be huge per Chapter 7, "Reagent." Note that the equations in this chapter use mass flows, whereas equations in mixing handbooks use volumetric flows.

RULE of Thumb *For a vessel with proper geometry, baffles, and axial patterns, the equipment dead time from mixing is approximately the turnover time.*

For continuous equipment, the effluent flow equals the summation of the influent and reagent flows by level control or overflow. For batch operation, the effluent flow is zero except at the end of the batch. Equation 5-3a is valid for batch and continuous operation except when the batch volume increases with time per Equation 5-3d. The agitator pumping rate can be estimated as the product of the agitator discharge coefficient, agitator speed, and the third power of the impeller diameter per Equation 5-3b. The discharge coefficient can be approximated by Equation 5-1c and usually ranges from about 0.4 to 1.4. Typical discharge coefficients for a hydrofoil and a pitched blade turbine impeller are 0.4 and 0.7, respectively. The impeller diameter in these equations is the diameter of the circle traced out by the tips of the propeller or turbine blades. These equations assume axial flow throughout the entire contents of the vessel. The lower limit on equipment dead time depends on the degree of agitation that starts air entrainment from the surface and the distance from the inlet to the exit nozzles. The dead time cannot be less than the shortest distance between the inlet and exit divided by the internal circulation velocity per Equation 5-3d. The internal circulation velocity rarely exceeds 1 ft/s. The largest of the equipment dead time values calculated using Equations 5-3a and 5-3e should be used as the equipment dead time. Note that this dead time does not include any transportation delays in the reagent piping or dip tube or from the equipment exit (discharge nozzle) to the *p***H** electrodes.

> **Key CONCEPT** The actual equipment dead time is often larger than the turnover time because of nonideal mixing patterns and fluid entry and exit locations.

For a vertical well-mixed vessel:

$$\theta_p = \frac{\rho \bullet V}{F_i + F_r + F_c + F_a} \tag{5-3a}$$

$$F_a = \rho \bullet N_q \bullet N_s \bullet D_i^3 \tag{5-3b}$$

$$N_q = \frac{0.4}{(D_i / D_t)^{0.55}} \tag{5-3c}$$

$$\theta_p = (L/U) \tag{5-3d}$$

where

D_i = impeller diameter (meter)
D_t = tank inside diameter (m)
F_i = influent mass flow (kg/min)
F_r = reagent mass flow (kg/min)
F_c = recirculation mass flow (kg/min)
F_a = agitation mass flow (kg/min)
L = distance between the inlet and outlet nozzles (m)
N_q = agitator discharge coefficient (0.4 to 1.4)
N_s = agitator speed (rpm)
ρ = average fluid density (kg/m^3)

θ_p = process dead time from mixing (min)
U = average bulk velocity (meters per min)
V = vessel liquid volume (m^3)

The equipment time constant, which is the time it takes the outlet concentration to reach 63% of its final value after the equipment dead time for a step change in the inlet concentration, can be approximated as the residence time minus the equipment dead time for the continuous mode of operation where there is a continuous effluent flow. Equation 5-3e is only true for well-mixed vessels and, thus, for small values of dead time relative to the residence time. The volume can be visualized as being split into a large back-mixed and a small plug flow portion. Note that this definition is for a linear change in outlet concentration. A *p***H** reading cannot be used to measure this time constant because of its nonlinear relationship to acid or base concentration.

RULE of Thumb *For a vessel with proper geometry, baffles, and axial patterns, the continuous equipment time constant is the residence time minus the dead time.*

The agitation term *mixing time* refers to the time required for the entire contents of the vessel to reach within a specified percentage of the final concentration for a pulse of reagent or influent and no effluent flow (batch mode of operation). This term cannot be related to and should not be confused with the negative feedback time constant for the continuous mode of operation. The mixing time can be used to estimate the equipment time constant for the fed-batch mode (simultaneous reagent and influent flows but zero discharge flow) if the percentage of uniformity reached is defined. Equation 5-3f shows that the batch equipment time constant is approximately equal to one-fifth of the mixing time to reach 99% uniformity minus the equipment dead time. The same agitator data sheet should be used to get the mixing and turnover times, and the turnover time should be used as the equipment dead time. It takes about 10 to 15 turnover times to achieve 99% uniformity for an aqueous solution and an axial flow turbine agitator. This translates to a dead time to time constant ratio of 0.3 to 0.5 for a well-mixed fed-batch vessel. Of course, a negative value for the equipment time constant is not physically possible and is a red flag for data inconsistency.

RULE of Thumb *For a vessel with proper geometry, baffles, and axial patterns, the fed-batch equipment time constant is one-fifth the mixing time minus the dead time.*

The agitation term *circulation time* is the interval between consecutive passes of a particle through the impeller. The circulation time is less than the mixing time. It is not a good estimate of the degree of mixing because a particle could be recycled back through the impeller region without mixing with the other regions.

For the traditional batch mode of operation with sequential feed and reagent phases, there is an integrating response because the concentration does not reach a steady state but ramps until a physical limit, such as tank overflow or a high-level interlock, is reached. The ramp slope divided by the change in the manipulated variable is the integrator gain. This gain depends on the choices of the manipulated and controlled variables and the units of both variables. Equation 5-3g gives the integrator gain for a controlled variable, the reagent acid or base weight fraction, and a manipulated variable, the reagent mass flow rate. Because the actual controlled variable is *p*H, the ramp will slow down and almost appear to stop on a very flat part of a titration curve. The effect of the nonlinear gain is addressed in the section on loop performance in Chapter 8 by translating a change in the abscissa to a change in the ordinate of the titration curve. For continuous operation, the abscissa is the ratio of reagent mass flow to influent mass flow. For batch operation, the abscissa simplifies to reagent weight fraction if the influent mass flow addition was completed before the start of the reagent addition and the reagent mass is small relative to the total mass.

> **Key CONCEPT** For a traditional sequential batch, there is no steady state, and the concentration ramp is described by an integrator gain rather than a time constant.

For continuous operation of a vertical well-mixed vessel:

$$\tau_p = \frac{\rho \bullet V}{F_i + F_r} - \theta_p \qquad (5\text{-}3e)$$

For fed-batch (simultaneous feed and reagent) operation of a vertical well-mixed vessel:

$$\tau_p = \frac{T_m}{5} - \theta_p \qquad (5\text{-}3f)$$

For traditional batch (sequential feed and reagent) operation of a vertical well-mixed vessel:

$$K_i = \frac{X_{rr} \bullet F_r}{\rho \bullet V \bullet \Delta F_r} \qquad (5\text{-}3g)$$

where

F_r = average reagent mass flow (kg/min)
ΔF_r = change in reagent mass flow (kg/min)

K_i = integrator gain (mass fraction per min per kg/min)
ρ = average fluid density (kg/m^3)
θ_p = process dead time from mixing (min)
τ_p = process time constant from mixing (min)
V = vessel liquid volume (m^3)
X_{rr} = reagent mass fraction in reagent stream

Even though the dynamic response of an agitated vessel is approximated as the combination of an equipment dead time and time constant, the actual response is S-shaped, as shown in Figure 5-3b for a step change in reagent flow with the controller in manual (open loop). No real process has the sharp transition shown at the end of the dead time. So, the theoretical definition of the time constant as the time to 63% of the change after the dead time, although accurate for field estimates, is better visualized via a tangent to the inflection point (the point where the slope starts to decrease), as shown in Figure 5-3c. The time from the start of the step change to the intersection of the tangent with the initial value is the dead time, and the time from the end of the dead time to the intersection of the tangent with the final value is the time constant. In an actual *p*H trend, the open-loop response is not just associated with the equipment but is also a combination of the equipment, valve, reagent, piping, and measurement dynamics, and the titration curve nonlinearity. The subscript $_o$ is used to designate the overall open-loop time constant seen on the trend to distinguish it from the closed-loop time constant used to describe the response of a controller in automatic to a set point.

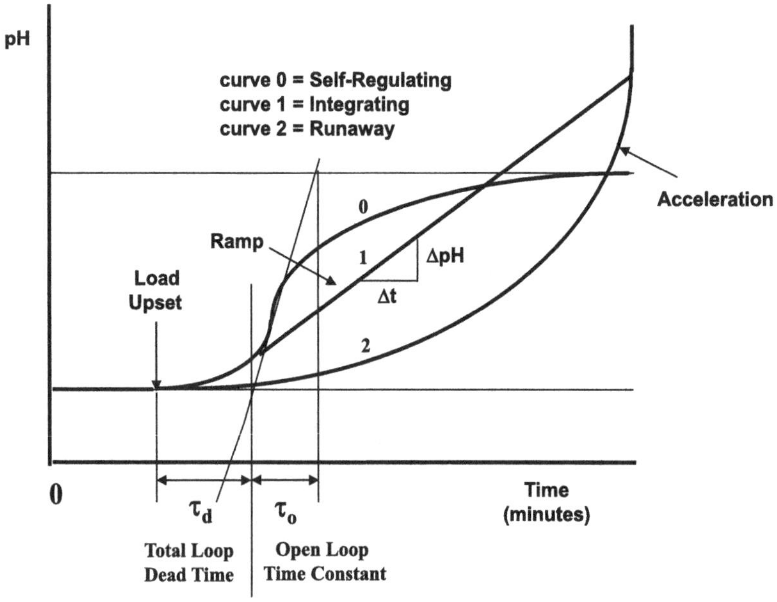

Figure 5-3c. Approximation of the dead time and open-loop time constant on a trend.

The entire contents of a vessel can be axial mixed only if the vessel height is equal to or slightly greater than the width. For a single impeller, the liquid height should be greater than 50% and less than 150% of the width. This type of vessel will hereafter be referred to as a *vertical tank*. Equation 5-3a for the equipment dead time is only valid for a vertical tank. If the ratio of the equipment dead time to the time constant is equal to or less than 0.05, the vessel is classified as a *vertical well-mixed tank*. The importance of this dead time to time constant ratio for controller tuning and control loop performance will be discussed in Chapter 8.

Horizontal tanks have a length much greater than the height. No matter how many agitators are installed, the complete volume cannot be considered as axially mixed. There will be regions of stagnation, short-circuiting, and plug flow. The result of stagnation is a region of fluid whose concentration is slowly dispersed by ion diffusion or migration. Short-circuiting results in a region where a change in the reagent or influent concentration bypasses the rest of the vessel contents and appears in the effluent prematurely. The result of plug flow is a region where the equipment time constant approaches zero, and the dead time to time constant ratio approaches infinity. Figure 5-3d illustrates a stagnant region in the far right corner of the horizontal tank from the lack of agitation or an entry flow, a short-circuiting region in the far left corner due to the proximity of the reagent and the effluent nozzles, and a plug flow region of influent to effluent in the middle of the tank caused by radial agitation. For a horizontal tank with proper flow entry and exit locations, the dead time as a percentage of residence time proportionally increases with the ratio of length to height [3]. For example, if the tank's horizontal length is four and eight times the tank's height, the dead time as a percentage of residence time would be 20% and 40%, respectively. Equations 5-3h and 5-3i approximate this relationship.

> **Key CONCEPT** Horizontal tanks are notorious for short-circuiting, stagnation, and plug flow that cause excessive dead time and an erratic *p*H response.

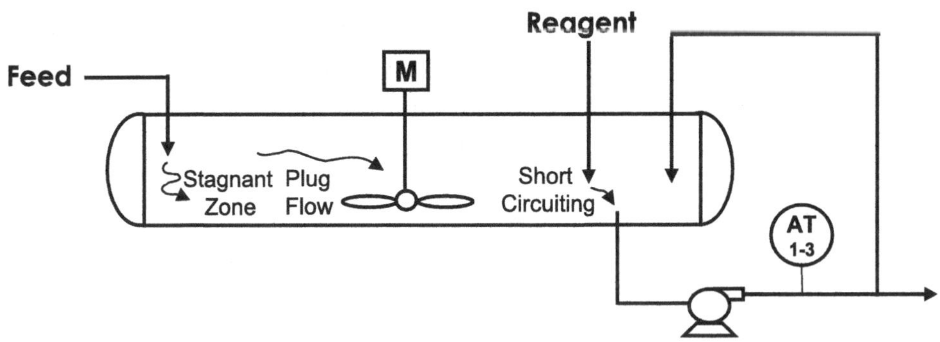

Figure 5-3d. Horizontal tank with short-circuiting, stagnant, and plug flow regions.

For continuous operation of a horizontal agitated tank with proper nozzle locations:

$$\theta_p = \frac{\rho \bullet V}{F_i + F_r} \bullet (L/H) \bullet K_z \quad (5\text{-}3h)$$

$$\tau_p = \frac{\rho \bullet V}{F_i + F_r} \bullet [1-(L/H) \bullet K_z] \quad (5\text{-}3i)$$

where

F_i = influent mass flow (kg/min)
F_r = reagent mass flow (kg/min)
H = liquid height (m)
K_z = geometric factor (0.05)
L = vessel length (m)
ρ = average fluid density (kg/m^3)
θ_p = process dead time from mixing (min)
τ_p = process time constant from mixing (min)
V = vessel liquid volume (m^3)

RULE of Thumb *For a horizontal agitated vessel, the dead time as a portion of the residence time is proportional to the ratio of the vessel length to the liquid height.*

Even if a vessel does not have a control loop, it can still play a critical role in attenuating oscillations. The time constant of a vessel is effectively the filter time in Equation 5-3j, which is an approximation of the amplitude ratio from the Bode plot analysis when the time constant is greater than the oscillation period [2]. Note that this equation is for the reagent concentration, not the *p***H**. Therefore, the input amplitude must be translated from *p***H** to the reagent-to-influent ratio, and the output amplitude must be translated from the reagent-to-influent ratio to *p***H** via the titration curve. For oscillations on the steep portion of the titration curve, the filtering effect on the *p***H** is not impressive, as shown in Figure 5-3e. If a volume is split into two volumes in series, the output oscillation from the first volume is the input oscillation to the second volume, and the effect is multiplicative by successively applying Equation 5-3j.

Key CONCEPT Back-mixed volumes attenuate oscillations in a concentration that must be translated to *p***H** via a titration curve to see the effect on the *p***H** trend.

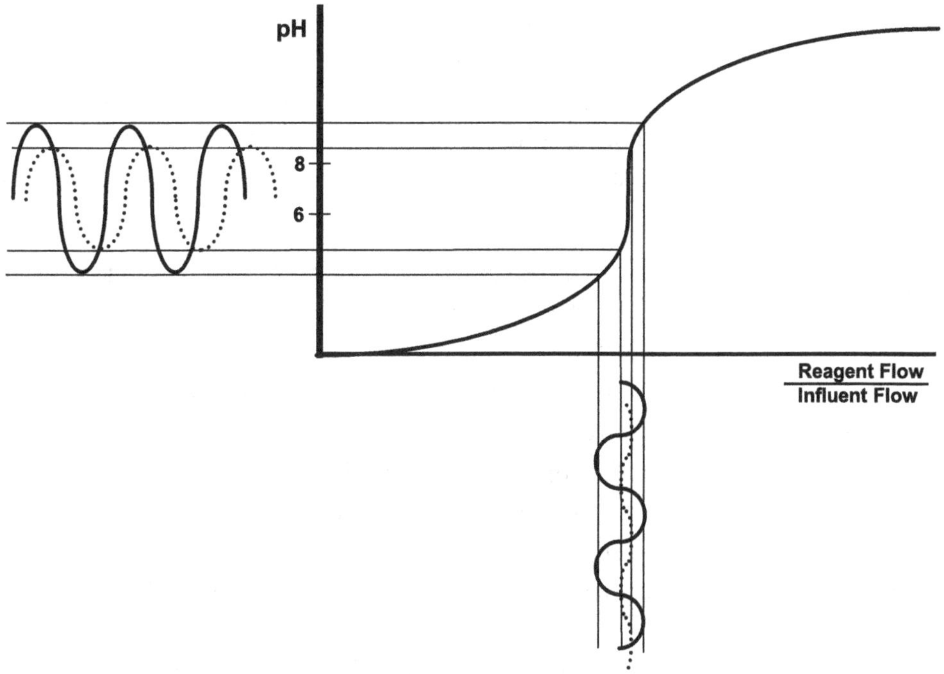

Figure 5-3e. Translation of oscillations between the axes of the titration curve.

For a single vessel volume:

$$A_o = A_i \bullet \frac{T_o}{2 \bullet \pi \bullet \tau_p} \tag{5-3j}$$

where

A_i = amplitude of input oscillation into the volume (reagent-to-influent ratio)
A_o = amplitude of output oscillation from the volume (reagent-to-influent ratio)
T_o = period of oscillation (min)
τ_p = process time constant from mixing (min)

If the vessel is upstream of a control loop, it can dramatically reduce reagent use by attenuating oscillations of influent concentrations and cross-neutralizing streams. If these are related to batch operations or the start-up and shutdown of unit operations, the vessel size required may be extremely large. For example, if the batch cycle time is 2 h, the effective oscillation period would be 4 h. However, the cost of operating the tank may be low because agitation may not be necessary if there is no layering of liquids or settling of solids from low velocities. Vessels or volumes with residence times of days do not need to be agitated to provide a beneficial filtering effect because there may be sufficient smoothing just from ion migration. For example, even if the back-mixed portion of the residence time is only 20% from feed and recirculation flow, the time constant for an unagitated vessel with a 24-hour residence time would be 4.8 h and would be able to noticeably filter the changes in a batch end point for a batch cycle time of 2 h. Thus, large tanks in a salvage yard may offer a big opportunity to reduce reagent costs.

> **Key CONCEPT** An unagitated vessel with a *p***H** control loop can be turned from a terrible loser into a big-time winner by moving the loop upstream or downstream.

If the vessel is downstream of a control loop, it can smooth out the oscillations from load upsets and stick-slip. For a reasonably well-tuned loop, the period for an oscillation with a decreasing (damped) amplitude is about twice the ultimate period or four to eight times the total loop dead time per equations presented in Chapter 8 on controller tuning. The period of the equal amplitude oscillations (limit cycle) for stick-slip is longer and increases as the controller is detuned in an attempt to kill it.

For a well-designed static mixer, the total loop dead time should be about 6 s, which means a time constant of just 24 to 48 s can smooth a damped oscillation noticeably and may be achieved by a wide spot in the line, such as an extra-large concentric pipe around the original pipe with holes to act as sparger and increase the back mixing. For a vertical well-mixed vessel with a total loop dead time of 1 min, a back-mixed time constant of 4 to 8 min downstream would appreciably attenuate a damped oscillation. Installing *p***H** measurements or implementing dynamic online *p***H** estimators on the discharge of downstream volumes can soothe concern and instill confidence in the system's performance by showing how much the oscillations have been smoothed. This is particularly important for static mixers because the *p***H** measurement will be noisy and have a propensity to oscillate severely because of the lack of back mixing. In fact, an in-line *p***H** loop that controls the filtered average of horrible, sustained oscillations for a strong acid and strong base system can draw a straight line in the discharge of a very large downstream volume. At a minimum, it can greatly reduce *p***H** constraint violations. For example, the historical trend of *p***H** on a vessel downstream of a static mixer can eliminate instantaneous Resource Conservation and Recovery Act (RCRA) limit violations.

> **Key CONCEPT** *p***H** measurements or dynamic online *p*H estimators on a downstream volume can reveal that the *p***H** loop is doing a great job.

RULE of Thumb *Use a large volume upstream to reduce reagent use and a relatively small volume downstream to reduce p***H** *constraint violations.*

5.4 Static Mixers

Static mixers have motionless internal elements that subdivide and recombine the flow stream repeatedly and cause rotational circulation to provide radial mixing of the stream but very little axial or back mixing. Consequently, fluctuations in *p*H over the cross section are smoothed, but fluctuations in *p*H with time, which are axial, show up unattenuated in the discharge, as shown in Figure 5-4a. The equipment dead time is about 80% of the residence time for a static mixer, and most manufacturers are working toward making their static mixers exhibit plug flow to reduce the residence time distribution [4]. Although this is beneficial for many chemical reactions, it makes the discharge *p*H more likely to spike and violate constraints.

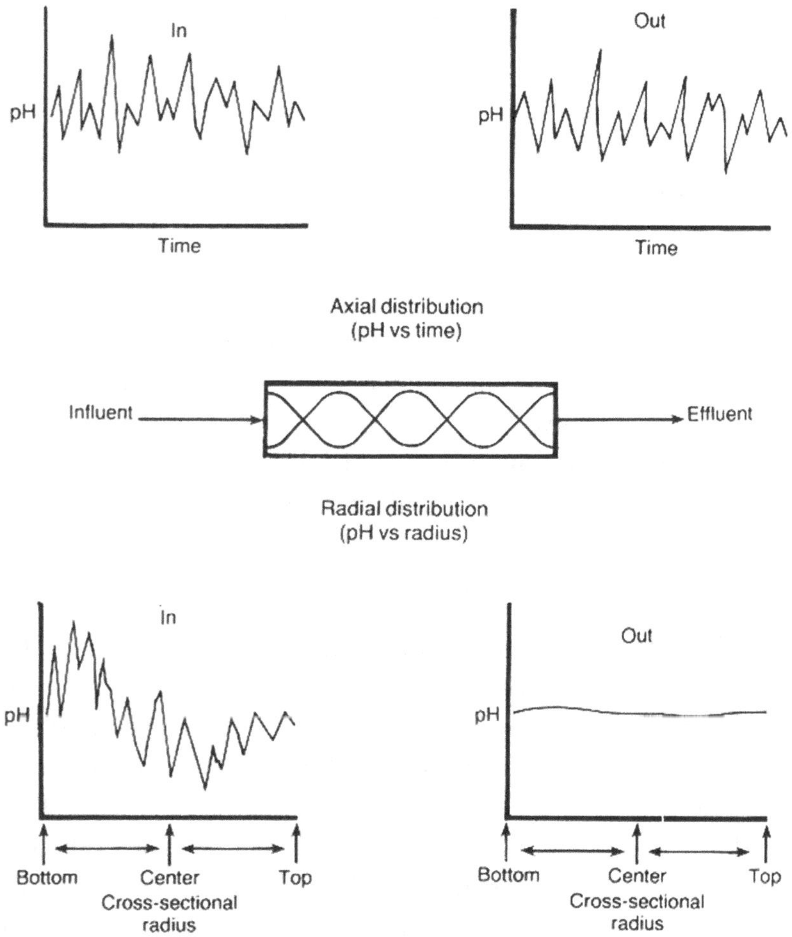

Figure 5-4a. Good radial mixing but poor axial mixing of a static mixer.

> **Key Concept:** Oscillations and noise will pass through a static mixer unattenuated, and the poor dead time to time constant ratio leads to more oscillations.

Flow pulses from a positive displacement reagent pump and drops associated with a high-viscosity reagent or low reagent velocity will not be back mixed and will cause a noisy *p*H signal. Bubbles from a gaseous reagent will also cause a noisy *p*H signal because the residence time is not sufficient for complete reagent dissolution. Using additional motionless elements beyond the normal number does not usually increase drop or bubble dispersion as much as an increase in stream velocity by a decrease in the static mixer diameter. Although a static mixer has a poor dead time to time constant ratio that tends to make a *p*H loop oscillate, it offers the significant advantage of a small magnitude of dead time and a small volume of off-spec material from a load upset. This is discussed in greater detail in the control system selection section of Chapter 8.

> **Key Concept:** A static mixer's extremely small residence time greatly reduces the magnitude of dead time and the volume of off-spec material.

About 100 equivalent pipe diameters of turbulent flow will provide about the same degree of radial mixing as a static mixer. However, the equipment dead time is several times larger because of the larger volume needed to get a sufficient number of diameters. The length of piping between the reagent injection point and the *p*H measurement point, and therefore the transportation delay, can be reduced by using fittings and valves that have a large number of equivalent pipe diameters.

5.5 Sumps, Ponds, and Lagoons

As the width or length of an agitated sump increases with respect to its height, the quality of agitation moves from that for a horizontal tank to that for a pond or lagoon. The equipment dead time will increase from 40% to 80% of the residence time. A pond or lagoon will have an equipment dead time that increases with size from 80% to 99% of the residence time. There will be stagnation areas that change with the amount of sunshine and wind. Submerged static mixers with air jets and submerged turbine agitators can provide localized back mixing [5]. Although the portion of the residence time that is dead time is decreased, the magnitude of the residence time, and therefore the dead time, is still enormous. The detrimental effects of such a large dead time on controller tuning and performance are detailed in Chapter 8. However, pond and lagoon volumes can significantly smooth oscillations by dilution and ion migration.

However, this is rarely considered in determining the RCRA violations of surface impoundments.

pH control loops should not be installed on sumps, ponds, and lagoons but on a static mixer or a vertical, well-mixed vessel upstream or downstream.

5.6 Best Practices

Equipment and the associated piping are the most frequent and largest sources of loop dead time. The lack of understanding of mixing profiles (and especially the fundamental concept of back mixing), the transportation delay of reagent into the mixture, and the transportation delay of changes in process *p*H appearing at the electrode are the biggest detriments to *p*H control system performance, assuming a healthy electrode and precise final control element. To minimize these problems, the distances to and from the mixture are minimized, the axial agitator pumping rate in vessels for back-mixing is maximized, and dead zones are minimized. Strategically employing vessels and static mixers in various *p*H control system designs and the importance of minimizing dead time are detailed in Chapter 8.

1. In equipment where a pH control system injects reagent, minimize the dead time contribution to the pH loop dead time from mixing and transportation delays to less than 6 s.
2. Use large volumes susceptible to large dead times upstream or downstream of the *p*H control system to average out *p*H changes for smoother *p*H control and less reagent consumption, particularly by upstream volumes where influent *p*H swings above and below the *p*H set point.
3. If large volumes exist downstream for smoothing *p*H fluctuations, a static mixer with a close-coupled reagent injection to the mixer inlet and electrodes in the pipeline about 20 pipe diameters downstream of the mixer outlet is an option for fast *p*H control to minimize new equipment cost.
4. Use static mixers in the feed or recirculation line of the vessel to premix influent and reagent added to the static mixer inlet to minimize mixing and injection delays.
5. Use impellers designed for axial agitation in vessels where a *p*H control system injects reagent. Keep the vessel diameter-to-height ratio between 0.5 and 1.5 and the agitator pumping rate to give a turnover time (mixing dead time) significantly less than 6 s.

6. In vessels where a pH control system injects reagent, use baffles in the well-mixed vessels to prevent dead zones and promote an axial agitation flow pattern.

References

[1] G. B. Tatterson, "Effect of Draft Tubes on Circulation and Mixing Times," *Chemical Engineering Communications* 19 (1982).

[2] Terrence L. Blevins, Gregory K. McMillan, Willy K. Wojsznis, and Michael W. Brown, *Advanced Control Unleashed – Plant Performance Management for Optimum Benefits* (Research Triangle Park, NC: ISA [The Instrumentation, Systems, and Automation Society], 2003).

[3] F. Greg Shinskey, *pH and pIon Control in Process and Waste Streams* (New York: John Wiley and Sons, 1973).

[4] T. Bor, "The Static Mixer as a Chemical Reactor," *British Chemical Engineering* 16, no. 7 (1971).

[5] Ralph L. Moore, *Neutralization of Waste Water by pH Control*, 2nd ed. (Research Triangle Park, NC: ISA [The Instrumentation, Systems, and Automation Society], 1995).

6 Control Valves

6.1 A Moving Story

Stick-slip occurs whenever the valve must move, even in the same direction. Even after the valve moves, it cannot move again unless the change in signal is greater than the stick. When the valve does move, it jumps or slips by an amount that is usually larger than the change in signal. Stick and slip generally occur together and have a common cause of friction in the actuator design, stem packing, and seating surfaces. Rotary valves with high-temperature packing and tight shutoff (the so-called high-performance valve) exhibit the most stick-slip. Rotary valves tend to also have shaft windup where the actuator shaft moves but the ball, disc, or plug does not move. It is much worse at positions less than 20% where the ball, disc, or plug is starting to rotate into the sealing surfaces. For sliding stem (globe) valves, the stick-slip increases below 10% travel as the plug starts to move into the seating ring. These problems are more deceptive and problematic in rotary valves because the smart positioner measures the actuator shaft position, not the ball, disc, or plug stem. Here is the rub: If there is stick-slip, the controller will never get to the set point and there will always be a limit cycle. The biggest culprits are low leakage classes and the big squeeze from graphite and environmental packing, particularly when the packing is tightened without a torque wrench. A bigger actuator may help but does not eliminate the problem. An undersized actuator can cause a huge additional slip. Stick-slip of 20% often occurs at breakaway from the seat of high-performance valves and for any valve at any position with graphite packing and no positioner. Even with a positioner, stick-slip of 4% has been commonly observed for high-friction packing [1].

RULE of Thumb *To provide isolation, use a separate on-off valve and avoid the specification of tight shutoff and high-performance valves for throttling reagent.*

Whether you can see them or not, the oscillations from control valves are everywhere in the process [2, 3]. It is just a matter of how large and how important they are. When the amplitude is less than the exception reporting setting of controllers, data highways, and historians, the oscillations do a disappearing act on operator displays and trends [1]. All control valves have a resolution limit determined by the amount of stick-slip, shown in Figure 6-1a as the staircase in valve position as the valve first starts to open. In the best throttling valves with air actuators, the stick-slip is typically about 0.1%, and its effect is not observable in the trend recordings of most control systems. This chapter discusses how this seemingly minor limitation in the control valve resolution will clearly show up as a large, sustained oscillation for a set point on the steep portion of the titration curve because of the high process gain. Thus, the extreme sensitivity of the *p***H** process requires a valve resolution that goes well beyond the norm, and *p***H** control systems are the best test for stick-slip in a control valve. The number of stages of equipment needed for neutralization may depend on the control valve's capability. It is difficult to use more than one control valve per stage effectively. To ensure the limit cycle from the valve resolution limit is within the control band, an extremely small and precise control valve is necessary. To achieve the large range of reagent addition and extreme precision required, several stages are used, with the largest control valve on the first stage and the smallest control valve on the last stage. This chapter explains how to estimate the rangeability and resolution required for a reagent valve based on the distance from the influent *p***H** to the set point and the slope of the curve at the set point.

Key CONCEPT When set points are on the steep portion of a titration curve, reagent control valve precision that goes well beyond the norm is required and offers the best test to determine a valve's actual stick-slip in installed conditions.

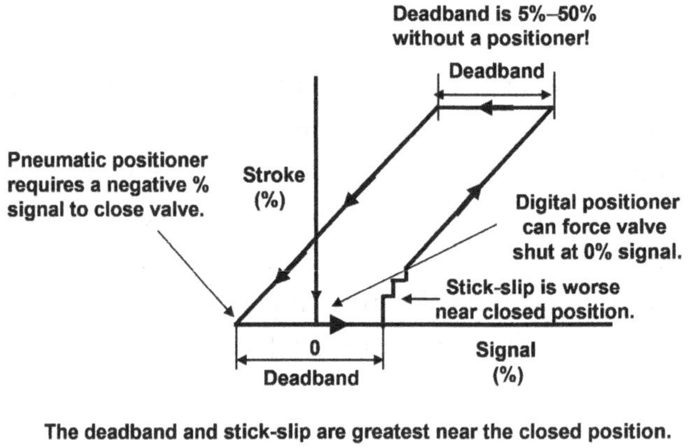

Figure 6-1a. Control valve stick-slip and deadband.

> **Key CONCEPT** Reagent valve stick-slip may determine the number of neutralization stages required, which has a huge impact on a project's capital cost.

The deadband in a control valve does not determine the resolution limit directly but adds dead time to the response, which is particularly significant for the slow *p*H response of large volumes and operating points on the flat portions of titration curves. Deadband occurs only when the valve must change the stroke direction. It is how much the signal must change direction to reverse the stroke direction. The official test for a full-scale stroke is performed in both directions, as shown in Figure 6-1, but deadband occurs for any stroke whenever the direction is reversed. It is caused by lost motion commonly associated with backlash from the actuator shaft and stem connections and linkages. Deadband is worse for rotary valves because of the gaps in the gear teeth in rack-and-pinion and the slot in scotch yoke actuators, key lock shaft-stem connections, and linkages to transfer vertical actuator shaft motion to the rotary ball, disc, or plug movement. Deadbands of 8% have been commonly observed for such valves even though they were outfitted with digital positioners. More actuator torque does not solve the problem. The time it takes for the controller output to work through the deadband is dead time that increases the errors from load disturbances. The problem is not seen for set point changes or step changes in the controller output that are much larger than the deadband. Thus, loop analysis or tuning based on large set point changes, open-loop step tests, or relay autotuner methods will not reveal the additional dead time [1]. If you consider that the peak and integrated errors for load upsets are proportional to the dead time and dead time squared, respectively, deadband is a hidden menace [4]. There is some consolation in that for deadband solely from lost motion, once the valve moves, it can respond to small changes in the signal in the same direction, and this deadband can only cause a limit cycle in a *p*H loop without a self-regulating response, such as batch *p*H control. However, Chapter 8, "Control System," explains how a limit cycle, which is a sustained oscillation of nearly equal amplitude, can also be caused by a *p*H loop driven between flat portions of a titration curve from too much controller gain or reset action.

> **Key CONCEPT** Deadband adds dead time to a *p*H loop that is particularly large for the slow response associated with a large vessel or a flat portion of the titration curve.

The control valve that responds best to small changes is a judiciously sized sliding stem (globe) valve with a digital positioner and a correctly sized diaphragm or electric actuator and properly tightened Teflon packing. It has negligible backlash and a stick-slip of less than 0.1%. A *p*H loop rarely needs the capacity of a rotary valve except for reactions or slurry reagents. If a rotary valve must be used, it is critical to avoid tight shutoff and high-friction packing and to use a diaphragm actuator with a short shaft and splined connections between the actuator shaft and the ball, disc, or plug stem. Make sure the ball, disc, or plug is cast with its stem; otherwise, the connection between the ball, disc, or plug and its stem is another source of backlash. If high-temperature or environmental packing must be used on any type of valve, increase the actuator size and positioner gain to help it better deal with the increased packing friction [1].

RULE of Thumb *Use sliding stem (globe) reagent valves with diaphragm or electric actuators, digital positioners, and properly tightened low-friction packing.*

A pump with a variable speed drive (VSD) is touted in the literature as a better solution than a control valve. Not realized is that the VSD installed flow characteristic is quick opening for high static heads, and both the rangeability and dynamic response are poor for cheaper inverter designs and excessive rate limiting and deadband introduced in the configuration [4].

Interestingly enough, valve specifications do not require that a control valve move. A response requirement should be added to the control valve specification that details the stick-slip, deadband, and response time for a small step in the throttle range [5]. It is ideal to use a ramp at the controller output's expected rate of change rather than a step to reveal the dead time from the deadband and the staircase from stick-slip [6].

Flow through a *p*H control valve is much lower than that through most other control valves, particularly for the last stage of a neutralization system. It corresponds to the last few drops added in lab titration. Extremely low reagent flows create a whole series of additional problems.

Small valves are more prone to improper sizing, irregular flow characteristics, greater stick-slip, and plugging. Here size does matter because most of these problems originate from extremely small Reynolds numbers, clearances, and stem diameters [1].

Key CONCEPT The extremely small reagent valves used in *p***H** control are prone to improper sizing, irregular flow characteristics, greater stick-slip, and plugging.

For a C_V less than 0.01 or extremely viscous fluids, the valve may operate in the laminar flow regime or in the transition region from turbulent to laminar flow. Liquid flow moves from being approximately proportional to the square root of the pressure drop toward being proportional to the pressure drop as the flow goes from being fully turbulent to completely laminar. For anything other than a 1 psi pressure drop, this can translate to an enormous sizing error. The result is an oversized valve that rides the seat where the high seating friction causes excessive stick-slip. Low loads, which decrease reagent demand, and low temperatures, which increase reagent viscosity, can cause a move from turbulent to laminar flow. The worst situation for operation occurs in the transition region, where the flow for a particular valve position has poor reproducibility because normally insignificant disturbances such as microscopic changes in roughness or small vibrations can trigger a switch between turbulent and laminar flow and an erratic installed flow characteristic [1].

> **Key CONCEPT** The extremely small reagent valves used in *p*H control may operate in laminar flow where the flow is proportional to pressure drop or, even worse, in the transition region where the flow response is erratic.

If you also consider the possibility of a significant distortion of the inherent flow characteristic caused by machining tolerances that are an appreciable portion of the clearances for such tiny trim sizes, the scene is set for an unknown and extreme nonlinearity. Tiny clearances can pose all sorts of problems because small particles and coatings cause plugging and sticking. The low flow velocities at the surface that are associated with laminar flow make the likelihood of coating much greater [1].

Finally, tiny stems are likely to be bent from normal handling before and after installation. Slight deflections of the stem can cause huge amounts of stick-slip. What good is a control valve if you can't drop or step on it [1]?

Is the situation hopeless? Not if you go with a control valve manufacturer that specializes in tighter machining tolerances and minimizing stick-slip. Stay away from tiny stems, insist on getting response and flow test results, and use computer programs that deal with laminar flow properly and offer an installed characteristic curve for your piping and operating conditions. Even though the smart digital positioner may be larger and cost more than the valve, it is more important than ever that it be used and properly tuned for the small actuator volume. The challenge here is for packaged platforms that are going for low bids not to get cheap where it hurts [1].

RULE of Thumb *Use control valves designed for small flows and industrial service with rugged shafts, tight machining tolerances, and minimum stick-slip.*

RULE of Thumb *Use sizing programs that deal with laminar flow properly and offer an installed characteristic for your piping system and operating conditions.*

If there is a tendency for plugging, pulse width modulation can be used if there is a sufficiently back-mixed volume to attenuate the pulses per Equation 5-3j in Chapter 5, "Mixing Equipment." It can also provide a linear flow characteristic and a flow large enough to be turbulent. The ratio of maximum to minimum pulse width establishes the rangeability. The maximum pulse width and cycle time determine the degree of variability that must be filtered and the additional dead time from the pulse-off time. Stroking time can be an issue because it is desirable to have the minimum pulse be as short as possible [1]. A globe control valve that is pulsed must be rugged enough to withstand the repeated hammering of the valve seat by the valve plug. A 1/4- or 1/2-inch piston-actuated ball valve is usually less expensive to replace and better able to withstand on-off service than a globe control valve is. Thus, a more practical arrangement for pulse width control is to pulse a small ball valve downstream of a globe control valve that is positioned to set the pulse amplitude. Even so, the life of the pulsed ball valve may not exceed six months. An installed spare ball valve in a parallel pipeline should be provided to minimize downtime.

RULE of Thumb *To prevent plugging and a transition to laminar flow for pH control on a well-mixed volume, use the pulse width modulation of a fast, cheap valve.*

Reagent control valves should be installed in the piping so that flow pushes up on the plug to help open the valve ("flow-to-open"). This avoids the vortex and instability of flow at low lifts, called the *bathtub stopper effect*, associated with "flow-to-close" valves. Valve trims can also develop a swirling action and vortices at particular lift positions, and actuators can develop an instability from inadequate spring stiffness and actuator force or improper positioner design and tuning. The flow becomes erratic and changes even though the controller is in manual. Also, the valve gain is nonreproducible and may even have local reversals (negative gains). Spool-type positioners should be avoided because the spool action tends to exasperate the potential for valve dither. If erratic flow or actuator and positioner behavior is suspected, request flow lab tests at your operating conditions over the whole throttle range from the valve manufacturer.

Key CONCEPT Some valve designs exhibit an erratic flow from swirling, vortices, and actuator instability for particular positions and operating conditions.

RULE of Thumb *Request a flow lab test at your operating conditions over the entire throttling range if erratic flow or actuator or positioner behavior is suspected.*

6.2 Resolution Requirements

Each loop has a control band that is the maximum desired positive and negative excursions of *p*H about a set point. If the width of this control band is marked on the ordinate of the titration curve and translated to a width on the abscissa, the result is the allowable error in the reagent-to-influent flow ratio. This error should be multiplied by the maximum influent flow, divided by the reagent valve capacity, and multiplied by 100% to express the error in terms of a percentage of the reagent control valve capacity. The reagent valve capacity should be about 25% greater than the abscissa at the center of the reagent error multiplied by the maximum influent flow to allow for data error. Figures 6-2a and 6-2b show how a steeper slope at the set point requires the reagent error to be smaller and, therefore, the control valve to be more precise. If Equation 6-2b for the maximum reagent valve capacity is substituted into Equation 6-2a for the allowable reagent error in percentage of capacity, the result is Equation 6-2c that has only the graph parameters A and B. Equation 6-2c shows the reagent error in percentage of capacity decreases for an increase in the distance A of the center of the reagent error band on the abscissa from the origin. Thus, a lower influent *p***H** requires a smaller allowable reagent error in percentage of capacity because the origin, which is the point of zero reagent flow, is shifted to the left. The equations are for the best case where the limit cycle from stick-slip is centered about a set point that is at the center of the control band. As a best practice, users should seek a valve whose stick-slip is about half the reagent error band (Equation 6-2d).

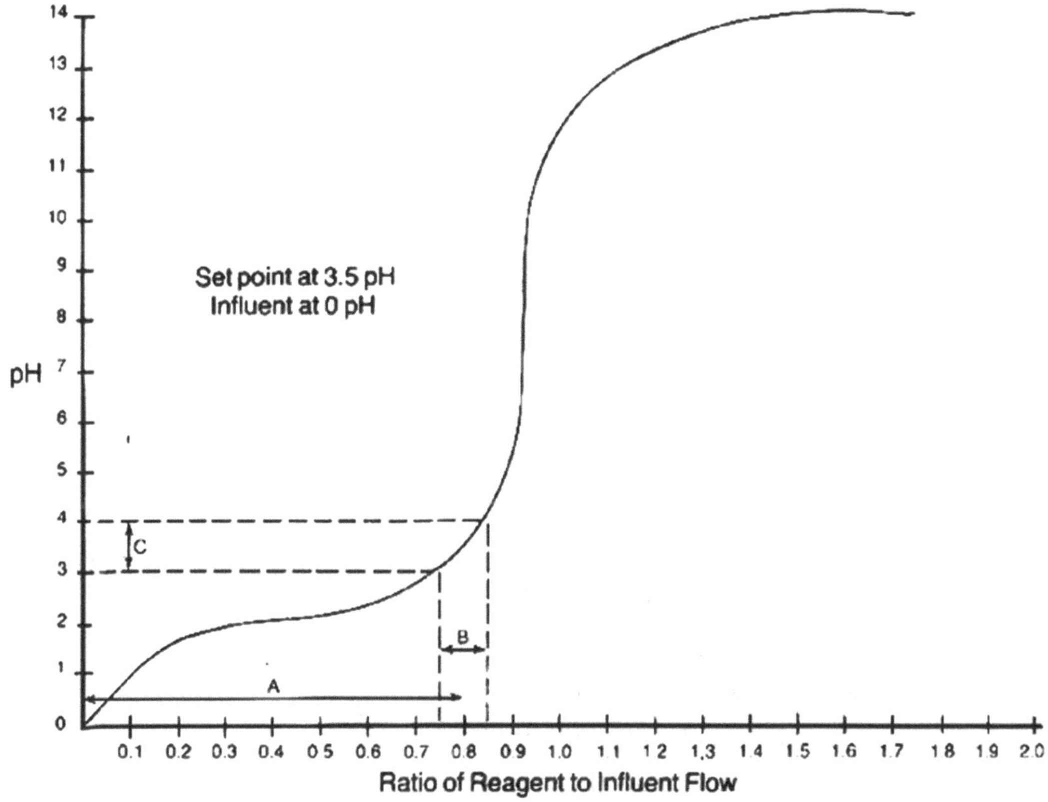

Figure 6-2a. Larger allowable reagent error on a flatter portion of the titration curve.

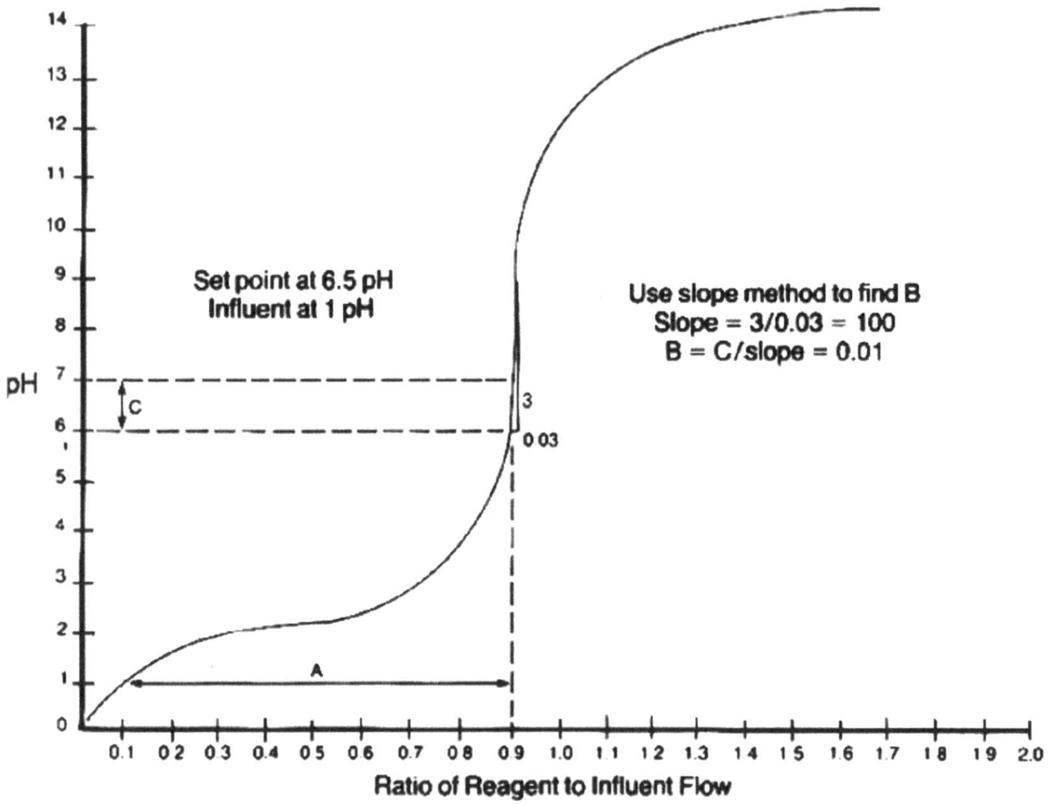

Figure 6-2b. Smaller allowable reagent error on a steeper portion of the titration curve.

$$E_r = 100 \bullet F_{i\max} \bullet \frac{B}{F_{r\max}} \tag{6-2a}$$

$$F_{r\max} = 1.25 \bullet A \bullet F_{i\max} \tag{6-2b}$$

$$E_r = 80 \bullet \frac{B}{A} \tag{6-2c}$$

$$S_s = 0.5 \bullet E_r \tag{6-2d}$$

where

A = distance of the center of the reagent error band on the abscissa from the origin (zero ratio) in Figures 6-2a and 6-2b

B = width of the reagent error band on the abscissa in Figures 6-2a and 6-2b

E_r = allowable reagent error (%)

$F_{r\max}$ = maximum reagent valve capacity (kg/min)

$F_{i\max}$ = maximum influent flow (kg/min)

S_s = allowable stick-slip (resolution limit) (%)

The number and type of stages of neutralization required are driven by the need to minimize the A parameter of the final stage and keep the B parameter small and fast from upstream stages so that it is effectively filtered out by the volume of the final stage or intermediate volumes. For a set point near the neutral point, this leads to the classical Rule of Thumb 1-2a from the 1960s that requires one vertical well-mixed vessel with *p***H** control for each 2 *p***H** units the influent is from the set point. An update to this rule takes credit for feedforward control and more precise control valves, as detailed in Chapter 8. The B parameter from the first stage should be substituted as the input amplitude in Equation 5-3j to estimate the effect of filtering the volume of the next stage. Inspecting the relationships espoused in this equation reveals that the first stage should be as fast and cover as much ground as possible. For a strong acid and strong base system, this translates to using a static mixer with a set point at the start of the bend of the overall curve (2 *p***H** in Figure 1-1a). The next stage would use a small well-mixed tank with a set point at the start of the bend of a first "zoom-in" (4 *p***H** in Figure 1-1b). The final stage would have a large well-mixed tank with a set point at the neutral point (7 *p***H**).

Key CONCEPT An oversized or sticky control valve can cause a limit cycle that is larger than the control band and necessitate an additional neutralization stage.

RULE of Thumb *The first and last stages should have the largest and smallest valve and the smallest and largest volume, respectively.*

6.3 Rangeability Requirement

Control valve inherent rangeability is the ratio of the maximum to the minimum controllable flow coefficient within which the "installed" flow characteristic does not exceed some specified limits. It is advisable to include the words *controllable* and *installed* because the allowable limits should be based on the magnitude of the change in slope (the valve gain) seen by the controller and because valve stroke precision near the closed position places a practical limit on the minimum controllable flow coefficient. While the flow coefficient of some valves becomes very irregular below a 5% opening, the change in slope is still usually less than the change in slope over the whole stroke for an equal percentage trim, a modified parabolic trim, or a varying pressure drop. Up to now, the emphasis has been on trying to define specified limits on the inherent flow characteristic, and the effect of stroke stick-slip on rangeability has been ignored in the literature. If stick-slip (which generally increases near the closed position) is ignored, the quick-opening inherent flow characteristic yields the best rangeability, the linear inherent flow characteristic yields the next best, and the equal percentage inherent flow characteristic yields the worst inherent rangeability. However, if the effect of stick-slip is included, the opposite is true. The equal percentage inherent flow characteristic gives a 50:1 inherent rangeability for a wide range of valve stroke resolution capabilities because of the flatness of the characteristic curve near the closed position. Most field instrument maintenance personnel have felt that equal percentage trim gave them the greatest rangeability despite what is in the literature. Also, for static mixers, the equal percentage trim characteristic, whose flow coefficient is proportional to the flow, helps compensate for a process gain that is inversely proportional to flow. For the well-mixed tank, a linear installed characteristic is desirable if the *p***H** controller is directly manipulating a reagent valve because the inverse relationship of process gain to flow is canceled out by an equipment time constant that is proportional to the residence time per equations in Chapters 5 and 8. However, most *p***H** loops on well-mixed tanks now manipulate a reagent flow controller set point because of the improvement in flowmeters, which isolates the nonlinearity of the control valve from the *p***H** controller [5].

> **Key CONCEPT** Most statements in the literature about valve rangeability are erroneous because they ignore the installed valve gain and stick-slip.

RULE of Thumb *Use a sliding stem valve with a digital positioner with equal percentage characteristics and minimal seating friction for maximum rangeability.*

If the pressure drop available to a *p***H** control valve decreases as the reagent flow increases because the frictional pressure drop in the rest of the system increases, a linear trim distorts toward a quick open installed valve characteristic, and an equal percentage trim distorts toward a linear installed valve characteristic. The rangeability is also adversely affected by the significant decrease in available pressure drop for the valve at high flows. However, for *p***H** control valves, the reagent flow is so small, unless consumed by a reaction, that the pressure drop in the piping system is negligible, and most of the pressure drop occurs across the control valve. In this case, the valve's pressure drop is constant, and the installed characteristic is the inherent trim characteristic.

> **Key CONCEPT** The reagent flow required for neutralization is usually so small that the installed valve characteristic is the inherent trim characteristic.

The rangeability required depends on the maximum and minimum influent *p***H** and flow. The maximum influent *p***H** corresponds to a minimum A distance, and the minimum influent *p***H** corresponds to a maximum A distance. The required rangeability is the ratio of maximum to minimum A distance multiplied by the ratio of maximum to minimum influent flow, as shown in Equation 6-3a, and then multiplied by 1.25 to provide a 25% margin for data error. The minimum A used should not be less than the reagent error band because the *p***H** is already within the control band. Equation 6-3b shows that the rangeability for this case where the minimum A is equal to the B parameter which is equal to the inverse of the allowable reagent error as a fraction (not a percentage) of the reagent valve capacity (per Equation 6-2a) multiplied by the influent flow ratio. Thus, if you have satisfied the resolution requirement, you usually have satisfied the rangeability requirement because the actual A ratio is typically much less than that assumed for Equation 6-3b. The literature to date has focused on the rangeability requirement, which is the need to deliver a small change in reagent flow only at low loads, and ignored the more stringent resolution requirement, which is the need to deliver a small change in reagent flow at all loads. Therefore, split-ranged valves solve only the rangeability problem, whereas sliding stem valves with digital positioners and control strategies for a coordinated coarse and fine reagent adjustment meet both requirements.

RULE of Thumb *Use precise valves and control strategies for a coordinated coarse and fine adjustment to meet both resolution and rangeability requirements.*

$$R_r = 1.25 \bullet \frac{A_{\text{max}}}{A_{\text{min}}} \bullet \frac{F_{i\text{max}}}{F_{i\text{min}}} \tag{6-3a}$$

For $A_{\text{max}} = A$ and $A_{\text{min}} = B$:

$$R_r = 1.25 \bullet \frac{A}{B} \bullet \frac{F_{i\text{max}}}{F_{i\text{min}}} \tag{6-3b}$$

where

A = distance of the center of the reagent error band on the abscissa from the origin

A_{max} = maximum distance of the center of the reagent error band on the abscissa from influent *p***H**

A_{min} = minimum distance of the center of the reagent error band on the abscissa from influent *p***H**

B = width of the reagent error band on the abscissa

$F_{i\text{max}}$ = maximum influent flow (kg/min)

$F_{i\text{min}}$ = minimum influent flow (kg/min)

R_r = required reagent rangeability

6.4 Split-Ranging

Although split-ranging control valves does not solve the precision problem, it is still required when both acidic and basic reagents are needed, or an inherent maximization of an inexpensive reagent is desired. For example, if the influent *p*H can be above or below the set point or overshoot correction is needed for *p*H control of a traditional batch, split-ranged acid and base valves are needed. Also, if using the waste stream as a reagent stream must be maximized, a split range can be set up to completely open the *p*H valve on the waste stream before the supplemental reagent valve is opened. Split-ranging must be done accurately to prevent overlapping strokes that alter the combined valve gain seen by the controller, increase the occurrence of stick-slip, and waste reagent by cross-neutralization or oscillations. The increased stick-slip at the closed position means that the loop will develop a limit cycle at the split-range point whose severity increases where an acid valve closes as a base valve opens, or vice versa. To minimize the propensity for a limit cycle, a split-range gap should be twice the stick-slip or the potential error in the positioner, whichever is largest. However, this gap adds a dead time equal to the gap divided by the rate of change of controller output. This dead time is quite large for a *p*H loop on a large volume or one that is detuned to deal with a steep titration curve.

> **Key CONCEPT** A split-range gap adds dead time, and a split-range overlap increases the severity of a limit cycle from slip-stick and increases reagent use.

RULE of Thumb *Use a split-range gap that is about twice the stick-slip or the potential error in the positioner calibration, whichever is largest.*

Usually, 50% controller output is used as the transition point for split-ranging an acid or base valve. Although this point is easy to remember for maintenance and operations, a better transition point for loop performance is one that equalizes the control valve gain seen by the controller over the whole signal range. The ratio of the portions of the signal range allocated for the acid and base should be equal to the ratio of the product of the reagent valve capacity and the reagent normality for the acid and base, as shown by Equation 6-4a. The same type of relationship holds for split-ranging different or identical normality acids or bases. (The normalities cancel out when equal.) For example, the split-range point with a 1% gap for a large linear trim valve whose capacity is 10 times that of a small linear trim valve on the same reagent should be at 9%. The small valve would stroke from 0% to 9% of the signal, and the large valve would stroke from 10% to 100% of the signal.

RULE of Thumb *Choose a split-range point to minimize the nonlinearity introduced by the valve and reagent so that the percentage of change in the controller output to either valve will cause about the same change in pH.*

$$S_1 = S_2 \bullet \frac{N_1}{N_2} \bullet \frac{F_{1max}}{F_{2max}} \tag{6-4a}$$

$$S_2 = 100 - G - S_1 \tag{6-4b}$$

If S_2 per Equation 6-4b is substituted into Equation 6-4a with $G = 1\%$, $X_1 = N_1 \bullet F_{1max}$, and $X_2 = N_2 \bullet F_{2max}$:

$$S_1 = 99 \bullet \frac{X_1}{X_1 + X_2} \tag{6-4c}$$

where

F_{1max} = maximum valve capacity of reagent 1 (kg/min)
F_{2max} = maximum valve capacity of reagent 2 (kg/min)
G = gap at the split-range point (%)
N_1 = normality of reagent 1
N_2 = normality of reagent 2
S_1 = portion of split-ranged controller output allocated to reagent 1 (%)
S_2 = portion of split-ranged controller output allocated to reagent 2 (%)

If acid and base valves are split-ranged, each has an opposite effect on the *p***H**. Because only a single *p***H** control action can be used, one of the valve actions must be the reverse of the other. If the two reagent valves have the same failure modes, and thus the same valve actions, the signal must be reversed for one of the valves. For a reverse *p***H** control action, the acid valve signal should be reversed. For a direct *p***H** control action, the base valve signal should be reversed if the control valves are both fail closed (increase to open). The opposite assignment should be used if the control valves are both fail open (increase to close). Some valve positioners and current-to-pneumatic (I/P) transducers have options for split-ranging and signal reversing, but for accuracy, flexibility, visibility, and maintainability, these functions are best handled by blocks in the control system (e.g., DCS) configuration.

RULE of Thumb *Split-ranging and signal reversing are better implemented in a Fieldbus function block than via a specially calibrated I/P or positioner.*

Various computations have been developed for split-ranging equal percentage characteristic valves. Although the nonlinear gain can, for some specific set of influent flow and concentration conditions, partially compensate for the nonlinearity of the titration curve, the assumption that these conditions always exist is tenuous. Using reset action in a *p***H** controller means that no one-to-one relationship exists between the measurement signal and the control valve signal. The *p***H** controller will drive its output to whatever value is necessary to reduce the error. Thus, nonlinearity in the process or a controlled variable cannot generally be compensated for by control valve trim nonlinearity for proportional-integral or proportional-integral-derivative (PID) controllers.

For more on the dynamic response problems of control valves, the consequences and methods of mitigation, see References 7 and 8.

6.5 Best Practices

Achieving a fast and precise final control element is critical for *p***H** control. This chapter's focus on control valve resolution and rangeability is an essential starting point, but there are many other considerations. Variable frequency drives (VFDs) are often touted as offering tighter control, but there are many design and installation choices that can make this expectation unrealistic. To help us all be aware of potential problems and recognized solutions, the following list of best practices is offered based on the content in Chapter 7 of the ISA book *Essentials of Modern Measurements and Final Elements in the Process Industry* [9].

1. Use sizing software with physical properties for the worst-case operating conditions.
2. Include the effect of the piping reducer factor on the effective flow coefficient.
3. Select the valve location and type to eliminate or reduce damage from flashing.
4. Preferably, use a sliding stem valve (size permitting) to minimize backlash and stiction unless crevices and trim cause concerns about erosion, plugging, sanitation, or the accumulation of solids (particularly monomers) that could polymerize. For single-port valves, use a "flow-to-open" trim design to eliminate the bathtub stopper swirling effect.
5. If a rotary valve is used, select a diaphragm actuator with a splined actuator-shaft-to-stem connection, integral cast ball or disk stem, and minimal seal and packing friction to minimize lost motion deadband and resolution limitation.

6. Use conventional Teflon packing. For higher temperature ranges, use ultra-low friction Teflon packing, avoid overtightening packing, and consider using a compatible stem lubricant.
7. Compute the installed valve flow characteristic for worst-case operating conditions.
8. Size actuator to deliver more than 150% of the maximum torque or thrust.
9. Select an actuator and positioner with threshold sensitivities of 0.1% or better.
10. Ensure that the total valve assembly deadband is less than 0.4% over the entire throttle range.
11. Ensure that the total valve assembly resolution is better than 0.2% over the entire throttle range.
12. Choose an inherent flow characteristic and valve-to-system pressure drop ratio that does not change the valve gain more than 4:1 over the entire process operating point range and flow range.
13. Tune the positioner aggressively (high proportional action gain) without integral action and with readback that indicates actual plug, disk, or ball travel instead of just actuator shaft movement.
14. Never replace positioners with volume boosters. Instead, put volume boosters on the positioner output to reduce the valve 86% response time for large signal changes with the booster bypass valve open just enough to assure stability.
15. Use small (0.2%) and large step changes (20%) to test the valve's 86% response time to determine if changes must be made to meet the desired 86% response time.
16. See ISA-TR75.25.02, Annex A [8], for more details on valve response and for relaxing expectations on travel gain and 86% response time for small and large signal changes, respectively.
17. Counterintuitively increase the PID gain to reduce the oscillation period and/or amplitude from lost motion, stick-slip, and poor actuator or positioner sensitivity.
18. Use external reset feedback, detailed in Chapter 8, with accurate and fast valve position readback to stop oscillations from poor precision and slow response time [10].

19. Use input and output chokes and isolation transformers to prevent electromagnetic interference (EMI) from the VFD inverter.
20. Use pulse width modulation to reduce torque pulsation (cogging) at VFD low speeds.
21. Use a VFD inverter duty motor with Class F insulation and a 1.15 service factor that is a totally enclosed fan-cooled motor with a constant speed fan or booster fan or a totally enclosed water-cooled motor in high-temperature applications to prevent overheating.
22. Use a National Electrical Manufacturers Association Design B motor instead of a Design A motor to prevent a steep VFD torque curve.
23. Use bearing insulation or path to ground to reduce bearing damage from electronic discharge machining (EDM). Damage from EDM is worse for the six-step voltage, older VFD technology.
24. Size the pump to prevent it from operating on the flat part of the VFD pump curve.
25. Use a recycle valve to keep the VFD pump discharge pressure well above the static head at low flow and a low-speed limit to prevent reverse flow for the highest destination pressure.
26. Use at least 12-bit signal input cards to improve the VFD resolution limit to 0.05% or better.
27. Use a drive and motor with a generous amount of torque for the application so that speed rate-of-change limits in the VFD setup do not prevent changes in speed from being fast enough to compensate for the fastest possible disturbance.
28. Minimize the VFD deadband introduced into the drive configuration (often set in a misguided attempt to reduce the response to noise) which causes delay and limit cycling.
29. For VFD tachometer control, use magnetic or optical pickup with enough pulses per shaft revolution to meet the speed resolution requirement.
30. For tachometer control, keep the speed control in the VFD to prevent cascade rule violations where the secondary speed loop is not five times faster than the primary process loop.
31. To increase rangeability to 80:1, use fast cascade control of speed to torque in the VFD to provide closed-loop slip control.

32. Use external reset feedback, detailed in Chapter 8, with accurate and fast-speed readback to stop oscillations from poor VFD resolution and excessive deadband and rate limiting in the VFD configuration [10].

33. Use foil-braided shield and armored cable for VFD output spaced at least 1 ft from signal wires—never crossing any signal wires—ideally via separate cable trays.

References

[1] Gregory K. McMillan, "What's Your Flow Control Valve Telling You," *Control Design* (May 2004).

[2] Michael Ruel, "Stiction: The Hidden Menace," *Control* (November 2000).

[3] Michael Ruel, "Control Valve Health Certificate," *Chemical Engineering* (November 2001).

[4] Terrance L. Blevins, Gregory K. McMillan, Willy Wojsznis, and Michael W. Brown, *Advanced Control Unleashed* (Research Triangle Park, NC: ISA [The Instrumentation, Systems, and Automation Society], 2003).

[5] Gregory K. McMillan, *Good Tuning: A Pocket Guide*, 4th ed. (Research Triangle Park, NC: ISA [International Society of Automation], 2015).

[6] *A Funny Thing Happened on the Way to the Control Room*, Reprint via ProQuest UMI "Books on Demand," 1989.

[7] Gregory K. McMillan and Pierce Wu, "How to Specify Valves and Positioners That Don't Compromise Control," *Contro*l (May 2016).

[8] ANSI/ISA-TR75.25.02, *Annex A - Valve Response and Control Loop Performance Sources, Consequences, Fixes, and Specifications* (Research Triangle Park, NC: ISA [International Society of Automation], 2023).

[9] Gregory K. McMillan, *Essentials of Modern Measurements and Final Elements in the Process Industry*, (Research Triangle Park, NC: ISA [International Society of Automation], 2010).

[10] ANSI/ISA-TR5.9-2023, *PID Algorithms and Performance* (Research Triangle Park, NC: ISA [International Society of Automation], 2023).

7

Reagent

7.1 Delivery Dilemmas

For good p**H** control, a small change in reagent dosage must be delivered precisely, reliably, and quickly into the middle of a mixture. This becomes an exceptional challenge for the extremely small flow required for most p**H** control systems and translates into specific demanding requirements for reagent selection and piping design.

For a reagent dose to be precise enough, the reagent's logarithmic acid dissociation constant (pK_a) should be close to the p**H** set point so that it tends to flatten the titration curve, reducing the effect of valve stick-slip. Weak acids, such as acetic acid, and weak bases, such as ammonia, have a pK_a close enough to 7 p**H** to reduce the precision requirement and possibly the number of neutralization stages. The reagent viscosity should also be low enough to ensure a fully turbulent flow which is important for consistent dosing. For the delivery to be reliable, the reagent should be free from solids and slime because of the tiny control valve trims used. A fast reagent delivery requires that the reagent viscosity be low enough for the dose to start quickly and mix rapidly with the effluent. An extraordinary time delay has been observed for starting the flow of 98% sulfuric acid through an injection orifice because of its high viscosity. It has been compared to getting ketchup out of a bottle [1]. Mixing an influent stream with a viscous reagent stream, such as 98% sulfuric acid or 50% sodium hydroxide that is about 40 times more viscous than the typical influent stream, is difficult and requires greater agitation intensity and velocity. For static mixers, it means more mixing elements and higher pressure drops. A highly viscous reagent dose tends to travel as a glob through the mixture. Lastly, the reagent should be in the liquid phase.

The neutralization reaction of liquid components is essentially instantaneous once reagent streams are mixed with influent streams. Gas or solid reagents take seconds and minutes, respectively, to dissolve and get to the point of liquid contact and mixing with the influent stream components. Reagent bubbles escape as a vapor flow when the bubble breakup time and gas dissolution time exceed the bubble rise time. Reagent particles travel downstream and accumulate on the walls of piping, equipment, and pump strainers when the solid dissolution time exceeds the shortest residence time. Unreacted ammonia and calcium hydroxide (lime) reagent have been the source of huge and largely undocumented extra reagent and additional emissions and solids disposal costs. Coating and plugging upstream control valves and downstream equipment from particles of unreacted reagent or precipitation of salts from reacted reagents can be so severe as to cause excessive equipment maintenance and downtime. Liquid reagents such as ammonia that can flash in the vena contracta of a control valve can choke the valve or cause cavitation damage in the valve and the piping immediately downstream. Particles in the reagent can erode the control valve trim. Some waste lime systems have rocks that can quickly tear up a valve seat and plug [1].

RULE of Thumb

To help ensure a precise, reliable, and quick reagent delivery, use low-viscosity, liquid weak acids and weak bases that contain no bubbles, particles, rocks, or slime and cause no excessive salt precipitation.

Key CONCEPT

Gaseous and solid reagents often have huge hidden costs from additional reagent consumption, emission, solids disposal, and maintenance costs.

If the reagent pipeline is partially filled or empty, or a dip tube or injection tube is backfilled with process fluid, the regent delivery delay becomes the biggest source of dead time in a *p*H loop for the small reagent flows commonly required for neutralization. Whenever a reagent control valve closes or a metering pump stops, the reagent continues to drain into the process, and process fluid can be forced back up into the reagent injection or dip tube. Even if hydraulics do not promote draining or backfilling the tubes, ion migration from high to low concentrations will proceed until an equilibrium is reached between the concentrations in the reagent tube and the process volume. As a result, the *p*H will continue to be driven by draining and migrating reagent after the control valve closes or the metering pump stops. If the valve is closed or the pump is stopped for a long time, when the valve opens or the pump starts, it must flush out process components in the tubes before the reagent gets into the volume. The worst-case delivery time delay is the volume of the backfilled tube divided by the reagent flow. Because dip tubes are designed to be large enough to withstand agitation and the design standard for normal flows is to take the dip tube down toward the impeller, the reagent delivery delay can be several orders of magnitude larger than the turnover time described in Chapter 6, "Control Valves." For example, if the reagent flow is 1 gal/h and the submerged reagent dip tube volume is 2 gal, it would take 2 h for the reagent to get into the vessel, which is almost 1000 times larger than the dead time from the turnover time in a well-mixed vessel. Although this is an extreme case that usually occurs only during start-up or batch operations, it is symptomatic of the reagent piping being a big potential source of dead time that is often overlooked during reagent delivery system design [1].

When a section of reagent pipeline is empty because of drainage and a siphoning effect, the time delay to fill the pipe is the volume of the empty pipe divided by the reagent flow.

If there is insufficient back pressure to keep the pipeline full, the delay is the time for a wave of process fluid to propagate down the horizontal sections and dribble down the vertical sections. If reagents are mixed or diluted in-line, the time delay is the piping volume from the mixing point to the entry point divided by the combined flow.

> **Key CONCEPT** The delivery delay from an empty or backfilled reagent pipe, injection tube, and dip tube is the largest source of dead time in a *p*H loop.

The best case is when the reagent piping and tubes are completely filled with a constant concentration of noncompressible liquid reagent. A change in valve position causes a pressure wave to travel at the speed of sound in the fluid from the valve that, on arrival at the entry point into the process, initiates a change in reagent flow within a second or two. Check valves can be used on the tips of injection and dip tubes to achieve this condition if they can be designed reliably enough for the process service and a reagent flowmeter is installed to verify reagent flow. Because the check valve is difficult to replace and its integrity is suspect, it is not a popular solution. An automated on-off isolation valve close-coupled to the process connection that closes when the control valve throttle position is below a reasonable minimum helps keep reagent lines pressurized and full of reagent. An isolation valve should be added to the pump suction of metering pumps to prevent the reagent from draining back through the pump [1].

> **Key CONCEPT** If the complete reagent delivery system can be kept pressurized with a constant liquid reagent composition, the delivery delay is negligible.

RULE of Thumb *Add an automated isolation valve close-coupled to the process connection and in the suction of a metering pump that closes when the control valve closes to prevent draining and backfilling reagent piping.*

The distance between the reagent control valve and the process connection should be minimized even when there is an isolation valve to minimize the transportation delay of changes in reagent concentration and the amount of reagent trapped between the valves. To protect against overpressurization from thermal expansion in a trapped section, a relief valve is installed with its discharge piped to a reagent storage or collection vessel.

The best method of reducing reagent delivery delay and mixing dead time is to inject the reagent flow into a feed or a recirculation line. The high feed or recirculation flow provides premixing, minimizes transportation delay for a properly selected injection point, and creates better reagent dispersion at the entry point into the process. Conventional dip tube design can be used with this method. However, localized high concentrations of reagent acids and bases may necessitate *p*H control and/or special construction materials between the injection and entry points into the volume. If the *p*H is controlled in a feed line and recirculation line, the injection point is usually into a static mixer and pump suction, respectively. If reagent addition is manipulated by a loop to control the *p*H in a vessel rather than in a pipeline, the injection point should be within 20 pipe diameters of the entry point into the vessel. A static mixer can improve the premixing, although the reduction in *p*H noise may not be noticeable for a large vessel.

RULE of Thumb *The best method of reducing reagent delivery delay and mixing dead time is to inject the reagent into a high feed or recirculation flow.*

7.2 Dilution

A common misconception is that the slope of the titration curve and, therefore, the system sensitivity can be decreased by reagent dilution. Reagent dilution has a negligible effect on the shape of the titration curve: the curve slope will appear larger if the same abscissa is used because only a portion of the original curve is displayed. The numbers along the abscissa must be multiplied by the ratio of the old to the new reagent concentration to show the entire original titration curve. For example, the abscissa values would have to be doubled if the reagent concentration were cut in half. While the allowable reagent error band is increased, the reagent control valve capacity requirement is also increased by the same factor, so the valve resolution requirement does not change.

Properly designed dilution systems offer a variety of performance benefits. Most chemists use diluted reagents to make titration easier and more accurate. Because this is true for the lab, you can imagine how much more important it is in the field fraught with nonideal conditions. Dilution can reduce reagent valve plugging, reagent transportation delay, and reagent viscosity. It can prevent laminar flow and partially filled pipes and dramatically improve dispersion in a mixture. It also decreases the freezing point and winterization problem and corrosion for sodium hydroxide and the plugging tendency of sulfur dioxide. However, sulfuric acid becomes more corrosive when diluted.

Key CONCEPT

Although the shape of the titration curve does not change much by dilution, dilution can dramatically decrease reagent delivery delays, improve reagent addition reproducibility by eliminating transitions to laminar flow, increase valve reliability by reducing plugging, and enhance mixing by increasing the reagent injection velocity and reducing the differences in viscosity and density between the reagent and the influent, which is a normally a dilute water stream.

If reagent dilution is used, the system must be carefully designed to prevent creating reagent concentration upsets and delivery delays. The *p***H** controller should throttle the diluted reagent. The mass flow of water should be ratioed to the mass flow of reagent, and a density controller should trim the ratio. Coriolis flowmeters should be used to improve the mass flow measurement reproducibility and provide an accurate density measurement for concentration control, as shown in Figure 7-2a. In addition, one could utilize the temperature from the Coriolis meter to compensate for the density, particularly if the feed temperature is not constant. If the *p***H** controller throttles the undiluted reagent or the water, the reagent delivery time delay is for the plug flow concentration response instead of a liquid flow response. The time delay for the concentration response is approximately equal to the residence time, whereas the time delay for a full pipeline liquid flow response is less than 1 s. If the titration curve is steep, the fast concentration disturbances from flow measurement inaccuracy and noise may cause *p***H** excursions outside the control band. For example, the single-step neutralization of 0 *p***H** hydrochloric acid with sodium hydroxide would require a flow measurement accuracy of 0.0001% for reagent dilution to achieve a 2 *p***H** control band about a 7 *p***H** set point. If a static mixer is used for *p***H** control, the fast concentration disturbances from the reagent dilution will create *p***H** measurement noise. Therefore, for steep titration curves and in-line *p***H** control, a storage tank for a diluted reagent with a high recirculation flow should be installed to smooth out the fast reagent concentration disturbances from the dilution.

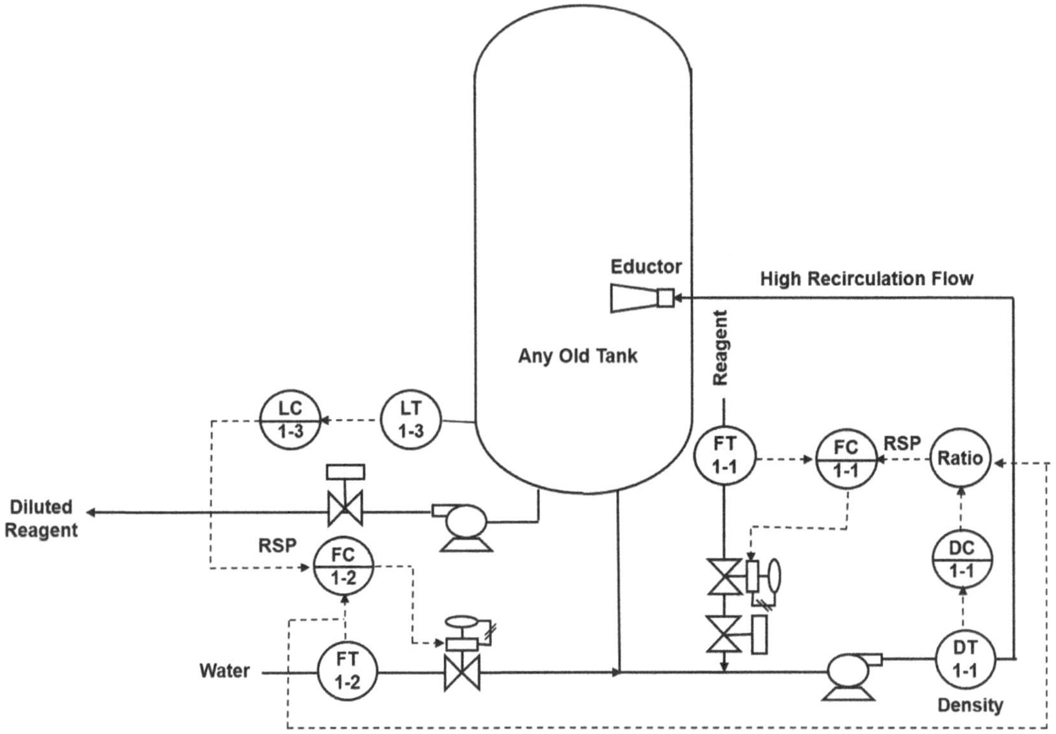

Figure 7-2a. Precision dilution control system in an upstream vessel. (High recirculation is used to provide mixing that smooths out reagent fluctuations and enables an approach to a new dilution set point.)

RULE of Thumb *Use Coriolis mass flowmeters to provide mass flow ratio control of water to reagent corrected by a Coriolis density loop that feeds a diluted reagent vessel volume sized to smooth out reagent concentration fluctuations to the* p***H*** *loop.*

7.3 Buffering

The buffering mechanism was described in Chapter 3, "Titration Curves." Adding a buffer with a dissociation constant *p***H** equal to the *p***H** set point can greatly reduce the slope and, therefore, the system sensitivity in the *p***H** control band. The buffer capacity, also known as the *buffer value,* is defined as the change in normality from adding a strong base or acid divided by the change in *p***H**, as shown in Equation 7-3a. The buffer capacity is a measure of the solution's ability to resist a change in *p***H**. Typical buffer capacities of buffer solutions used for *p***H** meter calibration range from 0.01 to 0.2. The equations in Chapter 2, "The Chemistry," can be used to convert from normality to weight fraction or a mass flow ratio of reagent to influent to increase the utility of the buffer capacity for system design.

$$Z = \frac{\Delta N}{\Delta p\mathbf{H}} \tag{7-3a}$$

where

ΔN = change in acid or base normality (gram-equivalents/L)

$\Delta p\mathbf{H}$ = change in solution $p\mathbf{H}$

Z = buffer capacity (gram-equivalents/L/$p\mathbf{H}$)

7.4 Dissolution

Dissolution is the act of dissolving. Whereas the *p***H** reaction time rarely exceeds a few seconds, the dissolution time of some solid or gaseous reagents can be several minutes. The dissolution time is equivalent to the batch conversion time and can be determined by adding a shot of reagent to an influent sample in a beaker and noting the time between the shot of the reagent and 99% conversion. The beaker must have the same degree of agitation as the field vessel. A *p***H** measurement can be used to note the 99% conversion point, but a very steep titration curve slope will probably result in a shorter than actual dissolution time measurement, and a very flat titration curve will probably result in a longer than actual dissolution time measurement. If the reagent is continuously added instead of pulsed, the *p***H** measurement may suddenly approach the final value and rapidly return to the previous value due to dynamic changes in the solution saturation. These "rebounds" add ambiguity to the dissolution time measurement.

A back-mixed vessel is much better than a plug flow vessel for control because it has a much lower dead time to time constant ratio. However, the residence time distribution has a much greater spread. Some of the material is in the vessel for a time much smaller than the residence time for continuous operation. Consequently, a greater residence time is required for a continuous back-mixed vessel to convert lime. The plot in Figure 7-4a shows that the residence time of a continuous back-mixed vessel must be 20 times the dissolution time for 99% conversion [2]. The agitator pumping rate must be scaled up in proportion to the vessel volume so the equipment dead time to residence time ratio remains the same. Because the horsepower required is proportional to the third power of the pumping rate, the additional energy cost may be greater than the savings from using lime instead of caustic as a reagent. The capital cost of the larger vessel and motor may also be significant. If the size of a single back-mixed vessel becomes prohibitive, several smaller ones in series can be used to achieve the desired conversion percentage. The effect of these vessels in series on conversion is multiplicative regarding conversion fractions, as shown by Equation 7-4a. For example, two vessels, each with a residence time that is twice the dissolution time, will have an individual conversion of 90% and an overall conversion of 99%.

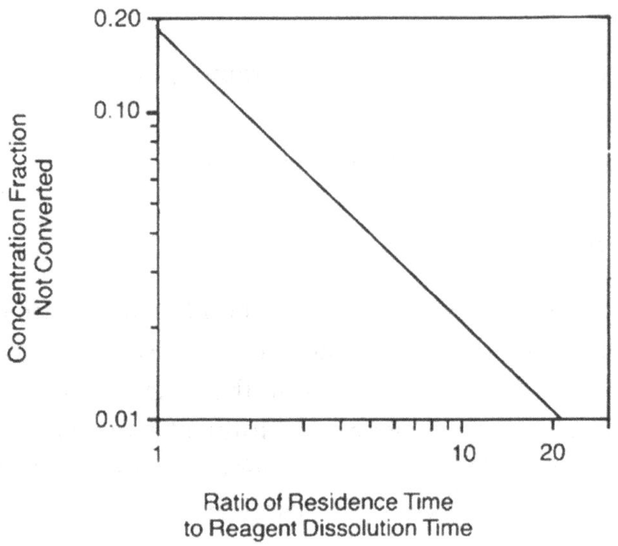

Figure 7-4a. Concentration fraction of lime not converted in a back-mixed vessel.

RULE of Thumb *For multiple stages, use lime on the first stage rather than on the last stage to decrease the amount of unconverted lime downstream.*

$$X_o = X_1 \bullet (1 - X_1) \bullet X_2 \tag{7-4a}$$

where

X_o = overall conversion fraction at the outlet of the last vessel
X_1 = conversion fraction at the outlet of the first vessel
X_2 = conversion fraction at the outlet of the second vessel

The dissolution time affects a control loop in a manner that is similar to a slow reagent control valve because the change in the quantity of dissolved reagent lags behind the change in the *p***H** controller output. The equivalent slow valve time constant is approximately equal to one-fifth of the dissolution time. Most of this time constant is converted to loop dead time, which increases the ultimate period of the *p***H** loop on a vessel. Chapter 8, "Control System," quantitatively shows this time constant's effect on the ultimate period and the controller mode settings. If the influent flow can be throttled, it is best to eliminate the slow, manipulated variable by base loading the lime feed rate and throttling the waste influent flow for *p***H** control, as illustrated in Figure 7-5c for lime feeders.

The dissolution time of limestone is too long for it to be used for feedback control, even when pulverized into fine particles. The main use of limestone is for manually treating acid spills by covering the spill area or diverting the spill to a limestone pit. The dissolution time of pulverized dry lime can be greatly reduced by slaking the lime and making it into a slurry. Figure 7-4b shows that the dissolution time for pulverized dry lime is about 32 min, and the dissolution time for lime slurry is about 8 min when added to a 1000 mL wastewater sample in a 1500 mL baffled beaker with a 100 rpm three-blade agitator. The aforementioned data is for "soft burned" lime. If the lime is heated in the kiln longer, the lime becomes "over burned" or "head burned" and takes longer to dissolve. Dry lime tends to lie on the surface unless wetted and forced below the surface by a water spray. Therefore, a water spray should be used for dry lime addition. The dissolution time of lime slurry increases with age due to the agglomeration of small particles into larger particles, even though the lime slurry storage tank is mildly agitated [3]. If the *p*H controller manipulates the lime feeder or water addition to the lime slurry storage tank, the equipment time constant of the slurry tank has the same effect as a slow valve time constant. To prevent adding this time constant to the loop, the lime feeder speed should be manipulated by a slurry tank level controller and the water addition rate ratioed to the lime feeder speed. The *p*H controller should manipulate the diluted lime slurry discharge flow that feeds the neutralization vessel. The diluted lime slurry discharge flow must be kept flowing by using a recycle loop to prevent settling and plugging in the reagent lines. A throttled globe control valve will plug. The liner of a pinch diaphragm valve will fail due to erosion. Pulse width modulation of an on-off ball valve is the best alternative because it has the fewest maintenance problems and is relatively inexpensive to replace periodically (see Chapter 6).

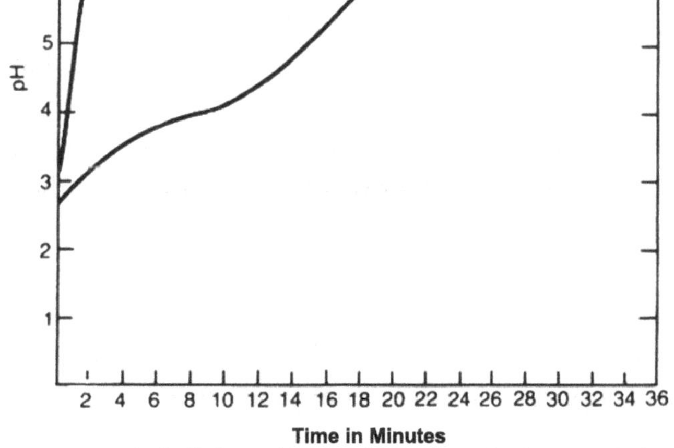

Time response of pH with lime added to a 1000 ml sample of wastewater in a 1500 ml baffled beaker with a 100 rpm, three-bladed agitator. Curve 1 records the response to 40 ml of 0.5 lime slurry; curve 2 is the response to 190 mg of pulverized dry lime.

Figure 7-4b. Time required for lime slurry and dry lime to dissolve.

Use "soft burned" lime slurry to reduce the dissolution time.

Precipitation can force the *p*H of the effluent to a level that may be outside the control band. For example, the precipitation of ferric and ferrous hydroxides acts as a reservoir of hydroxyl ions. An 11 *p*H upset for an hour will increase the reservoir of hydroxyl ions so much that the *p*H may remain in the 9 to 10 range for a day or more [4].

7.5 Special Strategies

A small reagent control valve for fine adjustment in parallel with a large control valve for coarse adjustment meets the resolution and rangeability requirements and enablesmaxmization of possibly cheaper coarse reagent use, but the two valves will interact. To reduce this interaction problem, the output of a separate proportional-only controller can be used to position the small valve, as shown in Figure 7-5a, or the output of a valve position controller can be used to position the big valve, as shown in Figure 7-5b. The measurement for the valve position controller is the small valve position and the set point is 50%. The valve position controller has the advantage of ensuring that the small valve precision is always available by keeping it near the middle of its stroke range by slowly adjusting the big valve. Because the valve position controller uses slow integral-only (I-only) action to reduce interaction between it and the *p*H controller, this configuration is sluggish for large load changes. External-reset feedback (ERF), detailed in Chapter 8, with set point limits on the analog output block can enable using proportional plus integral action for the valve position controller. ERF prevents the controller output from changing faster than the valve is responding. The set point rate limits are set to be slow when the small valve is in a good throttle range (e.g., 20% to 80%) and set to be fast when the small valve approaches the closed or wide-open position. Tuning a proportional-integral controller with ERF is generally simpler and more forgiving offering greater load rejection capability than a conventional I-only controller. Flow feedforward or ratio control should be implemented to proactively position the coarse valve for large feed upsets and the fine valve for small feed upsets using flowmeters with high rangeability (e.g., magmeters or Coriolis flowmeters). With the increased availability of model predictive control (MPC) integrated into control systems, an MPC matrix can be readily set up, as shown in Chapter 8, to do a better job of constraint and feedforward control because each controlled variable knows the effect of each manipulated and disturbance variable, and there are tuning adjustments for minimum move size, move suppression, and penalty on error.

Use a valve position controller or MPC to minimize the interaction of fine and coarse valves with feedforward control for feed upsets.

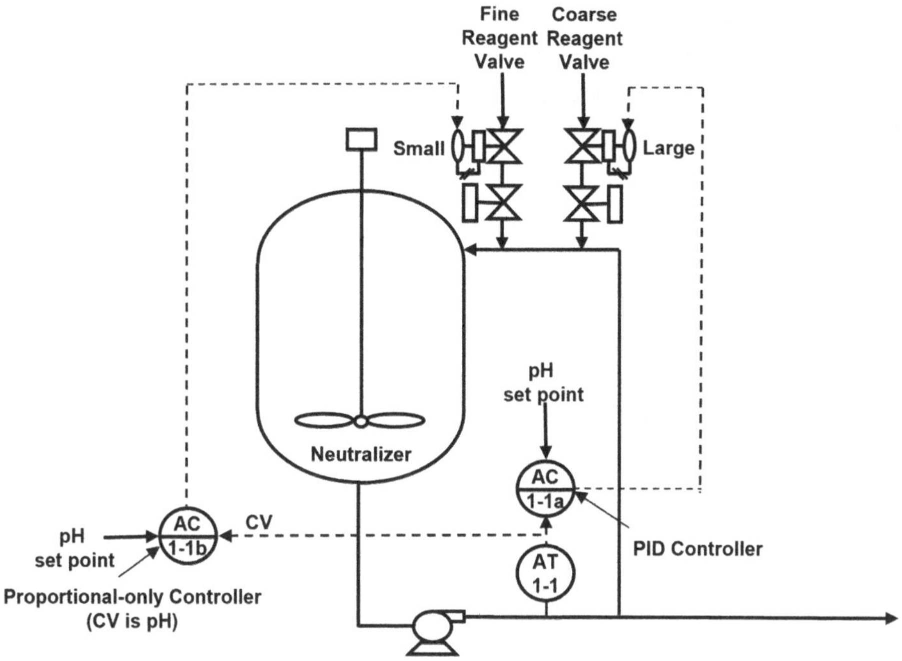

Figure 7-5a. Proportional-only *p*H control of a fine adjustment valve to reduce interaction with a coarse adjustment valve.

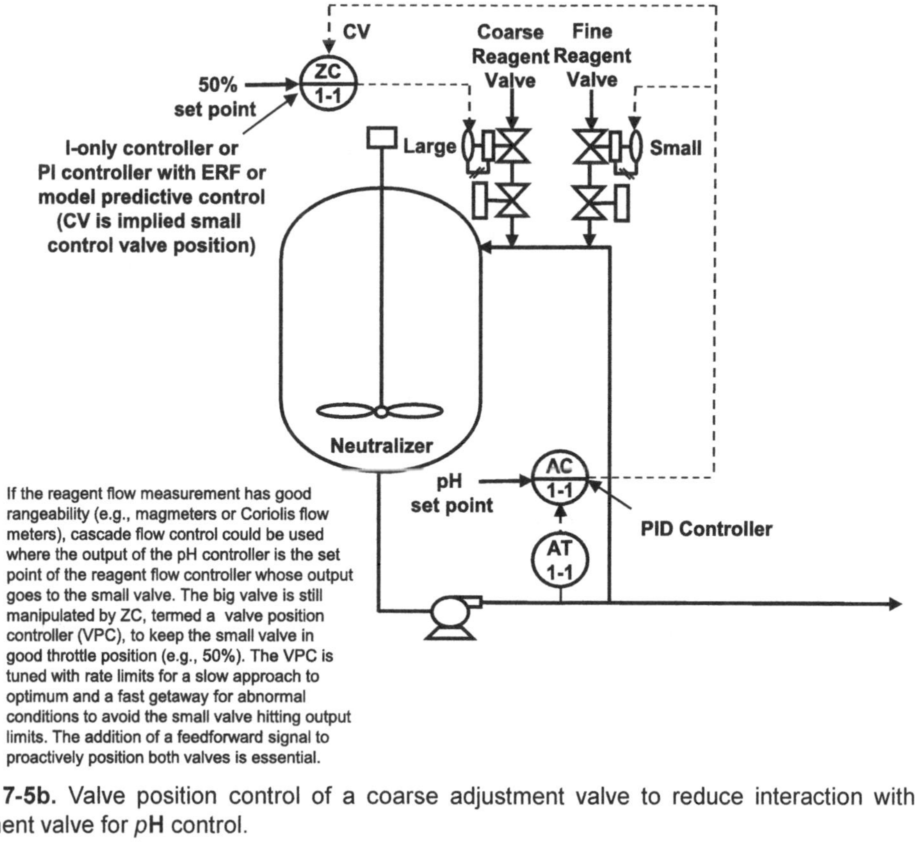

Figure 7-5b. Valve position control of a coarse adjustment valve to reduce interaction with a fine adjustment valve for *p*H control.

Metering or positive displacement pumps with pneumatically set stroke lengths typically have a resolution of 2% and a drop-off in flow at 5% to 10% of capacity (10:1 to 20:1 rangeability). The accuracy of the pneumatic calibration also deteriorates with time. Motor-driven metering pumps with electronically set stroke frequencies have better precision but about the same rangeability. Electronic metering pumps that use a solenoid and solid-state pulsing circuit instead of a conventional motor and gearbox assembly have a rangeability of 30:1 by combining stroke frequency and length adjustment. The pulsing flow and pressure from a metering pump create reagent injection and *p*H measurement noise problems for in-line *p*H control systems. Also, because the metering pump is usually located at grade level and is not closely coupled to a pipe or vessel nozzle, the reagent delivery delays can be enormous if the reagent drains back through the pump or into the vessel, or the process fluid backfills into the reagent delivery system.

Key CONCEPT **Metering pumps tend to have limited inherent or self-imposed rangeability, exhibit noise from pulsing, and have huge reagent delivery delays.**

Lime feeders have a transportation delay that is proportional to the length of the feeder divided by its speed. This transportation delay may be several minutes. The residence time of a vertical tank must be at least 20 times as large as the sum of this delay, lime dissolution time, and turnover time for the vessel to be classified as *well-mixed*. To eliminate the need to increase the vessel size and the agitation power, the lime rotary valve speed can be base loaded, and the *p*H controller can manipulate the conveyor speed or the influent flow. If the *p*H controller manipulates the waste flow, the dissolution time associated with an increased lime delivery rate is also eliminated. The level controller on the influent tank slowly corrects the lime rotary valve base speed if the waste inventory gets too high or low. A notch gain or error squared level controller or MPC is used to suppress control action when the level is in the midrange. A 50% output from the level controller corresponds to a zero correction of the base-loaded rotary valve speed. The operator can change the system capacity by changing the speed through a manual loader. Figure 7-5c shows this method of eliminating the reagent transportation delay and lime dissolution time by manipulating waste flow to control *p*H.

RULE of Thumb *To eliminate the transportation time delay and dissolution time associated with a lime feeder system, manipulate the waste flow for p**H** control.*

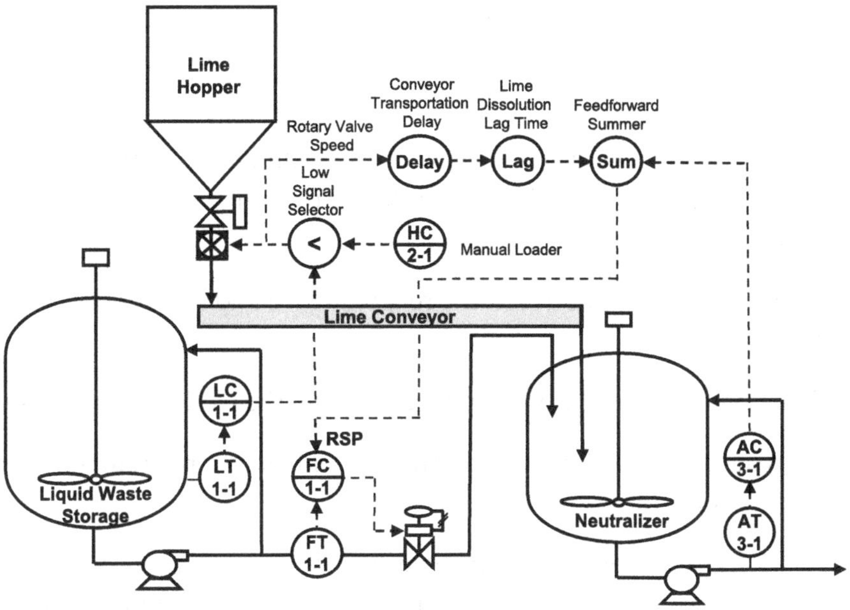

Figure 7-5c. Throttling the waste stream to eliminate lime feed transportation delay and dissolution time for *p*H control.

7.6 Best Practices

The time for solids to dissolve can be horrendous (minutes). Although the time for bubbles to dissolve is faster (seconds), the noise from fluctuations in concentration, besides the delay to complete dissolution, is problematic. For these reasons, using lime and ammonia for *p*H control is undesirable. Additional problems stem from highly viscous reagents, such as 97% sulfuric acid, that result in laminar flow in control valves and greater difficulty in mixing with process streams. Highly concentrated reagents cause extremely large reagent delivery delays due to low reagent flows (e.g., <1 gal/h). Strong acids and strong bases cause this same problem, and the additional difficulty of a steeper titration curve and the consequential need for more precise final control elements and possibly more neutralization stages. The following best practices are offered to provide more realizable reagents for *p*H control.

1. Avoid reagents that have solids or bubbles, take more than a second to react, are highly concentrated, or are strong acids or strong bases.

2. Increase the residence time in equipment and piping to provide complete dissolution of reagents and uniformity of reagents in the process stream.

3. If a conveyor is used, manipulate the motor speed of a gravimetric feeder rather than the flow dumped on the conveyor inlet to eliminate the conveyor transportation delay.
4. Use pressurized reagent pipelines with a coordinated isolation valve close coupled to the reagent control valve outlet to prevent process backflow into the reagent pipeline.
5. Minimize the volume between the reagent valve and its injection point into the process.
6. Avoid using dip tubes by injecting reagent into the feed or recirculation stream.
7. Dilute reagents using tight concentration control in a recirculation stream to a diluted reagent storage tank.
8. Use Coriolis mass flowmeters for accurate reagent flow and concentration measurement by using a high-resolution density measurement for reagent dilution and *p***H** flow ratio control.

References

[1] Gregory K. McMillan, *Process Control Case Histories* (New York: Momentum Press, 2010).

[2] Octave Levenspiel, *Chemical Reaction Engineering*, 3rd ed. (New York: John Wiley & Sons, 1998).

[3] Ralph L. Moore, *Neutralization of Waste Water by pH Control*, 2nd ed. (Research Triangle Park, NC: ISA [Instrument Society of America], 1995).

[4] Greg Shinskey, *pH and pIon Control in Process and Waste Streams* (New York: John Wiley & Sons, 1973).

8

Control System

8.1 Feedback Control

Nearly all control loops utilize some form of feedback control. The most prevalent type of controller is a proportional-integral (PI) controller, which basically provides a controller output that is the sum of a contribution proportional to the *p***H** error and the integral of the *p***H** error, where the error is the difference between the *p***H** set point and the measurement. If the *p***H** loop is on a vessel with a large time constant that provides a smooth transition in *p***H**, it is beneficial to include rate action to form a proportional-integral-derivative (PID) controller so there is a contribution to the controller output proportional to the rate of change of the *p***H** or the control error. If the set point is on a flat part or a bend in the curve, rate action is particularly effective at dealing with the drastic acceleration associated with an excursion to the steep portion of the curve.

Rate action on the measurement instead of the error is preferred to eliminate bumps to the controller output from step set point changes by the operator. However, for set point changes by model predictive or supervisory control and a batch sequence, rate action on error will help the controller recognize that the set point change is intended.

The Figure 8-1a loop block diagram details the dynamics that determine the PI and PID controller tuning settings and the loop performance. The reagent flow starts to correct for a disturbance, such as a change in influent composition or flow, only after it appears in the effluent, the controller sees it and reacts, and the reagent flow responds and enters the process at the same point as the disturbance. No matter where the upset enters, information must complete one cycle around the loop for a corrective action to

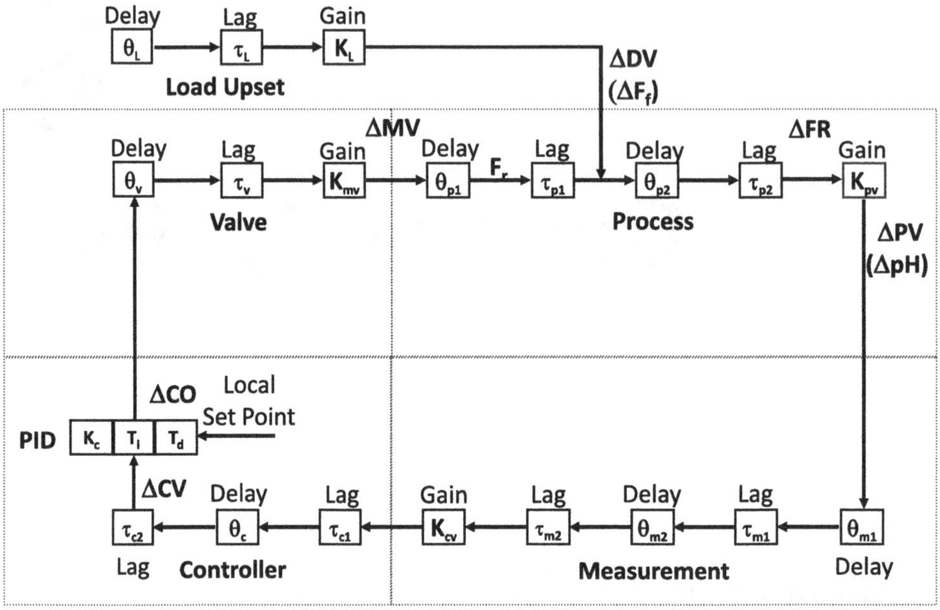

Figure 8-1a. Block diagram for *p*H with the major sources of dead time noted.

take effect. The time delay between the start of the disturbance and the arrival of the corrective action is the total loop dead time. If the dead time is zero, the *p*H will stay within the resolution limit of the measurement and the control valve at the set point. Because the *p*H hardly budges from its resting value on the titration curve, the variation in the slope of the curve is a moot point. Although it is not possible to have a zero dead time, it is more important in *p*H control than in any other loop to make this dead time as small as possible to reduce the effect of the extreme process nonlinearity typical for *p*H applications.

> **Key Concept** More so than for any other loop, it is important to reduce the dead time for *p*H control because it reduces the effect of the nonlinearity.

The total loop dead time is the sum of the valve dead time from deadband, stick-slip, and actuator pre-stroke delay and stroking time; reagent delivery delay; equipment mixing delay; and measurement dead time from transportation delay, sensor time constant, digital control cycle time, and filter time constants introduced into the transmitter or controller. The fraction of a small time constant that is converted to dead time is detailed in Section 8.11, "PID Tuning," but for now, small time constants for *p*H loops on vessels are effectively considered dead time because the sources of dead time are usually underestimated.

> **Rule of Thumb** *Add the pure time delays and the effective time delays from small time constants for one pass around the control loop to get the total dead time.*

8.2 Feedforward Control

Feedback control only takes action after an error has developed. The error must be large or persistent before feedback control makes a large correction for an upset. Feedforward control has the noble idea of calculating a corrective action to compensate for an upset before an error appears. The most common load upsets (disturbances) are influent feed and composition. Flow feedforward is widely used and is often referred to as *ratio control* because the reagent flow is ratioed to the influent flow. The ratio is the abscissa of the titration curve for the *p*H set point if the x-axis is converted to a ratio as recommended for *p*H system design.

The reagent and influent upset may enter the process equipment literally at the same point if the reagent is combined with the influent stream; otherwise, they take different but hopefully similar paths. There is a series of dead times and time constants in each path, as shown in Figure 8-2a. The feedforward signal is typically added to the *p*H controller output. If the combined signal goes directly to a reagent control valve, the feedforward calculation is complicated by the control valve's nonlinearity and sizing. Using a good reagent flow measurement, as described in the next section for cascade control, eliminates the difficulty of computing a valve position that corresponds to a desired flow. Although theoretically, a feedforward multiplier instead of a summer should be used on a *p*H controller output that directly manipulates a linear valve for flow feedforward for a static mixer, scaling issues make this undesirable.

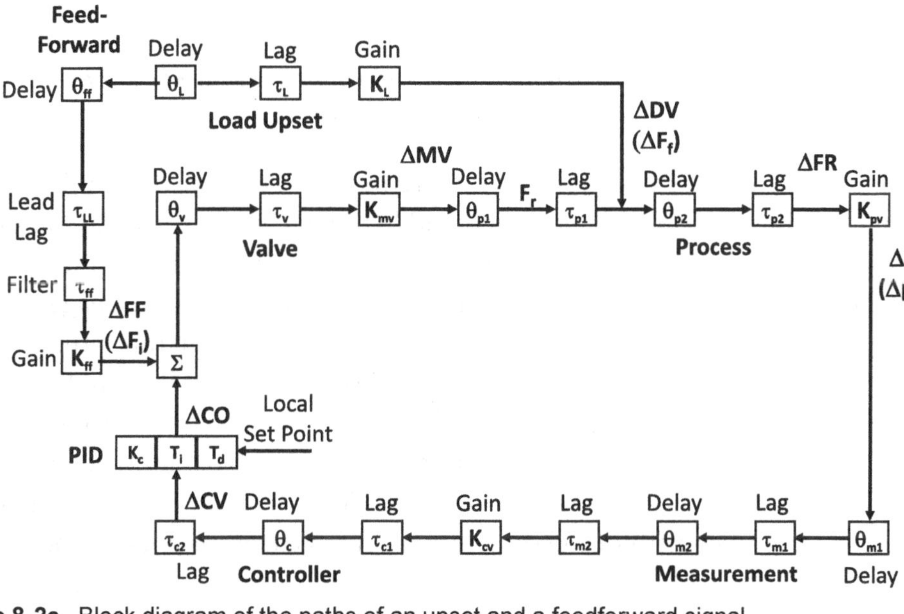

Figure 8-2a. Block diagram of the paths of an upset and a feedforward signal.

The effectiveness of feedforward control depends on the accuracy of the measurement of the upset and the calculation and timing of the corrective action. For there to be no error, the corrective action must arrive in the process at the same point and time as the upset with an opposite but equal value to the upset. The disturbance measurement is first filtered to eliminate noise. The timing (dynamic compensation) is then set via a lead-lag and a dead time function block. Nearly all books and articles on feedforward control setup and tuning take a simplified view where the disturbance is shown to enter the process output. The feedforward lead time is set equal to the lag in the path of the corrective action (reagent) and the feedforward lag time is set equal to the lag in the path of the disturbance (influent).

In reality, the corrective action (reagent) and disturbance (influent) are process inputs. Therefore, the approach commonly taken in the literature is wrong and results in large settings of both the lead and lag times. An error in one of these settings can lead to a serious timing error. In particular, a larger-than-needed lead setting will amplify noise and cause an inverse response by a feedforward signal that arrives too soon. An inverse response is where the initial direction of a change is the opposite of the final direction of the change. Needless to say, it confuses the heck out of the controller. Noise and inverse response can cause a feedforward control system to do more harm than good.

Key CONCEPT: The reagent correction from feedforward must arrive at the same point and time as the influent upset with an equal but opposite equivalent value.

RULE of Thumb: *Filter the feedforward signal to remove noise and make sure the corrective action does not arrive too soon and cause an inverse response.*

The lag times in the corrective action and disturbance path are usually similar enough to warrant excluding the calculation of lead and lag times per Figure 8-2b. Although a lead time might help kick a valve signal to get it to move, this is difficult to get right, and if it is too large, it creates additional slip. Besides, the stick-slip in reagent valves must be minimized because of the extreme sensitivity of the *p*H. In rare situations, the kicking action has helped break the trim free of solid buildup, but the burst of reagent flow causes overshoot. The one solid case where a lead would prove beneficial is in cancelling a lag time associated with dissolving calcium hydroxide (lime) or magnesium hydroxide particles. Here, the lead time should be set equal to the reagent lag time, which is about one-fifth of the dissolution time. Whenever a lead time is used, the lag time should be at least one-eighth of the lead time to prevent the excessive amplification of noise.

RULE of Thumb *If the reagent has particles, set the feedforward lead time equal to the dissolution lag time; otherwise, omit using a lead-lag in the feedforward.*

In a well-designed system, the reagent should arrive in the mixture at about the same time as the influent. However, as explained in Chapter 7, "Reagent," reagent delivery delays can be large. The dead time block in the feedforward calculation can delay the feedforward signal if it arrives too soon. Thus, to ensure that feedforward action does not arrive too late, it is important to follow the suggestions in Chapter 7 to increase the reagent flow rate as it enters the equipment by injecting the reagent into a feed or recirculation stream close to where it enters the volume or by precisely diluting the reagent in an upstream volume.

Key CONCEPT The effectiveness of feedforward control greatly depends on the ability to eliminate reagent delivery delays.

Flow feedforward is particularly helpful to get the reagent flow in the ballpark for start-up and batch sequences because the *p*H controller gain may be detuned to deal with the steep slope of the titration curve. This initialization of the reagent flow is so important that if feedforward is not possible because an accurate influent or feed flow measurement is not available, a head start should be implemented. In a head start, the controller output tracks the last reagent demand captured from the last batch or period of operation. The head start is held for one loop dead time before the controller is released for feedback control.

RULE of Thumb *If there is a reproducible influent flow measurement, use flow feedforward; otherwise, use a head start to initialize the reagent flow for the start-up.*

*p*H feedforward has not been used much to date because the influent *p*H measurement is typically on the tails of the titration curve. Here the electrode is more prone to errors because of acid and alkalinity errors, chemical attacks, and dehydration. It is also more likely to fail because of the harsh environment. Finally, the error in the calculated amount of reagent needed can be huge because of the flatness of the curve, as shown in Figure 8-2c. *p*H feedforward would work a lot better if the influent or feed *p*H was near the neutral point and the set point was on the extremities of the curve, but this is rarely the case.

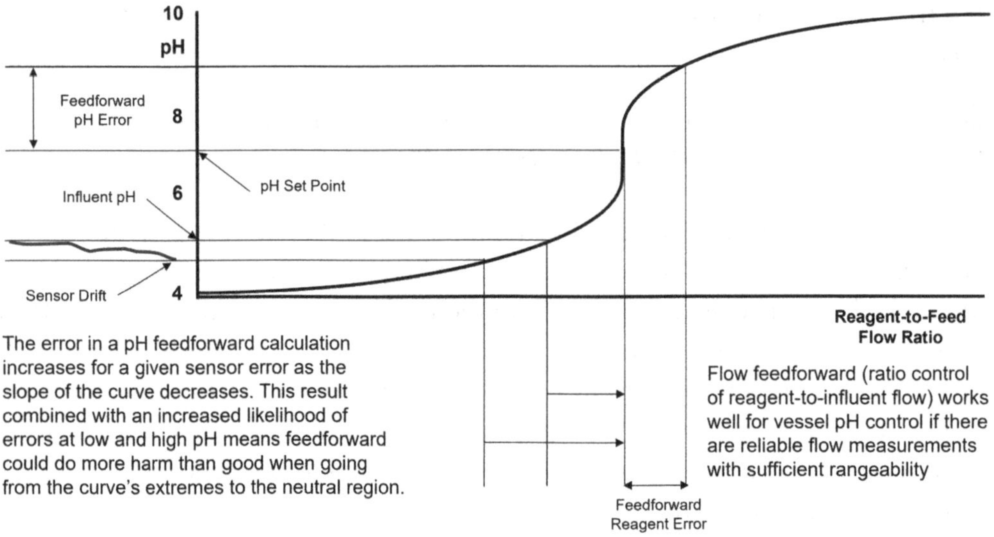

Figure 8-2c. Huge reagent demand error from *p*H feedforward measurement error.

> **Key CONCEPT** The reliability and error of a *p*H feedforward are unacceptable if the influent or feed *p*H measurement is on the extremities of the titration curve.

8.3 Cascade Control

In cascade control, the output of the *p*H controller becomes the remote set point of another controller instead of going directly to the reagent valve. The original *p*H controller is called the *primary* or *outer loop controller*, and the new controller with a remote set point is called a *secondary* or *inner loop controller*. A valve positioner, such as a digital valve controller, is a secondary controller that compensates for stick-slip and gets its remote set point from the process controller. The next most common cascade control system uses a secondary flow controller that compensates for pressure upsets and the reagent valve's nonlinearity. There were many failures of this usually sound concept of *p*H to flow cascade when orifice meters were used because the flow signal became erratic and caused more harm than good for low flow and transitions to laminar flow. Improvements in the technology of in-line flowmeters now make a cascade loop of *p*H control to flow control desirable to isolate the *p*H controller from changes in the valve gain and to facilitate ratio control. The first, second, and third choices in terms of rangeability and insensitivity to the Reynolds number and straight pipe runs are a Coriolis, magnetic, and vortex flowmeter, respectively [1]. The vortex meter is a distant third choice because of signal drop out at low flow and the sensitivity of the meter coefficient to viscosity. For *p*H loops there is a triple cascade where the *p*H controller output is the remote set point of a reagent flow controller whose output is the remote set point of a digital valve controller.

RULE of Thumb *Use a Coriolis or magnetic flowmeter for reagent flow control.*

To get the most benefit from cascade control, the secondary loop dead time and time constant should be five times faster than the primary loop dead time and time constant. If this is not the case, one of the controller gains and reset settings must be decreased and increased, respectively, to prevent excessive cycling from the interaction of the two loops. The time constants and dead times for the position of a small reagent valve and the flow in a pumped full pipeline are all less than 1 s, which on the surface, violates the cascade rule where the secondary loop must be five times faster than the primary loop. However, the digital valve controller has no integral action and a scan time that is more than five times faster than the flow, and the flow controller usually has a filter time and low controller gain setting to deal with the measurement noise and valve nonlinearity. In the days of pneumatic positioners and electronic analog controllers, the interaction problem with large valves was so severe a rule based on a simplified Nyquist plot analysis was generated specifying that boosters should be used instead of valve positioners on fast loops, such as a flow loop. This falsely legitimized rule was disastrous even before the advent of digital valve positioners and digital process controllers because the simplified Nyquist plot analysis ignored the nasty consequences of sticking, jumping, and wandering valve plugs and discs from improper bench settings, stick-slip, and the poor inlet and the high outlet sensitivity of boosters [2]. Every reagent valve must have a digital positioner (digital valve controller)—no ifs, ands, or buts.

RULE of Thumb *Every reagent valve must have a digital valve controller.*

Cascading *p*H to reagent flow on a static mixer also appears to violate the cascade rule because the dead time and time constant of a filtered flow loop are a couple of seconds, which is almost the same as the transportation delay through the mixer and the clean electrode time constant for a buffered system. However, the additional dead time of the reagent injection and the transportation delay from the mixer to the electrodes makes the *p*H loop dead time at least 6 s. For strong acid and base systems, the electrode time constant is at least 4 s, and a digital *p*H filter time of 12 s is often used to screen out axial noise and enable an averaging control of sustained oscillations [2]. The cascade control of mixer *p*H to reagent flow greatly improves the performance of a flow feedforward, so it is advisable except for very fast in-line mixers on buffered systems. Figure 8-3a is the block diagram for a cascade *p***H** control to flow control system. A digital valve positioner is not shown in the system diagrams, even though its use is essential.

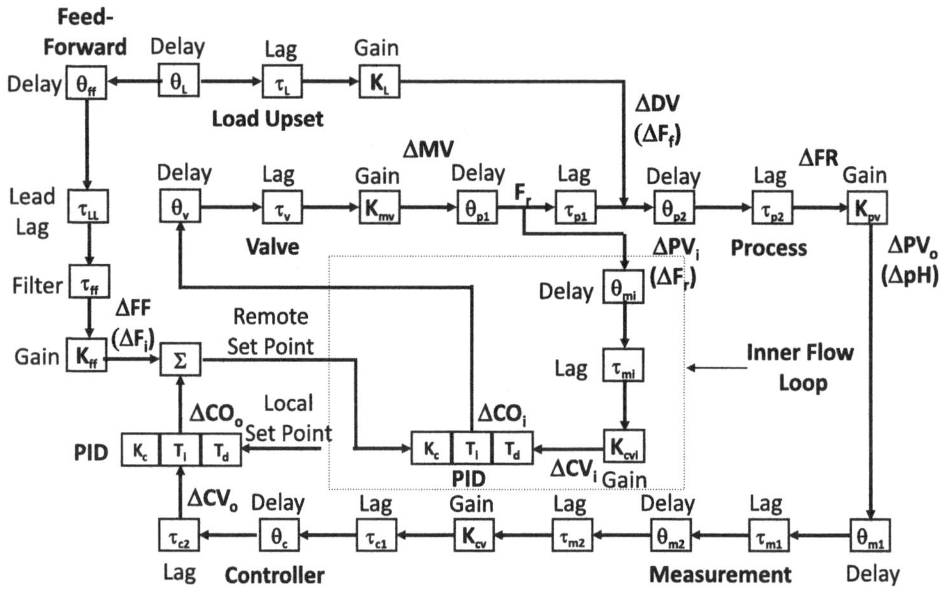

Figure 8-3a. Block diagram of the dynamics for cascade and feedforward control.

The cascade of a primary *p***H** control loop to a secondary *p***H** control loop, while less common, can still be important. The secondary loop is typically a *p***H** loop on an upstream static mixer that is actually a linear reagent demand loop, as described in Section 8.4. The primary loop is a *p***H** loop on a downstream vessel that trims the upstream loop set point. Even if the upstream loop is doing a great job, the average *p***H** from the mixer will not equal the average concentration and *p***H** in the vessel because of the nonlinearity and uncertainties of the titration curve. When the downstream *p***H** loop is on a poorly mixed volume commonly associated with the horizontal tank shown in Figure 8-3b, the primary loop uses slow integral-only control to gradually eliminate any long-term offsets [2]. Figure 8-6d shows the setup for a cascade of vessel *p***H** to recirculation *p***H** for advanced batch control.

8.4 Linear Reagent Demand Control

For a linear reagent demand (LRD) controller, the controlled variable is the x-axis of the titration curve instead of *p***H**. For convenience, the abscissa is labeled 0% to 100% reagent demand. For the first stage, 0% of scale is set at a reagent to influent ratio of 0%, and 100% of scale is set for a ratio that corresponds to the maximum expected overshoot in *p***H** expected. For successive stages, the 0% is set at the extreme of the control band for the upstream stage. The set point must also be converted to the percentage

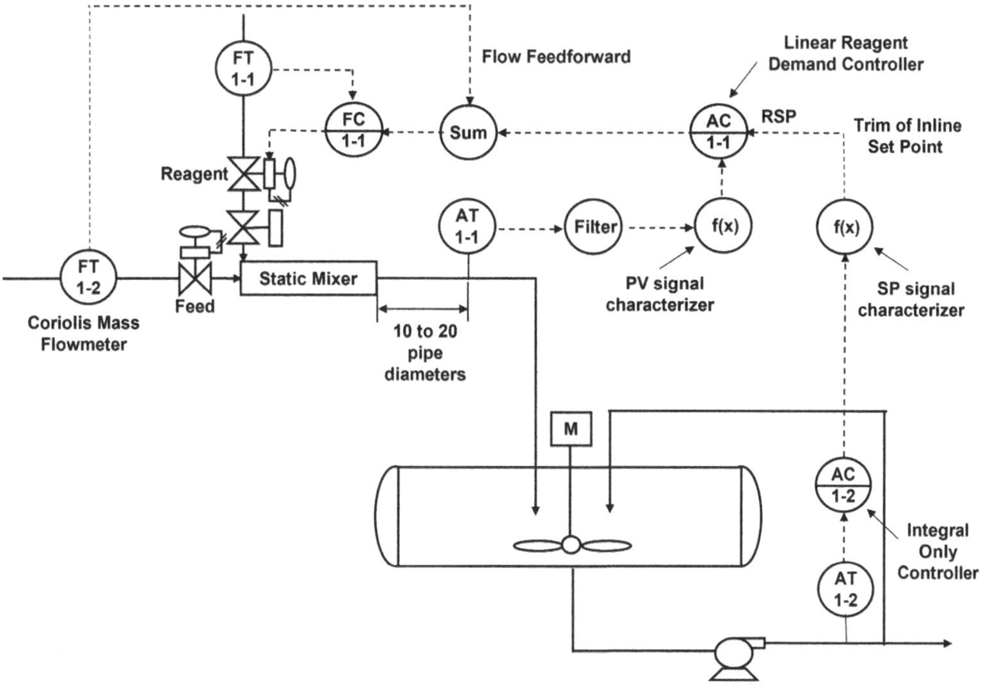

Figure 8-3b. Cascade control to eliminate downstream offset from an in-line system.

RULE of Thumb *Except for fast in-line buffered systems, use cascade control of pH to reagent flow to compensate for pressure upsets and enable flow feedforward.*

of reagent demand so it matches up with the new controlled variable. It is desirable to trend the percentage of reagent demand to show the benefits of using the LRD controller. If operations insist on a *p*H interface, the LRD controller can be kept as a detail display and a dummy *p*H controller presented to the operator. When the dummy controller is switched between the manual, auto, and cascade modes, the LRD controller mode is switched accordingly. In other words, the LRD controller mode tracks the *p*H controller mode. When the dummy controller is in manual and the operator changes the dummy controller output, the LRD controller output tracks it. However, when the dummy controller is in auto or cascade, the dummy controller output tracks the LRD controller output, and the LRD controller tracks the dummy controller set point after it has been converted to the percentage of reagent demand.

Users must be aware that the tuning settings of the dummy *p*H have no effect, and the LRD open-loop gain should be about one if the signal characterization curve is accurate.

Key CONCEPT The operator may prefer to use a dummy *p*H controller even though the LRD controller is doing the work behind the scenes.

A horizontal shift in the titration curve can be handled by simple bias to the LRD controller set point to make the *p*H controller measurement better match up with its *p*H set point. This can be automated by using an integral-only master *p*H controller on a downstream volume, as shown in Figure 8-3b, or by bypassing the *p*H through a big dead time and filter block to create a dummy version of the horizontal tank *p*H controller. In either case, the master real or dummy *p*H controller slowly adjusts a bias to the set point of the secondary dummy *p*H controller.

RULE of Thumb *Use a real or dummy master* p*H controller to bias a dummy secondary* p*H controller set point to compensate for horizontal shifts in the curve.*

A change in the shape of the titration curve requires either adapting the charge balance model or identifying the titration curve. Both can be done online, particularly for a static mixer where the dynamics are fast and relatively well-known. If a charge balance model is used to compute a curve online, the acid dissociation constant (pKa) closest to the *p*H set point can be slowly adjusted by model predictive control whose controlled variable is the model *p*H and whose target is the actual *p*H. If a static mixer and accurate mass flow measurements are installed for the influent or feed and reagent flows, the titration curve can be identified periodically online by ramping the reagent flow for a constant influent or feed flow, as shown in Figure 8-4a.

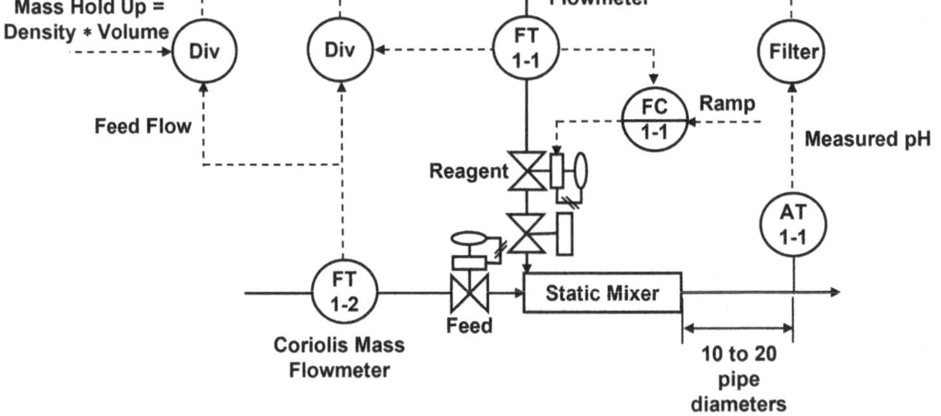

Figure 8-4a. Online identification of a titration curve.

RULE of Thumb

Adapt the charge balance model p***H*** *or use online titration curve identification to compensate for a curve distortion.ondary* p***H*** *controller set point to compensate for horizontal shifts in the curve.*

The match between the curve used for the LRD controller and the real curve does not have to be perfect. Any sort of linearization is generally better than none at all. An LRD controller on the steep portion of the titration curve will see less noise and less of a limit cycle from stick-slip and will see the original equipment time constant by eliminating the acceleration. Thus, the user gets much more value from investing in a vessel volume to provide a process time constant that slows down excursions. Also, the slope and gain for the process variable are now unity. The start-up and recovery from the flat portion of the curve are much faster because the LRD recognizes the true distance in terms of reagent requirement, and the controller gain is much higher because the steep slope has been eliminated and the time constant has been restored.

Key CONCEPT

Linear reagent demand can restore the time constant and capture the investment in well-mixed vessels, provide a unity gain for the process variable, simplify and improve controller tuning, suppress oscillations and noise on the steep part of the curve, and speed up start-up and recovery from the flat part of the curve.

The one case where an LRD controller may actually be detrimental is when the set point is on the flat portion of the curve and excursion to the steep portion must be avoided at all costs. In this case, the overreaction of a *p***H** controller to the acceleration as the *p***H** moves toward the neutral point is desirable, even if it means overcorrection and overshoot.

The most common method of generating the LRD-controlled variable is to use a signal characterizer block to provide a piecewise line segment fit whose input is *p***H** and whose output is the percentage of reagent demand. The number of data points should be increased in the control band to improve the fit in the region of greater interest. If third-order or higher polynomials are employed instead of line segments to fit a curve, the user must closely inspect the signal to make sure there are no dips and bumps that cause a localized gain reversal. Also, the user must prevent a calculation of a value below or above the 0% to 100% scale because polynomials can take off exponentially beyond the fit region and create bizarre results. There are similar concerns for neural networks. In the future, charge balance models will be used to provide a smoother and more exact fit.

8.5 Adaptive Control

Consciously or subconsciously, controllers are tuned based on a trade-off between performance and robustness. The capability to tightly control at an operating point is inversely proportional to the ability to weather changes in the behavior of the plant without becoming oscillatory. The operating environment for *p*H loops is stormy, and the last thing you want is for a control system to do more harm than good by increasing the variability when the controller is in automatic. Consequently, *p*H controllers are detuned (backed off from maximum performance) to provide a smooth response to weather the inevitable changes in the process dynamics [3]. Furthermore, because the changes in dynamics typically have not been identified, the controller tuning settings are backed off from the optimum much more than needed. In many cases, the loops end up in manual, and the operator attempts to chase upsets, which is a fruitless task because of dead time and the high sensitivity and nonlinearity of the *p*H response.

> **Key CONCEPT** Controllers are excessively detuned or put in manual because the changes in process dynamics that cause oscillations have not been identified.

In practice, three parameters are used to provide a first-order plus dead time (FOPDT) model of the process dynamics and to capture the essence of the process response. The most important of these is the loop dead time, which is the time delay between the change in a controlled variable (process output) after a change in a manipulated or disturbance variable (process input). If the dead time is zero and the measurement and valve resolution are unlimited, tuning is a nonissue and perfect control is possible. However, dead time always exists and is an accumulation of the pure delays from plug flow (transportation delays), valves (deadband and stick-slip), and digital devices (scan and execution time intervals) and the secondary lags from volumes in series, mixing, actuator, sensors, and signal filters [4]. The next most important parameter is *p*H process gain, which is the final change in *p*H for a given change in reagent flow. A high process gain (sensitivity) is desirable because it improves the inference of a process condition from a process measurement. However, a high process gain amplifies the stick-slip in valves and the noise from nonideal mixing. The third parameter is the process time constant, which is the time to reach 63% of the final change in a controlled variable after the process has started to change (after the dead time). In plants, this time constant is rarely constant. For a large back-mixed volume, it is essentially the residence time if the secondary lags take a back seat [1]. However, the time constant seen by the controller is greatly reduced by the acceleration of the *p*H response on the steep portion of a titration curve, even though the actual process time constant has not changed. The controller only knows what it sees. Using the model parameters for controller tuning and loop performance and the effect of the nonlinearity of the titration curve on these parameters are discussed in Sections 8.10 and 8.11.

Changes in the FOPDT model parameters reveal changes in the raw materials, process, equipment, valves, and sensors. The size, direction, and characteristics of these changes can provide a roadmap, knowledge of the terrain, and a route to more responsive controller tuning.

> **Key CONCEPT** An FOPDT model that consists of a process gain, time delay, and time constant can be used to describe the process response and calculate the controller tuning settings.

Nearly all the industrial adaptive controllers used to date in chemical processes require a rather long observation of changes in the process variable and show the results in terms of new tuning settings. The tuning rules are embedded and usually unknown. The more prevalent adaptive controllers use pattern recognition and will, if necessary, increase the controller gain to induce oscillations. The size of the transients or oscillations and the time required for identification can translate to significant process variability and an adaptation rate that is slower than the rate of change of the process parameters. In fact, most adaptive controllers are playing catch-up even if they have seen the same situation a thousand times before. At best, these controllers provide a snapshot of the current tuning requirements and no real process insight as to where the process has been or where it is going. Also, sudden unexplained shifts in the tuning settings or bursts of oscillations reduce the operator's confidence and decrease the likelihood that the controllers will run in the adaptive mode and be used in future applications.

> **Key CONCEPT** Nearly all the adaptive controllers used to date are playing catch-up and do not reveal the process model or the embedded tuning rules.

Recently, adaptive controllers have been developed to rapidly identify a process model and provide process model parameters that can be displayed, trended, and diagnosed. Changes in the model parameters can reveal changes in feed or reagent compositions and installed valve characteristics, as well as a coated sensor and plugged valve. Tuning rules can be selected based on user preference to match the process and the plant objectives. Furthermore, these controllers remember the results for similar conditions, eliminate repetitious identification, and take the initiative [3].

> Changes in the process model can be used to predict and analyze changes in the influent, reagent, valve, and sensor.

These controllers identify the dead time, process gain, and time constant for both manipulated and disturbance variables and save these as a function of a key variable, which in this case would be *p*H. The user can use the recommended tuning method or choose an alternative method to compute the current tuning settings for the current and memorized conditions. When the operating point has moved to a different portion of the titration curve, the tuning is then scheduled based on the process model saved in this region. The adaptive controller remembers the results from previous excursions and does not wait to recognize old territory. The adaptive controller takes preemptive action but still uses any excursions as an opportunity to refine its knowledge of the process [3].

> New adaptive controllers remember changes in the process model as a function of the operating point and preemptively schedule controller tuning.

One such adaptive controller computes the integrated squared error (ISE) between the model and the process output for changes in each of three model parameters from the last best values. All combinations of the three values (low, middle, and high) for three parameters result in 27 models. The correction in each model parameter is interpolated by applying weighting factors that are based on the ISE for each model normalized to a total ISE for all the models over the period of interest. After the best values are computed for each parameter, they are assigned as the middle values for the next iteration. This model switching with interpolation and recentering has been proven mathematically to be equivalent to least square identification and to provide an optimum approach to the correct model. The search is actually done sequentially, first for the process gain, then the dead time, and finally the time constant, which reduces the number of models to 9. Figure 8-5a shows the graphical concept and the equations of parameter evaluation and interpolation in the adaptation algorithm [3].

An in-line *p*H control system on the feed or in the recirculation flow to a large vessel volume can be used to automatically identify the titration curve online. The feed or recirculation flow is fixed, and the reagent flow is ramped over the course of a few minutes. The reagent to feed or recirculation flow ratio is computed and synchronized with the *p*H measurement by passing the ratio through dead time and filter blocks. The time delay of the dead time block is computed online as the in-line system volume divided by the total flow rate, and the time constant of the filter block is set equal to an estimated sensor lag time. The ramp should not start until the reagent pipe and injection lines are verified to be completely full of reagent. The ramp should go beyond the pH set point so the curvature is identified for overshoot. After the ramp is completed, the reagent flow should be initialized to the value identified to correspond with the desired pH set point to minimize the upset. If the ramp is fast compared to the size of the volume, the disturbance to the final pH should be negligible. The ramp can be scheduled periodically or initiated as a phase in a start-up or batch sequence. The accuracy of the curve is greatly improved by using mass flowmeters and linear control valves. Figure 8-5a shows the use of a static mixer and the dynamic compensation of the ratio to identify a titration curve online.

RULE of Thumb *Use in-line pH control, mass flowmeters, linear control valves, and dynamic compensation to automatically identify the titration curve online.*

The titration curve can also be identified during the normal pH adjustment phase of a batch operation. Here the ratio would be the totalized reagent flow divided by the batch mass. Synchronizing the ratio with the pH is more difficult because the dynamic compensation becomes a function of the reagent delivery and mixing time delays, and only the approach to the end point is identified because overshoot is not permissible.

The slope of the titration curve can be used to schedule the process gain and, therefore, the controller gain. Alternatively, the curve can be used to convert the controlled variable to the percentage of reagent demand discussed in Section 8.4. Either method frees up the adaptive controller to identify the changes rather than the whole shape of the curve. Finally, the curves and dynamic models can be used for online dynamic pH estimators, model predictive control, and real-time optimization of reagent consumption.

RULE of Thumb *Use gain scheduling or signal characterization based on the titration curve to free up an adaptive controller to identify the changes in the curve.*

8.6 Advanced Batch Control

In a traditional batch sequence, the reagent is charged based on a preset total. There may be a dribble flow rate near the end to help improve the accuracy at the cutoff. Often the batch is held for lab analysis and manual trim adjustments. The lab sample should not be taken until the last drops of reagent for the main charge and each trim charge have gotten into the batch and been thoroughly mixed. The hold time for reagent delivery and mixing delays, manual sampling, lab analysis, and trim charges is considerable and can exceed the main charge time. To reduce the time required for pH adjustment, five advanced batch control strategies are presented. These strategies reduce the pH overshoot that negatively impacts the batch quality, reagent costs, and batch cycle time.

RULE of Thumb *Batch samples should be taken only after all the reagent in the pipeline and dip tube has drained into the batch and been thoroughly mixed.*

Key CONCEPT Manual sampling, lab analysis, and trim charges can more than double the time required for *p*H adjustment.

If the titration curve is fixed, which implies the composition of acids and bases in the feed and reagent streams is constant, the reagent charge can be calculated based on a temperature-corrected titration curve. The x-axis of the curve, which is a reagent to feed mass ratio, is multiplied by the total batch mass to get the reagent charge totalizer setting. For sequential charges, where the reagent is added after the other charges are completed, the reagent charge valve is wide open until the reagent flow totalizer reaches the dribble set point. Then the charge valve is either throttled back or closed while a small trim valve in parallel with the main valve completes the charge. Using mass flowmeters greatly improves the accuracy of this method. If the reagent is charged simultaneously with other feeds that affect the *p*H (fed-batch operation), the reagent mass flow set point is the feed mass flow multiplied by the x-axis ratio that corresponds to the *p*H end point. Differences in the dynamics, noise, and tuning of the feed and reagent control loops lead to an error between the desired and actual running total of reagent flow. A master controller whose controlled variable is a filtered mass total error can be used, as shown in Figure 8-6a, to add a bias to correct the ratio of reagent to feed that is the set point of a secondary reagent flow controller. For end points on the steep portion of the titration curve, it may still be necessary to consider one of the next methods.

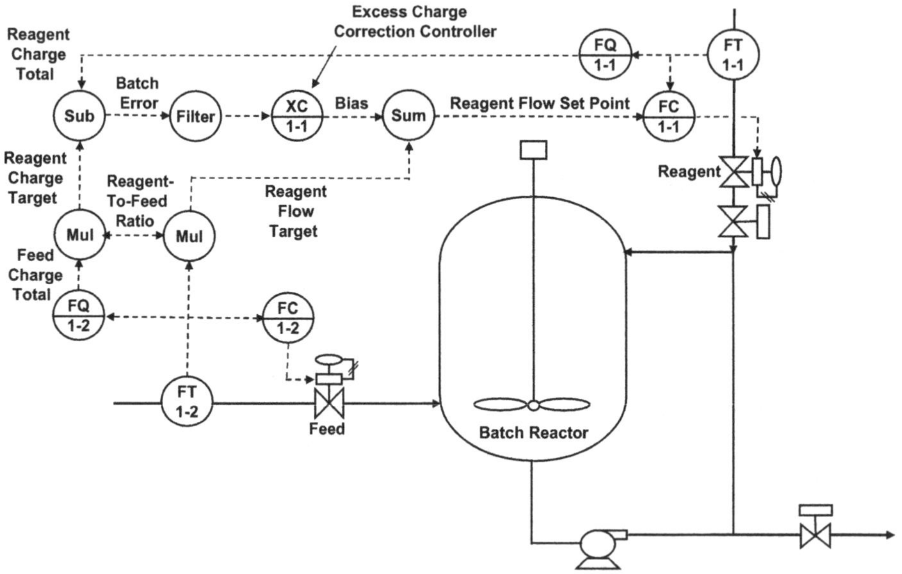

Figure 8-6a. Dynamic correction of the mass flow ratio for fed-batch operation.

RULE of Thumb

If there are mass flowmeters, the reagent charge can be calculated from the x-axis of a temperature-corrected, fixed-titration curve.pipeline and dip tube has drained into the batch and been thoroughly mixed.

The x-axis value that corresponds to the end point can be corrected for the next batch from the *p*H lab analysis of the last batch. This eliminates the hold time required for manual *p*H adjustment because the batch continues onto its next phase before the lab sample is taken. This method is adequate for dealing with slowly changing feed and reagent compositions, flowmeter and weigh tank errors, and upstream batch conditions. To eliminate chasing *p*H analysis errors, only half of the delta correction in charge is used.

RULE of Thumb

Half of the delta correction based on the p***H*** *analysis of the last batch can help correct for slowly changing errors and upstream conditions.*

For variable and steep titration curves and cases where large line sizes, high solids content, high temperatures, and special materials of construction preclude using mass flowmeters, online *p*H measurement is necessary. Using conventional *p*H feedback control is usually not advisable because the *p*H has an integrating response unless the reagent is consumed in the reaction. Reagent delivery and mixing time delays and reset action in a *p*H loop will cause considerable overshoot of the end point.

The fastest automated online *p*H adjustment consists of positioning the reagent valve wide open and shutting it when a predicted *p*H matches the batch end point. The *p*H measurement is velocity limited and filtered. The difference between the current value and a value passed through a dead time block is divided by the delay time (calculation time interval) of the dead time block to provide a rate of change of *p*H. The time delay is chosen to be large enough to improve the signal-to-noise ratio but small enough to not increase the system dead time by more than 10%. The predicted *p*H is the present *p*H plus the *p*H rate of change multiplied by the system dead time, which is the total of the reagent delivery, mixing, measurement, and rate of change calculation time delays. It is important that the reagent delivery time delay be minimized by injecting the reagent into a feed line or into a recirculation line just before it enters the vessel, as shown in Figure 8-6b.

RULE of Thumb

Use a wide open reagent valve that is shut based on a predicted online p***H*** *measurement to provide the fastest batch* p***H*** *adjustment.*

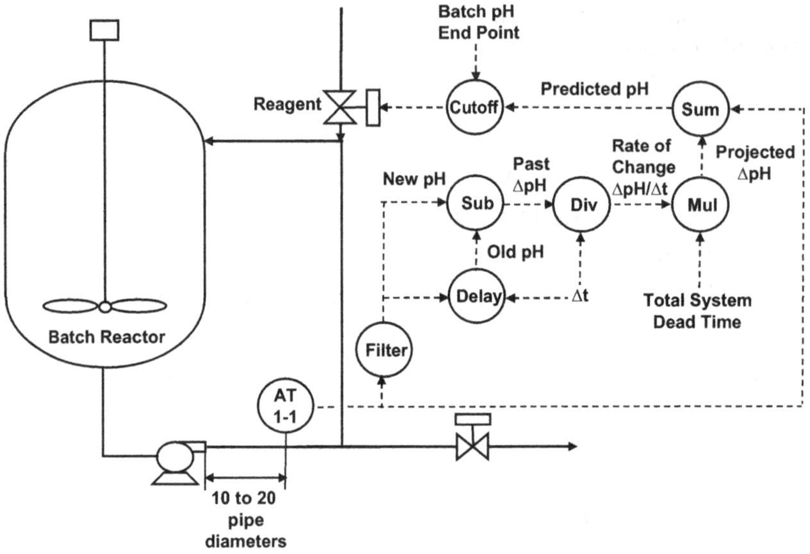

Figure 8-6b. Fast charge rate with an automated cutoff based on predicted *p*H.

A proportional- plus- derivative *p*H controller can be used for pulse width and amplitude modulation to mimic the lab titrations shown in Figure 8-6c where the drop size and frequency are decreased as the vessel *p*H approaches the end point. The pulse width and amplitude are proportional to the *p*H controller output. The pulse cycle time (time between the start of successive pulses) is set equal to the system dead time, which increases the total dead time by 50%. A minimum *p*H controller output, pulse width, and amplitude are selected based on a compromise between the need to minimize batch cycle time and prevent overshoot.

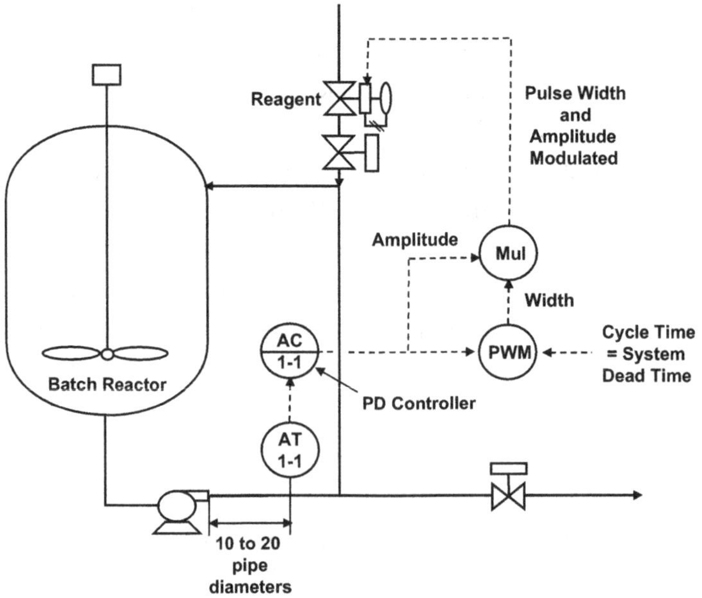

Figure 8-6c. Pulse width and amplitude modulation to mimic lab titrations.

RULE of Thumb *Use pulse width and amplitude modulation of a proportional-plus-derivative controller output to mimic lab titration for batch* p***H*** *control.*

For extremely variable titration curves, the titration curve can be identified online for linear reagent demand *p***H** control if the *p***H** extremes in the recirculation line do not cause quality problems. The master *p***H** loop uses a linear reagent demand proportional-plus-derivative controller to help it recognize the distance to the end point. Per Section 8.7, the future value of vessel reagent demand can be used as the controlled variable. The secondary *p***H** loop uses a proportional-plus-integral controller on a static mixer on the recirculation flow just before it enters the vessel. The remote set point of the secondary loop is the master controller *p***H** output with a bias to keep the reagent valve open as the vessel *p***H** approaches the end point. The titration curve is identified online at the static mixer, as shown in Figure 8-6d. The reagent valve is shut off when the master *p***H** controller output equals the bias. The output limits of the master controller must match the set point limits of the secondary controller that minimize quality and corrosion problems in the short section of the line between the static mixer and the vessel.

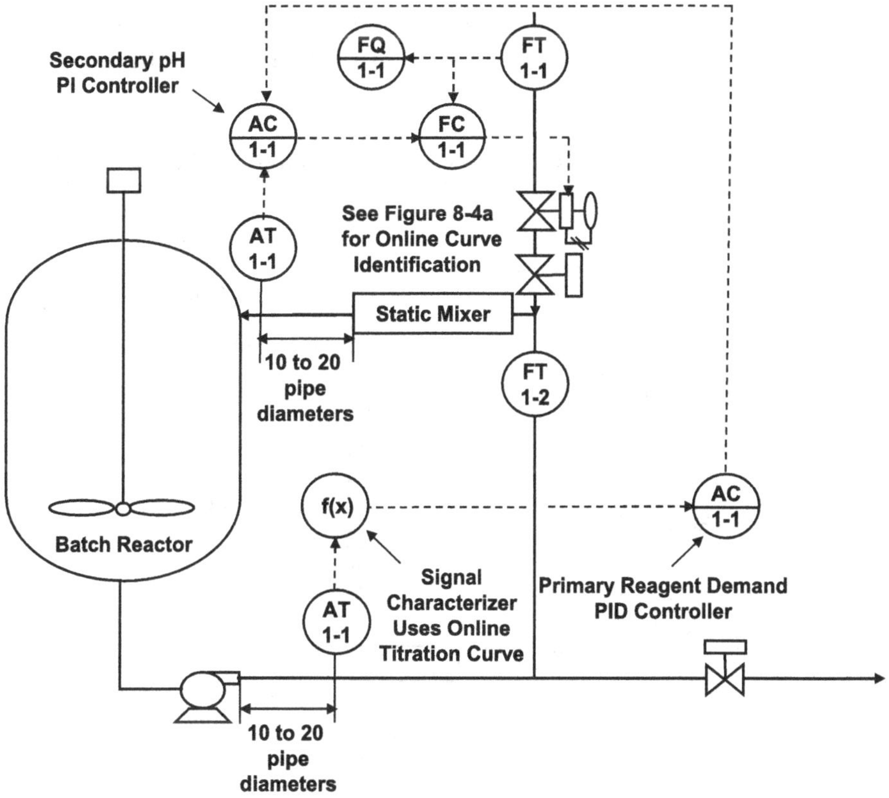

Figure 8-6d. Online titration curve identification and linear reagent demand cascade recirculation line *p***H** control.

RULE of Thumb *Use online titration curve identification and linear reagent demand pH control for extremely variable and sharp titration curvature.*

8.7 Online Dynamic pH Estimators

If the titration curve is fixed, identified, or computed online and mass flowmeters are used, an online dynamic *p*H estimator can be readily constructed to provide a faster, smoother, and more reliable *p*H measurement. The ratio of reagent to feed mass flow is filtered and converted to a future *p*H via the titration curve generated by a signal characterizer or a charge balance model. Note that in this case, the y-axis value is computed from an x-axis data point, whereas for signal linearization, the x-axis value is computed from a y-axis data point. The future *p*H value is sent through dead time and filter blocks to synchronize it with an online measurement. The time delay and filter time are set equal to the open-loop dead time and time constant, respectively. A fraction of the difference between the measured and synchronized estimated *p*H is used to provide a bias correction [1]. If the *p*H measurement is suspected of having failed, the correction freezes.

The future *p*H value can be used for continuous control, as shown in Figure 8-7a, or in conjunction with the advanced batch *p*H control shown in Figure 8-6b if it has a 0.5-minute filter and 0.2-minute scan time to provide the proper cascade

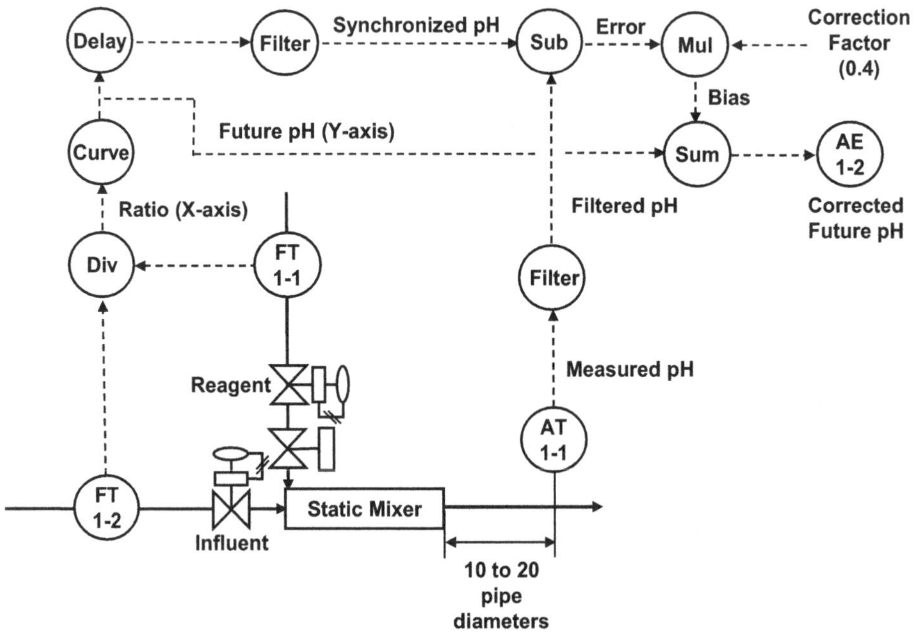

Figure 8-7a. An online dynamic *p*H estimator for future value *p*H control.

dynamics. This can provide an important advantage for vessel *p*H control by predicting steady-state *p*H values 20 min or more sooner. Note that the system dead time and time constant must be accurately known. New model-based adaptive controllers or the process identification software of model predictive controllers can be used to identify the FOPDT model. Reagent delivery delays and poor mixing should be minimized because these create an unknown variation in system dead time. Transportation delays should be computed from volume and flow rate and updated in the model.

RULE of Thumb

Use an online dynamic pH estimator to provide a much faster, smoother, and more reliable pH value if the open-loop dead time and time constant are known and there are feed and reagent mass flowmeters.

8.8 Model Predictive Control

If the total loop dead time approaches or exceeds the open-loop time constant, the *p*H loops interact with concentration or density control loops, or there is a need for constraint control, then model predictive control (MPC) should be used instead of PID control. The controlled variable should be linear reagent demand instead of *p*H because MPC is based on linear superposition [5] and the titration curve drastically changes the process gain and severely distorts the process time constant, as discussed in Section 8.10 on dynamics and performance. Online dynamic *p*H estimators can be used, but the future value should be the original reagent flow to feed flow ratio converted to the reagent demand value percentage as described in Section 8.4. Industrial MPC software has built-in disturbance variables for feedforward control and optimization variables that can be used to maximize the cross-neutralization of waste streams or low-cost reagent while keeping the control valve for the more expensive supplemental reagent in a useable throttle range per the matrix shown in Figure 8-8a. If the feed is free to float within limits, the feed can be maximized based on low cost or waste reagent availability per the matrix in Figure 8-8b.

A more common effective use of MPC is for simultaneous coarse and fine reagent adjustments shown in Figure 8-8c. In this case, the controlled variables are *p*H and small valve position, the manipulated variables are small and large valve position, and the disturbance variable is feed flow. The move suppression for the manipulated variables is dramatically increased for the big valve compared to the small valve. The penalty on error for the small valve as a controlled variable is considerably smaller compared to the penalty on error for *p*H. This provides an optimization that slowly moves the big valve to keep the small valve in a good throttle range. MPC has the advantage of being able to act simultaneously and incorporate knowledge of the future

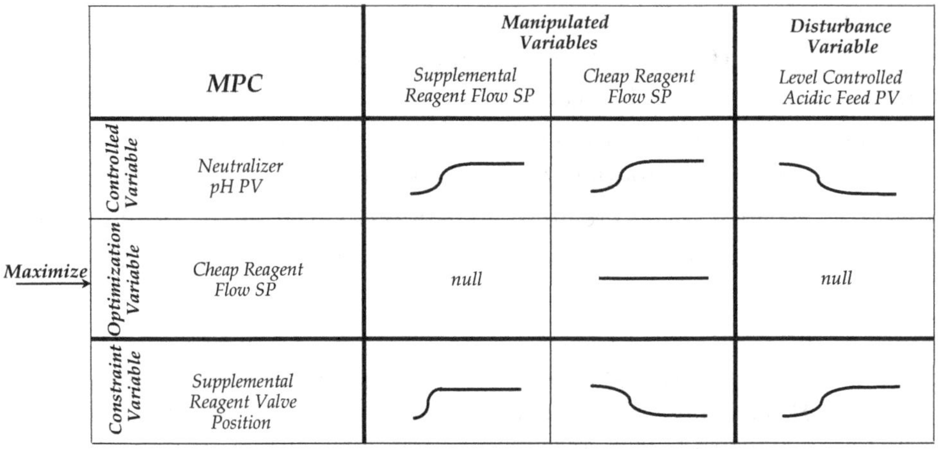

Note that the cheap reagent flow is maximized for a given feed flow to keep the supplemental reagent flow valve at the minimum throttle position for minimum stick-slip.

Figure 8-8a. Matrix for minimizing reagent costs for level-controlled feed.

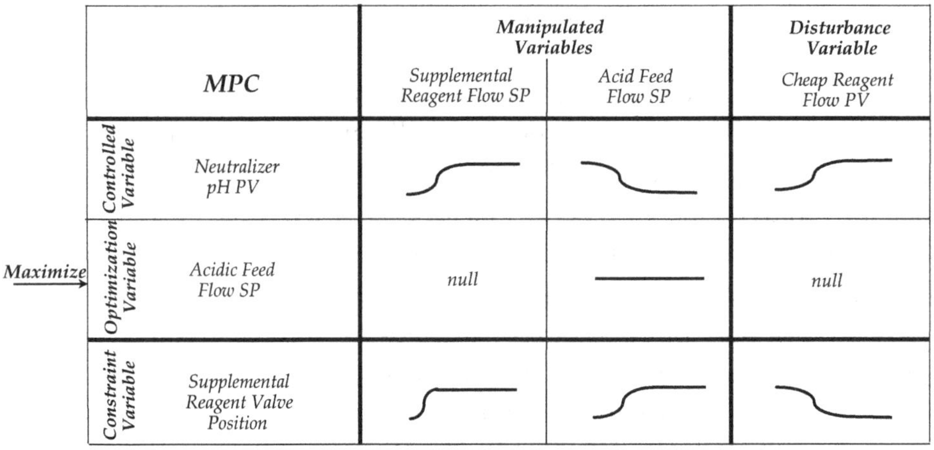

Note that the cheap reagent valve is wide open, and feed is maximized to keep the supplemental reagent flow valve at the minimum throttle position for minimum stick-slip.

Figure 8-8b. Matrix for minimizing reagent costs for maximized feed.

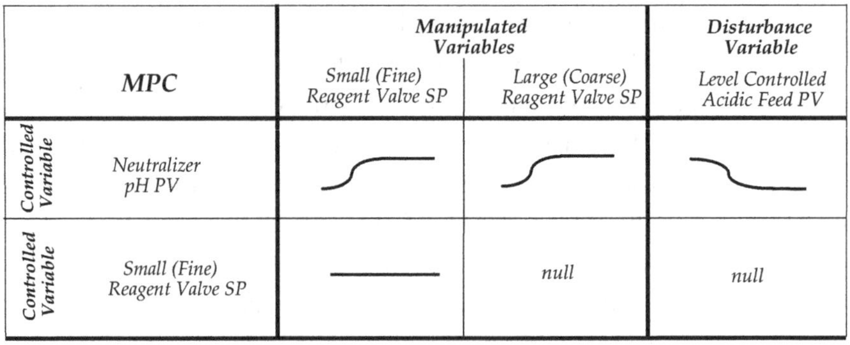

Figure 8-8c. Matrix for simultaneous coarse and fine reagent adjustments.

consequences of a valve movement and a feed upset. It can also be used for a nonintrusive online adaptation of the charge balance model *p*H for linear reagent demand control and real-time optimization [5].

RULE of Thumb *Use linear reagent demand MPC for dead time dominant or interacting systems and constraint or valve position control.*

8.9 Real-Time Optimization

If the titration curve can be identified or computed online from a charge balance model, a *p*H set point can be computed that minimizes reagent consumption based on reagent flows upstream and the variability in *p*H at various points on the titration curve from valve stick-slip and measurement noise. The *p*H set point is then positioned so the projected variability does not cause excessive violations of the *p*H constraint limit. Because the constraints are usually on a much flatter portion of the titration curve, the savings in reagent can be enormous, as shown in Figure 8-9a.

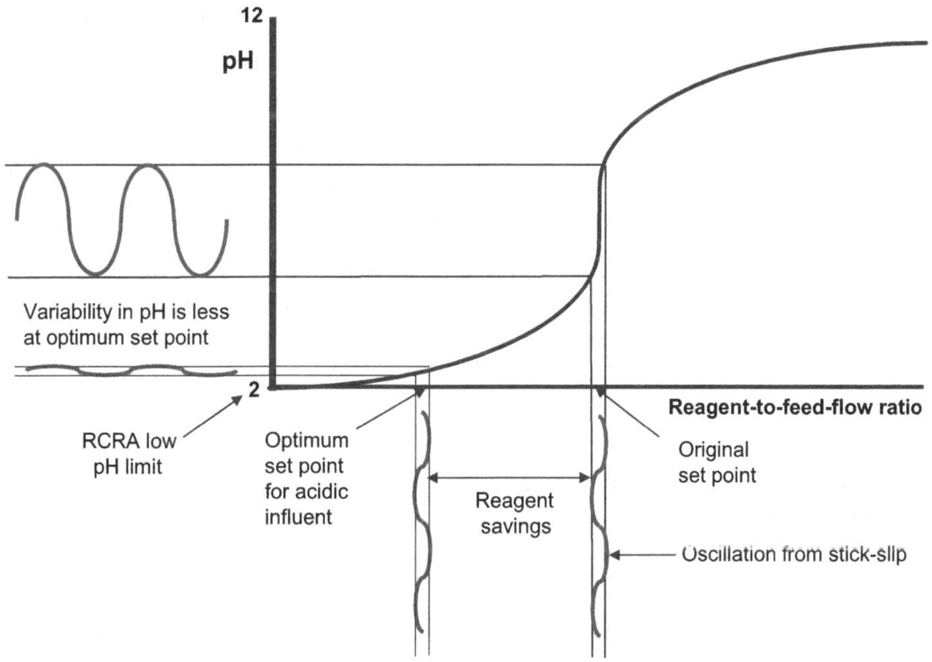

Figure 8-9a. Real-time optimization of reagent use based on variability projections.

8.10 Dynamics and Performance [6]

Important aspects affecting dynamics and thus performance are discussed for temperature in addition to *p*H control because poor temperature control can cause poor *p*H control because the actual process *p*H changes with temperature, as discussed in

RULE of Thumb *Use real-time optimization and variability projections to minimize reagent use if the titration curve can be identified or computed online.*

Chapter 2, "The Chemistry," and Chapter 4, "Electrodes." Both types of loops can have large secondary time constants due to sensor design and fouling. A runaway response can also be apparent because of the acceleration of the process variable from exothermic reactions in temperature loops and an approach to the neutralization point in strong acid and strong base *p*H systems. Derivative action can be critical for both.

A large primary process time constant is beneficial because it slows down the changes in the process output from upstream load disturbances, enabling a higher PID gain and giving the PID more time to catch up to and correct for the load disturbance. This is true for self-regulating processes where the primary process time constant is from negative feedback and for runaway processes where the primary process time constant is from positive feedback. For integrating processes, a large primary process time constant is equivalent to a slow integrating process gain. A large secondary process time constant or a time constant anywhere in the automation system (valve, measurement, or PID) is detrimental because it creates an equivalent dead time that reduces the maximum permissible PID gain and delays the correction. Derivative action may be used to help compensate for a secondary process time constant or automation system time constant. A measurement time constant that is much larger than the process time constant and total loop dead time is particularly confusing because it enables a higher PID gain and gives historian trend plots with smaller oscillations due to the filtering effect. This is a common problem in *p*H control with fouled electrodes or users who just get carried away with signal filtering. Equation 8-10pp at the end of this section enables users to back-calculate the actual process oscillation amplitude from the measured oscillation amplitude.

Key CONCEPT A large primary process time constant is beneficial because it slows down the disturbances and filters the oscillations in the process that affect *p*H.

The complexity of the automation process and its dynamics often distract us from the fundamentals, most notably the key role of dead time. If the dead time is zero, perfect control would be possible regardless of other dynamics because the PID could immediately see and correct for any load disturbances. The PID gain could be set at the maximum available setting, and the integral time could approach zero, assuming there is no noise. The dead time is, of course, never zero, and its identification and minimization are key to knowing and achieving the best performance. Dead time knowledge is important for control strategy design, feedforward feasibility and implementation, valve or variable-frequency drive (VFD) and measurement selection and installation, controller tuning, dead time compensation, process simulation, the dynamic compensation of online process metrics, and using neural networks and data analytics. Fortunately, dead time is the easiest dynamic term to identify, but its many sources are rarely recognized and properly minimized.

> **Key CONCEPT** A large measurement time constant is detrimental because it slows being aware of and correcting disturbances, fooling the PID and humans with an attenuated version of *p***H** from actual oscillations within the process that have already affected the *p***H**.

> **Key CONCEPT** Dead time is the most detrimental dynamic because the PID is unaware of what is happening during this time and it delays starting the correction.

Assuming healthy electrodes, the largest source of dead time is often associated with transportation delays from piping and mixing delays from vessels with improper agitation. The cumulative effect of dead times from all the automation system components can be significant, and aged or fouled electrodes can create a huge dead time.

Equations for estimating the dead time from various automation system components are provided in the following sections. Equations for estimating the effect of total loop dead time on the ultimate limit to the peak and integrated errors for load disturbances are also provided. The effect of the execution interval (scan time) can be seen in terms of the additional dead time being simply one-half of the execution time interval. To prevent the minimum peak error from increasing by more than 10%, the maximum execution time interval should be less than 20% of the total open-loop dead time from all other sources.

> **Key CONCEPT** Although transportation delays (particularly of reagent injection) and equivalent dead time from fouled or aged electrodes are the largest individual sources of dead time, the cumulative effect of other sources—most notably in the valve or VFD, transmitter, and controller response—can be significant.

Equations are also provided to show the effect of the execution interval on the practical peak and integrated errors for load disturbances that include tuning settings. The effect can be larger when derivative action is used and results in a smaller integral time. The execution interval can also affect the degree and smoothness of derivative action on an update. However, there are serious words of caution about any guidance based on existing tuning settings and the closed-loop response, such as the rise time for set point changes and the arrest time for load disturbances. The rise time is greatly affected by PID structure set-point lead-lags and tuning. The arrest time is mostly affected by tuning, which for reasonable robustness and best performance, typically has a nonoscillatory response with minimal overshoot. Note that there is always a trade-off between robustness and performance. Most controllers are tuned for excessive robustness and consequently achieve much less than the best performance. If the execution interval is set based on the tuning and closed-loop response, subsequent improvements in valve and measurement response and better tuning and performance will be unnecessarily limited by the additional dead time from an unnecessarily large execution interval. The excessive robustness can be equated to an implied dead time that is larger than the actual dead time. For example, if the integral time is doubled and the controller gain is halved compared to what is possible for reasonable robustness, the implied dead time is doubled and the integrated error is quadrupled for processes that are not dominated by a total dead time much larger than the open-loop time constant. For these reasons, it is better to set the maximum execution interval as a fraction of the observed minimum total loop dead time that often occurs at a maximum production rate. Also, one-half of the existing execution interval should be subtracted to get the minimum dead time.

Key CONCEPT

The peak and integrated errors for load disturbance depend on the tuning settings, which in turn depend on the open-loop dynamic response and making sure the tuning is not too sluggish due to seeking excessive robustness that equates to an implied dead time much larger than actual dead time.

There are many different metrics, objectives, and methods for improving loop performance that are a result of many different types of products, processes, dynamics, markets, and PID capabilities. Typically, the goal is to improve process performance and profit, which should dictate the required process responses, such as good load regulation or minimizing the propagation of variability using nonaggressive, nonoscillatory tuning. Here the focus is on the effect of PID algorithms, structures, and tuning.

Load Disturbances and Goals

Measured and unmeasured changes in process inputs—most notably in stream composition, flow, pressure, temperature, and physical properties—are prevalent in the process industry. There are also changes in equipment conditions that affect mass transfer and energy transfer, reaction rates, and changes in phase, such as fouling surfaces and deactivating the catalyst. These changes are classified as load disturbances. The stream conditions shown on a process flow diagram are rarely constant or exactly match what is happening in the process. Load disturbances can be extended to include any change in process parameters for the material and energy balances shown in Appendix F. Every PID controller can deal with these differences; otherwise, automatic control would not be necessary, and simply setting the PID controller at a fixed manual output for a given production rate would be sufficient. The streams may or may not be entering the process at exactly the same point as the manipulated variable, as illustrated in Figure 8-10a. Also, the effect may be seen in terms of changes in utility streams, such as utility temperature, and changes in environmental conditions, such as ambient conditions—particularly cold rainstorms that affect heating and cooling process equipment or piping. The load changes may affect the controlled variable much faster and sooner than the manipulated variable, but load changes still work through the material and energy balances in the process that determines the controlled variable.

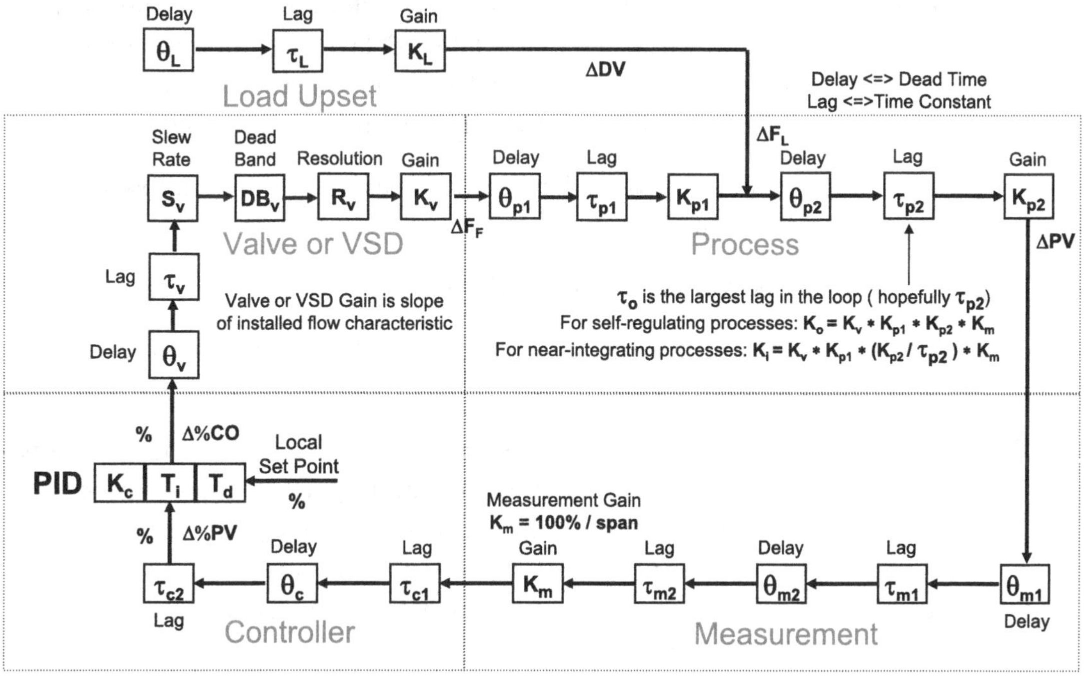

First Order Approximation: $\theta_o \cong \theta_v + \theta_{p1} + \theta_{p2} + \theta_{m1} + \theta_{m2} + \theta_c + Y*\tau_v + Y*\tau_{p1} + Y*\tau_{m1} + Y*\tau_{m2} + Y*\tau_{c1} + Y*\tau_{c2}$

Y is fraction of small lag that becomes equivalent dead time in a logarithmic function of the ratio of the small to largest lag (Y = 0.28 and 0.88 for ratios = 1.0 and 0.01, respectively).

Figure 8-10a. Block diagram for automation and process simulation dynamics.

> **Key CONCEPT** It is important for control system design, analysis, and tuning that (contrary to what is often shown in the literature) disturbances in industrial processes do not enter at a process output but arrive as a process input working through the material and energy balances and are classified as load disturbances to help alert practitioners.

Recognizing load disturbances offers the following advantages for the process industry:

1. Understanding the benefit of a larger primary process time constant and a smaller integrating process gain in reducing the impact of load disturbances. This is brought about by slowing down and averaging disturbances plus smoothing out changes in PID output. This enables aggressive gain and rate action, including overshoot of the final resting value needed to deal with diminished or missing process self-regulation. For example, better back mixing and a larger associated volume can greatly attenuate the load disturbance where the primary time constant is residence time.
2. Reducing the impact of load disturbances by adding control loops and using better valves and through cascade control.
3. Better timing of the manipulated stream by a better entry point into the process to avoid large and variable transportation delays ranging from minutes to hours in dip tube and piping volumes between the control valve and the point of injection for *p*H control due to extremely low and variable reagent flows.
4. Better timing of ratio control by having flows arrive at the same point in the process at the same time.
5. Better timing of feedforwards for half decoupling.
6. Recognizing the value of feed composition measurements by adding analyzers and inferential measurements.
7. Recognizing confusing oscillations from the existence of the window of allowable controller gains in near-integrating, integrating, and runaway processes. The problem originates from the user not realizing an aggressive gain is possible and needed to deal with load disturbances in non–self-regulating processes. The confusion comes from the expectation that oscillations are caused by a PID gain that is too high or too low.
8. Realizing that the PID should first be tuned for load response. Load response tuning can be tested by making a change in a process input that affects the controlled variable. Ideally, the tuning is tested by changing the actual load

disturbance, for example, the stream flow controller set point. If the PID being tuned can be put in manual for several seconds without disrupting the process, put the PID in manual, make a PID output change to create a load disturbance, and put the controller back in automatic to observe its load response (the set point should remain the same during this test). If this is not possible, the PID output may be configured to have a bias that can be changed while the PID stays in its highest mode. It is better if these tests are automated for greater consistency and faster results. After getting the desired load response, a set point lead-lag or two degrees of freedom (2DOF) structure is used to get the desired set point response that is tested by set point changes. A set point feedforward signal with dynamic compensation via a lead-lag or procedure automation can be used to help make the set point response faster.

9. Recognizing the load disturbances originating from problems at the split-range point that are often due to the changes in the installed valve characteristic, process dynamics, and valve resolution. This load disturbance is identified by changes in the PID output at the split-range point and possibly minimized by the intelligent split-range point (described in Section 8.12 on external reset feedback and Section 8.5 on adaptive tuning) and more linear and precise valves.

10. Recognizing that controllers transfer variability from the process variable to the manipulated variable, and in some loops, the process performance is improved by minimizing the variability of the manipulated variable rather than the process variable.

In this book, and generally throughout the literature, process responses are defined for process input step changes to provide a common basis for identifying, comparing, and discussing process dynamics. However, in the process industry, process input step changes are rare because of the response time of control valves and variable-frequency drives, primarily due to rate limiting and time constants and because integral action slows down changes in the PID output that cause changes in streams. Slowing down a load disturbance can significantly reduce its impact. Therefore, using throttling valves manipulated by a PID instead of on-off valves set by a sequence on a stream is much less disruptive to loops affected by the stream.

Key CONCEPT

Control loop responses are defined for step inputs even though most changes to a loop's process variable are fortunately not steps but are more gradual because of valve and measurement lags and reset action in controllers, giving time for the controller to react before it sees the total effect of the change.

For a step input, the time constant for a self-regulating process response can be visualized as the time it takes the output from the start of its response to reach the final value if the maximum rate of change observed at the beginning of the response is conceptually sustained. Thus, the time constant is the final change in output divided by this maximum rate of change of the output for a step input. For a higher order (multiple time constants) plus dead time response to a step input, the maximum rate of change occurs substantially after the output has started to respond, which is termed the *inflection point* (the point where acceleration stops and deceleration begins). When fitting an FOPDT model to a high-order response for a step input, the equivalent dead time can be approximated as the intersection of a line tangent to the inflection point with the original value. The time constant can be approximated as the time from the intersection of the tangent line with the original value to its intersection with the final value of the output. This approach is consistent with the Ziegler-Nichols reaction curve tuning method where L is the equivalent dead time and R is the slope of the tangent line in response to a step input. For processes where the time constant is more than four times the equivalent dead time, the slope of the tangent line is a near-integrating process gain.

The effects of automation system dynamics and process dynamics are important for meeting performance goals, as evident in the metrics. To help one understand and appreciate these effects, the figures for the process responses use an open-loop gain, open-loop time constant, and total loop dead time. The open-loop gain is the product of the manipulated variable gain, process gain, and measurement gain. The open-loop time constant is the largest time constant in the loop, hopefully in the process of self-regulating responses and necessarily in the process for runaway responses. The total loop dead time in a first-order response is the sum of all the pure time delays and equivalent dead times from time constants smaller than the largest time constant in the process and automation system, as noted in Figure 8-10a. The simple ordinary differential equations (ODEs) associated with the figures do not include the automation system dynamics. The ODEs presented were developed in Appendix F and are focused on the primary process response to help in understanding the effect of material and energy balances.

PID metrics particularly depend on the type of process response. The major types of process responses are self-regulating, integrating, and runaway. The open-loop response of a self-regulating process will reach a steady state for a given change in PID output because of negative feedback within the process, provided there are no disturbances during the test. In an FOPDT approximation, the response is characterized by a total loop dead time (θ_o), an open-loop time constant (τ_o), and an open-loop self-regulating process gain (K_o). If the total loop dead time is much greater (approximately >4X) than the open-loop time constant, the process is classified as *dead time dominant*. If the total loop dead time is about equal to the open-loop time constant, the process is termed *balanced*. If the total loop dead time is much less (approximately <1/4) than the open-loop time constant, the process is termed *lag-dominant* and possibly classified as *near-integrating*. The open-loop response of a true integrating process is a ramp without a steady state due to a lack of negative feedback within the process. In an FOPDT approximation, the response is characterized by a total loop dead time and an open-loop integrating process gain (K_i). Often a secondary time constant (τ_s) is included in identifying near and true integrating process responses because of its dramatic effect on performance and the benefit of cancellation by rate action. The subscript $_p$ denotes a process, $_o$ denotes an open-loop, $_I$ denotes integrating action, $_d$ denotes derivative action, $_v$ denotes a valve or VFD, $_m$ denotes a measurement, and $_a$ denotes an analyzer.

> **Key CONCEPT** The open-loop response of a self-regulating process will eventually reach a steady state due to the existence of negative feedback within the process if there are no additional disturbances or changes to the manipulated variable.

> **Key CONCEPT** The open-loop response of an integrating process will continue to ramp because of a lack of negative feedback within the process until the loop is closed or the process variable reaches an equipment or measurement limit.

Self-Regulating Process Response

The self-regulating process will reach a steady state provided there are no other disturbances or changes in the controller output because of negative feedback in the process shown in Figure 8-10b. This self-regulating action is due to negative feedback occurring in the process, as shown in the following Equation 8-10a expressed in terms of a primary process time constant (τ_p), steady-state process gain (K_p), manipulated variable (MV), and disturbance variable (DV). The negative feedback is evident when the PV is subtracted on the right side of the Equation 8-10a. Integrating the ODE in the next equation gives the exponential time response to changes in the manipulated and disturbance variables with continual deceleration after the dead time. As a result of other time constants in the loop, there is an inflection point where acceleration moderated by these time constants is exceeded by deceleration from the primary time constant. The PV, MV, and DV are in engineering units because the manipulated variable gain

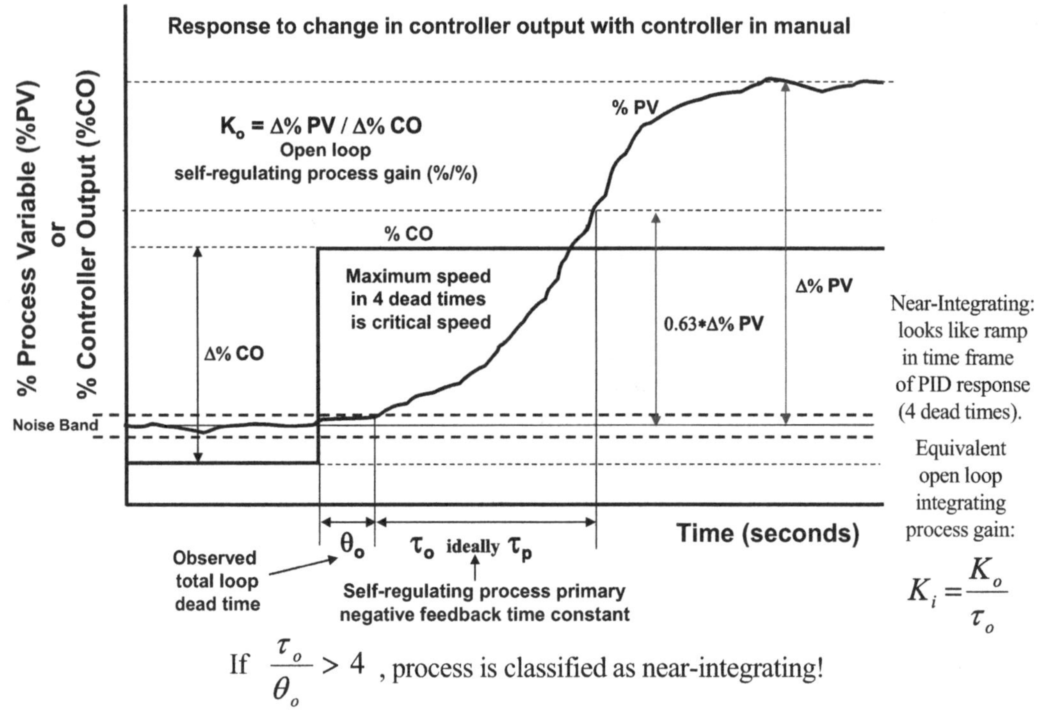

Figure 8-10b. Self-regulating process open-loop response approximated by a primary process negative feedback time constant and the process gain plus dead time model.

(e.g., valve gain) and measurement gain are not included, which would make the gain dimensionless.

$$\tau_p \bullet dPV/dt = K_p \bullet (MV - DV) - PV \tag{8-10a}$$

$$\Delta PV = K_p \bullet \left(1 - e^{-t/\tau_p}\right) \bullet (\Delta MV - \Delta DV) \tag{8-10b}$$

The open-loop response depends on the automation system dynamics and is described by an open-loop time constant (τ_o), total loop dead time (θ_o), and an open-loop gain (K_o). The open-loop time constant, also known as the *primary time constant,* is the largest time constant in the loop. Hopefully, it is in the process so it can slow down disturbances on the process input. In flow, pressure, and many in-line loops (e.g., static mixer *p***H** and concentration control), the largest time constant is in the measurement. The second largest time constant, known as the *secondary time constant* (τ_s), is sometimes identified; otherwise, a fraction of it, as with all other smaller time constants, is taken as an equivalent dead time (θ_i). The fraction (Y_i) for each smaller time constant (τ_i) varies from 0.28 to 0.88 as the ratio (X_i) of the small to largest time constant varies from 1.0 to 0.01 per the following equation:

$$X_i = \frac{\tau_i}{\tau_o} \quad (8\text{-}10c)$$

$$Z_i = \frac{1}{1 - X_i} \quad (8\text{-}10d)$$

$$Y_i = 1 + \frac{1}{X_i} - \frac{1}{X_i^{Z_i}} - \frac{\ln(X_i)}{1 - X_i} \quad (8\text{-}10e)$$

$$\theta_i = \sum_i Y_i \bullet \tau_i \quad (8\text{-}10f)$$

Multiple time constants and sources of dead time in the process and automation system noted here for self-regulating processes also exist for all types of process responses. The Figure 8-10a block diagram shows many of the gains, lags, delays, and valve or variable speed drive dynamic response limitations that are useful for analyzing and improving performance by minimizing dead time, nonlinearities, and filtering action.

If there are equal noninteracting time constants in series (θ_j), the fraction (Y_j) of the total that becomes dead time increases and approaches 1 for a large number (N) of these time constants in series. This methodology was used for simulating dead time in mathematical analysis and in analog computing modules that did not have a pure dead time capability. The total loop dead time is the sum of these dead times from time constants and the sum of pure dead times ($\Sigma\theta_k$) from digital updates (e.g., analyzer cycle times, execution rates, scan times, wireless update rates), the time for a signal to get through resolution limitations, and deadband and process dead times (e.g., transportation and mixing delays). What is left of the smaller and equal time constants not taken as equivalent dead time is summed and added to a lone primary time constant to get the open-loop time constant. Note that data compression and slow historian update times can hide the true dead time in historian data and trend plots.

$$\theta_j = Y_j \bullet N \bullet \tau_j \quad (8\text{-}10g)$$

$$Y_j = A_o + A_1 \bullet [\ln(N)]^1 + A_2 \bullet [\ln(N)]^2 + A_3 \bullet [\ln(N)]^3 + A_4 \bullet [\ln(N)]^4 \quad (8\text{-}10h)$$

$$\theta_o = \theta_i + \theta_j + \Sigma\theta_k \quad (8\text{-}10i)$$

$$\tau_o = \sum_i [(1 - Y_i) \bullet \tau_i] + \sum_j [(1 - Y_j) \bullet N \bullet \tau_j] + \tau_p \quad (8\text{-}10j)$$

> **Key CONCEPT** Separate equations can approximate the equivalent dead time from noninteractive time constants in series seen in all automation systems and interactive process time constants because of bidirectional flow between volumes (e.g., distillation column trays).

Appendix I explains how to estimate noninteractive time constants for two interacting volumes in series. A step change can provide a first-order approximation of a single time constant and effective dead time, but more accurate analysis and tuning settings are possible by identifying the two equivalent noninteracting time constants. This is particularly important for getting the derivative (rate time) setting, which is often set equal to the second largest noninteractive time constant.

The open-loop self-regulating process gain (K_o) for a control loop manipulating a valve or VFD is the product of the valve or VFD gain (K_v), process gain (K_p), and measurement gain (K_m). This open-loop gain is dimensionless for controllers working in percent signals, which is the case for most industrial PIDs. The valve or VFD gain is the slope of the installed flow characteristic for small changes in controller output. The measurement gain is simply 100% divided by the measurement span in engineering units. The gains originating in the automation system (shown in Equation 8-10k for self-regulating processes) also apply to other types of processes.

$$K_o = K_v \bullet K_p \bullet K_m \tag{8-10k}$$

$$K_v = \frac{\Delta F_v}{\Delta \%CO} \tag{8-10l}$$

$$K_m = \frac{100\%}{PV_{max} - PV_{min}} \tag{8-10m}$$

For composition, temperature, and *p***H** control, the process gain has a factor not commonly recognized that is revealed in the ODEs for the material and energy balance. The hint to this factor is that most plots of composition, temperature, and *p***H** (e.g., titration curves) have an abscissa that is the ratio of the manipulated flow to the feed flow. Consequently, the process gain consists of two components. The first is the change in the ratio (ΔR) with a change in the manipulated flow (ΔF_v), which simplifies to being the inverse of the process feed flow (F_p). The second component is the slope of the plot of the process variable versus the ratio.

$$K_p = K_{p1} \bullet K_{p2} = \frac{\Delta PV}{\Delta F_v} \tag{8-10n}$$

$$K_{p1} = \frac{\Delta R}{\Delta F_v} = \frac{\left(\Delta F_v \big/ F_p\right)}{\Delta F_v} = \frac{1}{F_p} \tag{8-10o}$$

$$K_{p2} = \frac{\Delta PV}{\Delta R} \tag{8-10p}$$

Key CONCEPT The open-loop process gain for a self-regulating process seen by a PID controller manipulating a valve or VFD is the product of a valve or VFD gain, process gain, and measurement gain with a flow ratio gain in the process gain that is inversely proportional to feed flow for composition, *p***H**, and temperature control.

If a primary controller is manipulating a secondary set point (e.g., cascade loop), the valve or VFD gain is replaced by the set-point gain that is computed from the set-point limits and controller output limits. Note that the output limits are in percentage values that correspond to the set-point limits, taking into account the set-point scale because industrial PID algorithms work with percent signals.

$$K_{SP} = \frac{SP_{max} - SP_{min}}{\%CO_{max} - \%CO_{min}} \tag{8-10q}$$

Key CONCEPT For the primary loop in cascade control, the valve or VFD gain is replaced by a simpler and linear set-point gain based on the secondary PID set-point limits and units and the primary PID output limits.

The process gain simplifies to the change in *PV* (e.g., primary loop) for a change in the manipulated *SP* (e.g., secondary loop). If the engineering units are the same for the *SP* and *PV*, there is no longer a flow ratio factor. For cascade control, where a primary vessel *p***H** loop is manipulating a secondary *p***H** loop on feed, the process gain may be about 1 because a 1 *p***H** change in feed *p***H** may correspond to a 1 *p***H** change in vessel *p***H** at a steady state. In this case, the vessel *p***H** loop is linear, and all the nonlinearity is relegated to the in-line feed loop, which has a much smaller dead time and response time, enabling it to recover from the adverse consequences of uncompensated nonlinearities much faster.

$$K_p = \frac{\Delta PV}{\Delta SP} \tag{8-10r}$$

Key CONCEPT

If the secondary loop *PV* in cascade control has the same units as the primary *PV* (e.g., cascade control of the vessel pH to static mixer pH), there is no longer a flow ratio gain factor and the process gain greatly simplifies to just the change in secondary *PV* divided by the change in the secondary *PV* set point, which is 1.0 for the common case of the same secondary *PV* and set-point scales.

An increase in the process gain provides an increase in sensitivity, amplifying the effect of changes on the process input. Increased process sensitivity is generally beneficial because this translates for a given measurement repeatability and accuracy to a correction of smaller disturbances. For example, if a resistance temperature detector sensor and pH electrode have a repeatability of 0.04°C and 0.04 pH, respectively, process changes that result in a temperature or pH approaching this are not well controlled. Strong acid and strong base pH systems exhibit extraordinarily large process gains and process sensitivity to changes in actual hydrogen ion concentration because of the logarithmic relationship. However, the mixing requirements (e.g., uniformity) to reduce noise and the resolution requirements of the valve are beyond what is normally possible, requiring multiple stages of pH control to gradually approach the neutral region.

Near-Integrating Process Response

Near-integrating processes are actually self-regulating processes with a large process time constant to dead time ratio that appear to have an integrating response because the response to a step change resembles a ramp in the normal PID response time (e.g., four dead times). The ramp rate ($\Delta\%PV$/s per $\Delta\%CO$) (1/s) is the integrating process gain (K_i) that can be estimated as the open-loop self-regulating process gain divided by the open-loop time constant. Rules for tuning and methods of analysis for true integrating processes can often be beneficially applied. Switching tuning rules from a self-regulating to an integrating process for a process time constant greater than four times the total loop dead time provides better load disturbance rejection. Also, near-integrating processes behave like true integrating processes in terms of developing extremely slow and large, nearly undamped process oscillations for a PID gain less than the inverse of the product of the integrating process gain and reset time. Because the literature emphasizes that a high PID gain causes oscillations, a frequent reaction is to decrease the PID gain, which makes the problem worse.

$$K_i = \frac{K_o}{\tau_o} \tag{8-10s}$$

Key CONCEPT

Self-regulating processes with an open-loop primary process time constant greater than four times the total loop dead time are approximated as near-integrating for better analysis and tuning, with the open-loop integrating process gain being the open-loop self-regulating process gain divided by the primary time constant.

True Integrating Process Response

A *PV* has a true integrating process response if the open-loop response is a ramp whenever there is an unbalance between process inputs and outputs. Because a perfect balance never occurs when a PID is in manual, the process will eventually ramp as seen in Figure 8-10c. There is no steady state. Without feedback correction, the ramp does not stop until a physical equipment limit is reached, a safety instrumentation system (SIS) is activated, or a relief device opens. An integrating response exists when there is no discharge flow or when a discharge flow is weakly dependent on the *PV.* The term *non-self-regulating* is used to denote the fact that a controller provides the missing negative feedback in the process seen in the omission of the *PV* on the right side of Equation 8-10t. Integrating the ODE from an initial steady state gives a ramp, as shown in Equation 8-10u, that is simply the integrating process gain multiplied by the time in the response to step changes after the delay from dead time. If the *MV* exactly equals the *DV,* the ramp stops. The *DV* is often termed a *load.* All integrating processes have a load, or else the process can only move in one direction. The *MV* and *DV* are flows to give an integrating gain process gain (K_{ip}) with inverse time units, as seen in more detailed Equations 8-10w through 8-10z for level.

$$dPV/dt = K_{ip} \bullet (MV - DV) \tag{8-10t}$$

$$\Delta PV = K_{ip} \bullet t \bullet (\Delta MV - \Delta DV) \tag{8-10u}$$

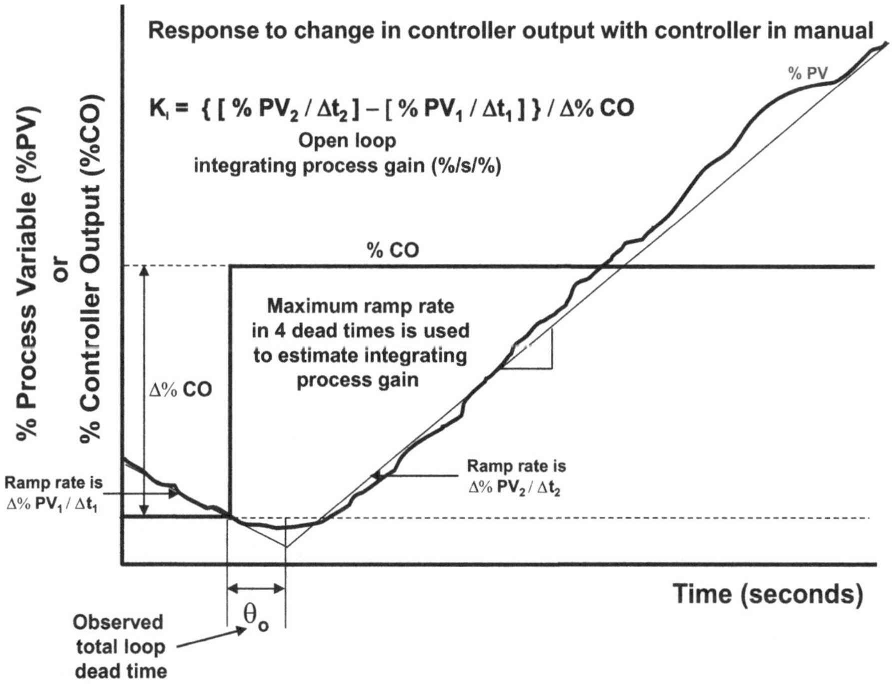

Figure 8-10c. Integrating process open-loop response approximated by an integrating process gain plus dead time model.

For a PID to reject a disturbance or reach a new set point, the controller output moves beyond its final resting value that matches the load. For a fast response, the overshoot of the final resting value should be large; this is best achieved by the proportional mode providing immediate negative feedback. In fact, there is a window of allowable controller gains, where too low or too high of a controller gain leads to oscillations. As the reset time is decreased, the low limit to the PID gain should be increased. The oscillation amplitude and period are much larger for a PID gain that is too low, making it more difficult to filter out in downstream volumes. Users and operators tend to be wary of large PID gains that are possible on these processes and are focused on reset action eliminating small differences on digital displays of *PV* and *SP* (often beyond sensor repeatability). Often the reset time is one or two orders of magnitude too small, making the gain requirement extraordinarily large. Operators do not like sudden and large movements of controller outputs. Section 8.11 will explain how external reset feedback (ERF) can solve the problem. The maximum ramp rate in four dead times can be used to estimate the integrating process gain.

> **Key CONCEPT** Integrating processes require the controller output to significantly overshoot its final resting value to correct for load disturbances or to reach a new set point.

The following equation shows how to calculate the low limit on the PID gain to prevent oscillations from starting in loops where the total dead time is not a major factor.

$$K_c > \frac{4}{K_i \bullet T_i} \qquad (8\text{-}10v)$$

> **Key CONCEPT** Integrating processes have a window of allowable PID gains where too low of a PID gain is commonly violated because of the tendency to use more integral and less proportional action, causing slow rolling oscillations that can take hours to die out.

The primary example of an integrating process is the level where the discharge flow from the bottom is pumped, and the amount of flow pumped is not greatly affected by the pump inlet pressure created from the level in the equipment. Tight level control is needed for residence time control of reactors and material balance control in evaporators where changes in discharge flow must be immediately translated to changes in feed flow for concentration control. Feedforward control can help in these cases, but tight feedback control helps to deal with unmeasured disturbances. Where tight level control is detrimental, absorbing variability as the *PV* changes can be maximized by minimizing the controller gain and making sure the low limit is not violated by maximizing the reset time.

For integrating processes, the open-loop self-regulating (steady state) gain (K_o) is replaced with an open-loop integrating process gain (K_i) with inverse time units (1/s) that result from the process gain term units not canceling out the mass flow (F_v) time units as shown in Equations 8-10w through 8-10z for level (L) in engineering units. For level and other integrating processes, we can eliminate the ratio gain term in the open-loop gain for a self-regulating process, giving an equation for the integrating process gain with the distinction of no longer being dimensionless but having units of 1/s. For level, the process gain is the inverse of the product of the cross-sectional area (A) and fluid density (ρ), revealing the process gain vulnerability to fluid composition and horizontal tanks. The measurement gain, as with most loops, is 100% divided by the measurement span in engineering units.

$$K_i = K_v \bullet K_p \bullet K_m \tag{8-10w}$$

$$K_v = \frac{\Delta F_v}{\Delta \%CO} \tag{8-10x}$$

$$K_p = \frac{\Delta L}{\Delta F_v} = \frac{1}{A \bullet \rho} \tag{8-10y}$$

$$K_m = \frac{\Delta \%PV}{\Delta L} = \frac{100\%}{L_{max} - L_{min}} \tag{8-10z}$$

Composition, temperature, and *p*H control loops of batch processes have an integrating response because the bottom discharge valve is closed during the batch cycle when these loops are active. The same situation can occur during the start-up of continuous processes when the discharge valve on a volume is kept closed until normal operation resumes. The integrating process gain for composition and *p*H control tends to be larger for low liquid volumes if the reaction rate is not a consideration. Often reactions are not started until volumes are relatively large. The effect of liquid volume on temperature may be largely offset by the effect of a smaller portion of the heat transfer area covered, especially for heating or cooling by a jacket instead of coils.

While fed-batch processes are sometimes termed *semicontinuous*, they do not have a steady state unless the controlled variable is a reactant concentration and the reaction rate consumes the reactant roughly in proportion to its concentration.

> **Key CONCEPT** Batch processes and the start-up of continuous processes have an integrating response for composition, temperature, and *p*H control because the liquid discharge valve on the volume is closed.

In batch processes, changing the controlled variable to the slope of a batch profile can provide, in most cases, a steady state with the ability to decrease besides increasing the controlled variable, which is often not possible when the original controlled variable is intermediate or the final product concentration. The increasing accumulation of a component in a batch conflicts with the need for PID to be able to both decrease a controlled variable that is the component concentration by changing the manipulated variable. If the controlled variable is the slope, the PID can decrease besides increase the controlled variable (rate of component accumulation). This technique is used to control the cell growth and product formation rates in fermenters and bioreactors. It is also used for highly exothermic reactors where the temperature should be increased cautiously as the batch progresses to ensure there is no excessive acceleration leading to a runaway.

There are additional variations of the integrating response that may be useful to identify, including integrator plus first-order lag, inverse response, and lead. These additional dynamics are important for PID tuning and feed forward settings.

Rise Time

The set-point rise time (T_r) is a set-point response metric that is the time it takes for the process variable to go through a specified amount of the set-point change, such as the time to go from 10% to 90% of the set-point change ($\Delta\%SP$). If the rise time is specified to be from 0% to 100%, as shown in Figure 8-10d, it is the total loop set-point response rise time, including dead time. This total rise time is particularly useful for batch cycle time analysis and improvement. For a non-self-regulating process with an integrating process gain (K_i) and an optional set-point feedforward gain (K_{ff}) dominated by proportional action via PID controller gain (K_c) rather than integral action, which is often the case for set-point changes for integrating processes, the 10% to 90% rise time ($T_{r(10-90)}$) can be estimated as follows:

$$T_{r(10-90)} = \frac{0.8 \bullet \Delta\%SP}{K_i \bullet Min\left(\left|\Delta\%CO_{max}\right|, \left(K_c + K_{ff}\right) \bullet 0.8 \bullet \Delta\%SP\right)} \tag{8-10aa}$$

For a maximum available output change ($\Delta\%CO_{max}$) larger than the step from the proportional mode plus feedforward, the changes in the set point in the numerator and denominator cancel out, yielding:

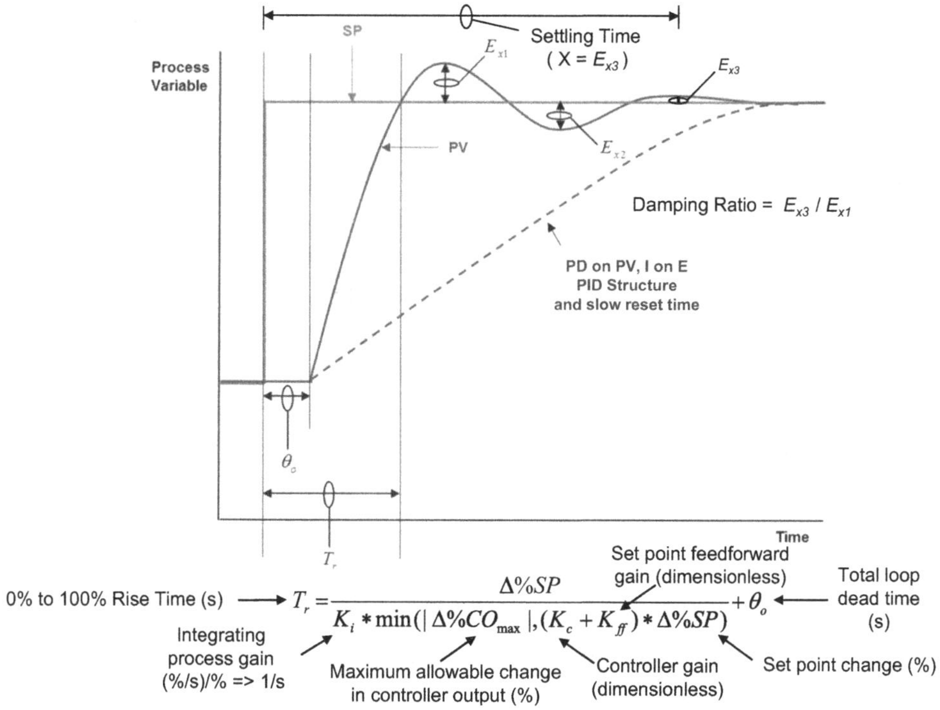

Figure 8-10d. Set point response metrics.

$$T_r = \frac{1}{K_i \bullet \left(K_c + K_{ff}\right)} \qquad \text{(8-10bb)}$$

> **Key CONCEPT** The rise time to reach a new set point is inversely proportional to the product of the open-loop process gain and PID gain and can be minimized by simply adding a set point feedforward which effectively increments the effect of the PID feedback control gain by the feedforward gain.

For the lag-dominant "near-integrating" process response often seen in vessel and column temperature loops and large-volume gas pressure loops where the primary process time constant is significantly larger than the total loop dead time, the integrating gain is estimated as the open-loop gain (K_o) divided by the open-loop time constant (τ_o), which is the process time constant, yielding:

$$T_r = \frac{\tau_o}{K_o \bullet \left(K_c + K_{ff}\right)} \qquad \text{(8-10cc)}$$

A process is generally classified as *near-integrating* if the process time constant is greater than four times the dead time. If a dead time is not perceptible, a minimum dead time is one-half the sum of the input/output (I/O) scan time and controller execution times. Realize that data historian compression and update times may hide dead time. Switching from a self-regulating process to integrating process tuning rules for near-integrating processes enables tuning settings that do not overestimate the integral time. It also enables analyzing the challenges associated with true integrating processes, such as the need for the product of the PID gain and reset time to be large enough to prevent slow oscillations with a poor damping ratio.

A faster set-point rise time metric reduces the time to reach a new set point to increase process performance during start-up and for changes in production rate and operating points, transitions to different products, procedure automation of continuous processes, and steps and phases in batch and fed-batch processes. Process capacity increases because of the shorter time to reach a set point and an increase in process efficiency from less off-spec material generated for the same feed rate, particularly for start-up, transitions, and optimization. If the feed rate is reduced to match the decrease in off-spec material, the improvement is a decrease in operating raw material cost from the increase in process efficiency.

For balanced and dead time dominant self-regulating processes, the 63% step response time is generally a more appropriate metric for the set-point response time. An approximate rise time can be computed if the dead time is identified separately and a first-order approximation is valid.

63% and 86% Step Response Times (T_{63} and T_{86})

The step response time is the time it takes to reach a specified percentage of the final response to a step change. If the percentage is not specified, the metric is meaningless. Consequently, the percentage of the final response included as a subscript is helpful. The final value can be the open-loop steady-state value after a step change to the input of a part of the control loop, such as a valve, process, or measurement. The final value can also be the set point for a closed-loop response time to a step change in the set point. The step response time includes the process dead time. For a first-order response, the 63% and 86% responses include a single and a double time constant, respectively. The closed-loop 63% response time is used by some tuning methods for self-regulating processes. The open-loop 86% response time is helpful for representing and diagnosing the response of control valves and *p***H** electrodes because it is more representative of the effect on loop performance and is not distorted by a much slower late approach to the final value, which occurs particularly for *p***H** electrodes. Unfortunately, many citations of the step response time for measurement devices are for 98% of the final response without the percentage stated.

> **Key Concept** The response time metric is meaningless unless the final percentage change is specified, preferably in the parameter subscript.

> **Key Concept** The 86% response time (T_{86}) provides a more practical measure of the dynamic response of control valves and *p*H electrodes.

Process Variable and Controller Output Overshoot

Process variable overshoot is the maximum excursion of the process variable past the set point after crossing the set point for a set-point change in either direction or returning to the set point for a disturbance from either direction. Overshoot can be a positive or negative excursion depending on the direction of the set-point change and disturbance. Overshoot can translate to a loss of process efficiency and capacity in processes operating close to constraints or sensitive to operating conditions. Temperature and *p*H overshoot in bioreactors can cause cell death. Temperature overshoot in highly exothermic reactors can trigger an acceleration in temperature so fast it becomes a point of no return leading to a shutdown. Processes that respond in only one direction to the manipulated variable (e.g., batch *p*H rise by base reagent or temperature rise by heating) cannot reduce an overshoot. Consequently, PID controllers for these unidirectional processes tend not to have integral action (e.g., proportional on error, derivative on PV, and no integral).

Controller output overshoot is the maximum excursion of the controller output past the final resting value (FRV), defined as the constant output the controller ends up with in response to a given set-point change or disturbance. Minimizing FRV overshoot is important in minimizing the interaction between loops, particularly in unit operations where there is a lot of heat integration. However, in the overzealous pursuit of minimizing FRV overshoot, it is sometimes forgotten that non-self-regulating processes must have FRV overshoot to reach or return to a set point. For example, a level loop manipulating a discharge flow must momentarily increase the discharge flow to be greater than the feed flow to achieve a lower level set point before the discharge flow FRV ends up being the feed flow. For tight level control for reactor residence time control, the FRV overshoot should be fairly large.

Damping Ratio

The damping ratio is the ratio of the second peak to the first peak in the same direction of an oscillation in response to a set-point change or disturbance. In the early days of PID controllers, tuning methods such as the Ziegler-Nichols ultimate oscillation method provided aggressive tuning with a quarter amplitude damping ratio to minimize the peak and integrated error from disturbances. However, such oscillations indicated a lack of robustness and posed a disruption to other loops. There is a tradeoff between PID tuning robustness and performance. Because control loop dynamics are nonlinear—varying with the operating point, direction, rates of change, and operating conditions with many unknowns—there is a considerable risk that the damping ratio could increase and the loop could become unstable with such aggressive tuning. Also, the oscillations can be a particularly disruptive disturbance to other loops and be amplified by resonance when the oscillation periods of the loops are similar. Smooth and gradual responses are favored by operators; pose the least disruption to other loops, particularly those without back mixing (e.g., in-line systems and sheet lines); enable better ratio control; and are the primary objective in some processes (e.g., pulp and paper). However, highly exothermic reactors need extremely aggressive temperature control, often resulting in a damped oscillation for a set-point change or a large, fast disturbance. The resulting oscillations are not a problem if they are effectively filtered out by a downstream volume.

Oscillations with a constant amplitude and, thus, a damping ratio of 1 are termed *limit cycles*.

Settling Time

The settling time is the duration of time after a set-point change or disturbance-induced deviation for the process variable to enter and stay within *X*% of the set point, where *X* is a percentage of the set-point change or a percentage of the maximum deviation from the set point after a disturbance (e.g., 5%). For processes sensitive to variability, the need for *X* and settling time to be minimized is greater. The minimum *X* is set by several factors including measurement noise and the automation system resolution that depends on the sensor, signal I/O bits and span, and final control element. The control valve resolution is multiplied by the final control element gain (e.g., the slope of installed flow characteristic) and process gain and converted to a percentage of the set point.

For limit cycles, the minimum *X* is set by the limit cycle amplitude. Note that the limit cycle is usually not centered exactly around the set point due to disturbances, offsets, and nonlinearities. Limit cycles develop from resolution limitations if there is integrating action anywhere in the process and control loop (e.g., process, positioner, secondary PID, primary PID). Limit cycles can develop when the final control element has a significantly longer response time for small amplitude controller output signal reversals in loops with integral action observed in some positioners. Limit cycles develop from lost motion deadband if there is integration action in two or more places anywhere in the control loop. The only control loops without limit cycles would be on self-regulating processes without any integral action in the positioner or controller(s). The appearance of limit cycles may be hidden by noise, disturbances, and historian data compression.

The limit cycle amplitude (A_o) from a resolution limitation is independent of controller tuning and is simply the resolution limitation (R_v) multiplied by the open-loop gain (K_o) for a self-regulating process:

$$A_o = R_v \bullet K_o \tag{8-10dd}$$

The limit cycle period (T_o) from a resolution limitation increases as the reset time (T_i) increases, and the controller gain decreases for self-regulating processes:

$$T_o = 4 \bullet T_i \bullet \left\{Max\left[2, 1/\left(K_o \bullet K_c\right)\right] - 1\right\} \tag{8-10ee}$$

Key CONCEPT

Limit cycles due to resolution limitation have an amplitude proportional to the product of resolution and open-loop process gain and a period proportional to the PID reset time and inversely related to the PID gain.

The limit cycle amplitude from lost motion deadband (DB_v) is inversely proportional to controller gain:

$$A_o = DB_v / K_c \tag{8-10ff}$$

The limit cycle period from lost motion deadband increases as the reset time increases and the controller gain decreases, as exemplified by the following equation for integrating processes:

$$T_o = 5 \bullet T_i \bullet \left[1 + \left(2/K_c^{0.5}\right)\right] \tag{8-10gg}$$

Key CONCEPT

In limit cycles due to lost motion deadband, the amplitude is proportional to the ratio of deadband to PID gain with a period proportional to the PID reset time and inversely related to the PID gain.

These equations showing the effect of PID tuning on the limit cycles from resolution limitation and lost motion deadband were developed for a single PID controller with integral action but not derivative action. The effect of a positioner is not included. If there is no integral action in the positioner, the resolution and lost motion seen in the positioner response may be used in the equations.

The equation for limit cycles due to a resolution limitation is for a self-regulating process that is not lag-dominant (not near-integrating). Thus, there is only one point of integration, which is in the PID controller. A limit is added to the original equation to protect against a negative result for a PID tuned aggressively for a process with an open-loop time constant greater than the total loop dead time.

The equation for limit cycles due to lost motion is for an integrating process and a single PID controller, so there are two points of integration. For a self-regulating process with integral action in the positioner and in a PID, a limit cycle will develop from backlash or other sources of lost motion. This can also occur for a cascade control loop with integral action in both the primary and secondary controllers.

Key CONCEPT

Limit cycles due to resolution require one point of integration, whereas limit cycles due to lost motion occur when there are two points of integration. The points of integration can occur in the process, positioner, and primary or secondary PIDs.

Although Equations 8-10dd through 8-10gg are for a single PID controller, no positioner, and a specified process type, they can be used to alert the user to the development of limit cycles and the relative effects of tuning for more complex loops, providing a means of diagnosing and mitigating the consequences of limit cycles. For example, reducing a limit cycle amplitude enables it to be more easily filtered by downstream volumes. It is also a motivation for not using integral action in positioners, which creates limit cycles from lost motion and makes the amplitude larger from a smaller positioner gain needed to deal with integral action.

The limit cycles of the PID PV are a square wave, and the PID output is a sawtooth for a self-regulating process whose open-loop time constant is less than the total loop dead time. Some rounding of the oscillation peaks occurs as the time constant increases above the dead time. For near-integrating and true integrating processes, the PV oscillation becomes a sawtooth, and the output oscillation is rounded.

Limit cycles can also originate from process instabilities and extreme nonlinearities where the process response is so large and/or fast that it overwhelms the controller, such as the 4 to 10 *p*H oscillations in a strong acid and base system.

Integrated Error

The integrated error (E_i) is the result of integrating the deviation between the process variable and the set point after a disturbance, assuming the process variable was originally resting at the set point. This can be viewed as the net total area of the error on a trend recording, considering that the area of plus and minus excursions cancels out to some degree. In processes with back mixing, the cancellation may occur from attenuation, resulting in an integrated error corresponding to the net amount of off-spec material downstream. If the oscillations do not cross the set point, the integrated error is the same as the integrated absolute error (IAE), which is the more common metric discussed in the literature. The advantage of the integrated error is the ability to estimate the practical limit from controller tuning settings, controller execution rate (Δt_x), signal filter time (τ_f), and the ultimate limit (best theoretically possible metric) from the loop dynamics. The following equation to estimate the integrated error practical limit is obtained from the original equation by Shinskey by converting the change in percent controller output required to reject a disturbance to the percent open-loop error (E_o), which is the error from a disturbance if the loop is open (the controller is in manual).

$$E_i = \frac{T_i + \Delta t_x + \tau_f}{K_c \bullet K_o} \bullet E_o \qquad (8\text{-}10hh)$$

The integrated error cannot be less than what is possible based on the total loop dead time and the rate of change of the excursion determined by the open-loop time constant. The equation is for a step disturbance. The effect of disturbance lag can be approximated by applying a filter to the open-loop error with a filter time equal to the disturbance lag time.

> **Key Concept** The practical limit to the integrated error for a step load disturbance is proportional to the sum of the PID reset time, execution time interval, and filter time.

The integrated error from a controller response increases with the ultimate period, which varies from two to four times the total loop dead time as a self-regulating process moves from dead time dominant to lag-dominant. The ultimate period of an integrating process starts out as four times the dead time but increases considerably as the secondary time constant increases. Tuning methods try to minimize this detrimental effect by canceling it with a rate time setting equal to the secondary time constant.

The following equation for the ultimate limit to the integrated error is based on the fact that any correction for a disturbance does not occur until one total loop dead time has passed. The maximum possible rejection of a step load disturbance to a self-regulating process is two right triangles back-to-back to form an equilateral triangle, as shown in Figure 8-10e. Each right triangle has an altitude that is the peak error and a base that is the dead time. Approximating the integrated error as the area of this triangle works best for a lag-dominant process. The result is the ratio of the total loop dead time squared to the open-loop 63% response time. For integrating processes, the open-loop time constant is approximated as the inverse of the integrating process gain.

$$E_i = \frac{\theta_o^2}{\theta_o + \tau_o} \bullet E_o \tag{8-10ii}$$

Key CONCEPT The ultimate limit to the integrated error for a step load disturbance is proportional to the ratio of the dead time squared to the 63% response time (T_{63}).

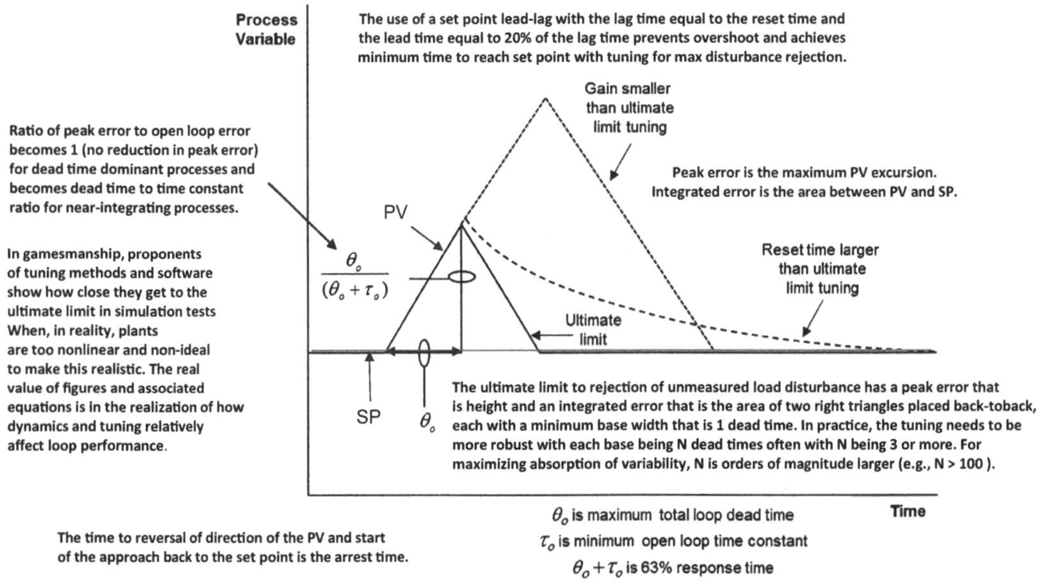

Figure 8-10e. The ultimate limit to integrated and peak errors.

For limit cycles, there is an IAE in addition to what is due to disturbances. This additional IAE can be conservatively estimated as the total area of the limit cycle set by its oscillation amplitude and period over the analysis time frame. There is some cancellation of the limit cycle IAE by the disturbance IAE resulting from the sign of the disturbance being the opposite of the sign of the limit cycle due to different oscillation periods or an opposite sign of the effect of the disturbance source.

The ISE normalized by integration time duration is indicative of the process variance, and the square root is the long-term standard deviation.

Peak Error

The peak error is the maximum excursion in the direction caused by a load disturbance. A large peak error can cause unsafe operation, damage equipment, and trigger undesirable process responses. Process onstream time is adversely affected by a peak error that activates the SIS, opens the relief valves, and/or blows rupture disks. Peak errors in waste treatment *p***H** can cause reportable violations of Resource Conservation and Recovery Act (RCRA) limits.

The peak error for a PID controller can be approximated using the following equation for self-regulating processes, realizing that the second term in the denominator is usually negligible, making the peak error mostly a function of controller gain. If there is no derivative action, the numerator should be increased from 1.25 to 1.5. The equation is for a step disturbance. The effect of disturbance lag can be approximated by applying a filter to the open-loop error with a filter time equal to the disturbance lag time. For near and true integrating processes, the peak error can be estimated by multiplying the ramping open-loop error ($\Delta E_o/\Delta t$; ramping PV in Figure 8-10c) by the arrest time (time to stop the initial excursion from a disturbance proceeding in the same direction) that can normally range from one dead time for aggressive control to five dead times for robust control. The arrest time is much larger when it is maximized for maximum absorption of variability (e.g., surge tank level control).

$$E_x = \frac{1.25}{K_c \bullet K_o \bullet \left(1.0 + \dfrac{0.5 \bullet \theta_o}{(T_i + \Delta t_x + \tau_f)}\right) + 1.0} \bullet E_o \qquad \text{(8-10jj)}$$

> **Key CONCEPT** The practical limit to the peak error for a load disturbance is inversely proportional to the PID gain and is only slightly decreased by a smaller reset time.

The ultimate limit to the peak error is the altitude of the right triangle in Figure 8-10e, which is the ratio of the total loop dead time to the open-loop 63% response time as discussed for the ultimate limit to the integrated error. The approximation works best for a lag-dominant process. For dead time dominant processes, the peak error approaches the open-loop error. For near-integrating processes, the open-loop time constant can be approximated as the inverse of the integrating process gain.

$$E_x = \frac{\theta_o}{\theta_o + \tau_o} \bullet E_o \qquad \text{(8-10kk)}$$

> **Key CONCEPT** The ultimate limit to the peak error for a load disturbance is proportional to the ratio of the dead time to the 63% response time (T_{63}) and becomes essentially the open-loop error for dead time dominant processes.

Automation System Response

Most of the dead time in flow, level, and pressure loops typically comes from automation system response, with notable exceptions occurring when heat transfer surfaces are involved in the action of the manipulated variable. The automation response is also the major source of dead time in composition and *p*H control on well-mixed vessels if you include the effect of lost motion and resolution limitations, slow electrodes, and transportation delays to the sensor and manipulated flows. Measurement transportation delay is the sum of each volume from the process to the sensor divided by the respective flow. The transportation delay should be minimized by high recirculation line flows and locating the sensor near the exit of the recirculation line from the process. The manipulated flow transportation delay can be negligible for pressurized completely full lines with the same composition as the manipulated flow. However, for a nonuniform line composition caused by backfilling process fluid from the injection point or by blending streams, the transportation delay is the volume divided by the manipulated flow. For *p*H control, where the reagent flow is a few gallons per hour and a 2-inch minimum dip tube size extends down toward the agitator per mixing guidelines, the transportation delay after mixing reagents or a sustained closure of the reagent valve can be several hours. Also, the reagent may continue to leak into the vessel for several hours after the reagent valve is closed. For this and other reasons, the reagent is injected into a recirculation line right before entry into the vessel. For conveyors, the transportation delay is obvious if the feed rate is changed by manipulating the amount added to the conveyor. Speed control, particularly with a gravimetric feeder, can eliminate this transportation delay and provide more precise mass feed flow manipulation.

The dead time from noncontinuous updates of digital devices (θ_d) is one-half of the update rate (Δt_{dx}) plus the latency (Δt_{dz}), which is the time from the start of a digital update time interval to when a result is communicated as an output. Wireless PID inputs have a variable latency. The published periodic messages from the device travel through the mesh from the source to the destination. The various routes a packet may take through the mesh are not all the same in hops. There are also retries. In one study for a wireless update rate of 4 s, the signal took about 0.4 s to arrive at the PID. Many signals arrived in 10 ms (one time slot), with a few taking up to 4 s to arrive, making the average latency about 0.4 s. The latency for a PID module can be considerable if complex calculations are done in the module which delays the output signal update.

$$\theta_d = 0.5 \bullet \Delta t_{dx} + \Delta t_{dz} \quad (8\text{-}10ll)$$

Writes from the control system to a wireless actuator (valve) were designed in the standard to also be periodic (windowed or change of state). However, the most common method in use does not support this. So, an output is sent as a request/response. Studies reported these downstream messages taking an average of 1 to 4 s. There were a few changes that took up to 15 s.

> **Key CONCEPT** The dead time from digital devices is one-half of the update rate plus the latency that is usually negligible, except for complex calculations or wireless routes.

The latency for an analyzer is the cycle time because the analysis result is available at the end of the cycle time. Thus, the dead time from the analyzer is one and one-half times the analyzer cycle time (Δt_{ax}). The time the analyzer takes to take another sample from a given sample point is the multiplex time. The total analyzer dead time (θ_a;) is increased by one-half of the multiplex time (Δt_{az}).

$$\theta_a = 1.5 \bullet \Delta t_{ax} + 0.5 \bullet \Delta t_{az} \quad (8\text{-}10mm)$$

> **Key CONCEPT** The dead time from analyzers is 1 one-halftimes the cycle time plus one-half of the multiplex time.

The dead time from a measurement sensor resolution or threshold sensitivity limitation (θ_{mx}) can be approximated as a measurement sensitivity limitation percentage (S_m) divided by the rate of change of the process variable ($\Delta\%PV/\Delta t$). For fast and large disturbances, this dead time is negligible.

$$\theta_{mx} = \frac{S_m}{\Delta\%PV/\Delta t} \quad (8\text{-}10nn)$$

> **Key CONCEPT** The dead time from measurement resolution for sensitivity limitation is approximated as the limitation divided by the percentage rate of change of the PID process variable.

Similarly, the dead time from the control valve or VFD resolution limitation (θ_{vx}) can be approximated as a valve or VFD percentage resolution limitation (R_v) divided by the rate of change of the controller output ($\Delta\%CO/\Delta t$). The equation also applies to deadband from lost motion or deadband in VFD settings for reversing the signal direction. For large or fast changes, this dead time is negligible.

$$\theta_{vx} = \frac{R_v}{\Delta\%CO/\Delta t} \tag{8-10oo}$$

Key CONCEPT

The dead time from a valve or VFD resolution and lost motion deadband is the resolution or deadband divided by the percentage rate of change of the PID controller output.

The time lags from composition and thermowell measurements can be quite large for low velocities and fouling. *p***H** electrodes are especially sensitive to fouling and aging, where a nearly imperceptible coating on or aging of the glass electrode can increase the electrode time constant from 3 to 300 s. A temperature measurement time constant can also dramatically increase because of a ceramic protection tube or a loose fit in a thermowell, or a sensor tip not touching the bottom of a thermowell. Unless there are abrasion concerns, a process velocity of about 5 ft/s is used for electrodes to minimize response time and fouling. A tight-fitting, spring-loaded stepped thermowell is used to minimize temperature response time. The thermowell insertion length is made large enough to minimize error from heat transfer in the thermowell wall from process to thermowell connection. An uninsulated connection or steam jacket can increase the emersion length requirement. An elbow connection offers greater immersion lengths in small-diameter pipelines and is generally installed with a tip pointing upstream for more uniformity.

To minimize noise and get a representative measurement, the sensor is located far enough downstream of an in-line equipment outlet to have a good concentration and velocity profile, with the tip in the center of a pipeline and not exposed to bubbles in liquids or droplets in vapors or gasses. In vessels, the sensor tips extend past baffles and are not near sparge lines or stream entry points.

Transmitter damping (time constant) settings and system signal filter time constant settings are minimized to be less than 10% of the total loop dead time and less than 10% of the reset time for tight control. If the measurement time constant becomes the largest time constant in the loop, the measurement amplitude (A_m) is an attenuated version of the process amplitude (A_p) that can be approximated by the following equation for a measurement damping or signal filter time constant (τ_m) larger than the oscillation period (t_o).

$$A_m = A_p \bullet \frac{t_o}{2 \bullet \pi \bullet \tau_m} \tag{8-10pp}$$

Key CONCEPT The amplitude of a *PV* oscillation seen by the PID is proportional to the ratio of the oscillation period to the measurement damping or signal filter time constant.

Key Diagnostics

A proper response of the control systems to the process variable statuses or control system failures should be considered, analyzed, and configured carefully. The analog output (AO) or digital output (DO) behavior should be properly addressed when there is a communication failure with the central processing unit (CPU) (possibly caused by a CPU failure, I/O backplane failure, or network failure with distributed I/O), SIS action when the condition contains a variable with bad status, and a loss of one controller that contains a process variable read by another controller (a typical case that arises based on process complexity, I/O density, and/or geographic dispersion). Here are some examples of key diagnostics with definitive explanations.

- Excessive time by a PID (e.g., more than 10 dead times for a constant set point) at the output limit is indicative of excessive fouling, excessive reset action, wrong valve action or PID action, the wrong manipulated variable, and/or a valve or pump-sizing problem.
- Excessive time by a PID not in the best mode is a symptom of a lack of operator training, bad diagnostics, poor tuning, and incorrect control strategy.
- Bad PV can be caused by incorrect instrument selection or installation and signal wiring problems.
- A faltering PV response in approaching the set point can be the result of a PID gain, a rate time setting that is too large, or the positive feedback in a hidden loop due to a negative relative gain.
- Limit cycles can be caused by resolution if there are one or more integrators and by a deadband if there are two or more integrators. Limit cycles can occur when the response time of the final control element becomes large compared to the normal loop response time for small magnitude reversals of the controller output. Limit cycles can also be caused by negative resistance, such as the buzzing associated with dips in compressor characteristic curves and steam jet curves and the extreme reversals of flow. Strong acid and strong base *p*H systems can develop something similar to a limit cycle from the oscillations through the steepest part near 7 *p*H to the flat parts of the titration curve at 2 and 12 *p*H.
- Power spectrum peaks are indicative of oscillating loops that could be increasing PID loop variability if loops are upstream or are associated with utilities,

recycle streams, or heat integration. If the peaks are larger when the PID is in automatic, resonance is occurring.

- An underdamped oscillation period relative to dead time offers possible clues. An oscillation period of about 3 times the dead time is indicative of a PID rate time that is too large. An oscillation period of about 4 times the dead time is indicative of a PID gain that is too large. An oscillation period of about 10 times the dead time is indicative of a PID reset time that is too small. An oscillation period of about 40 times the dead time for a near or true integrating process or runaway process is indicative of the product of PID gain and a reset time that is too small.
- Oscillations in manual are indicative of measurement noise and oscillating loops affecting PV.
- A slow valve response for small changes (e.g., <0.25%) can be caused by poor positioner or actuator sensitivity, lost motion (deadband), and resolution. For large changes (e.g., >25%) the slow response can be caused by insufficient positioner and air regulator flow capacity relative to the actuator size. For pneumatically actuated valves, high gain positioners and diaphragm actuators on sliding stem valves offer the best response to small signal changes. If the response to large signal changes must be faster, volume boosters on the positioner output with the bypass valve slightly open, larger air regulators, and larger tubing and actuator connections are used. Substituting a booster for a positioner on a diaphragm-actuated butterfly valve can cause positive feedback and erratic action. Rack-pinion, scotch yoke, and link-arm piston actuators and spool positioners have the least sensitivity, resolution, and deadband.

8.11 PID Tuning [1]

The incredible range and impact of the size of the primary time constant relative to the dead time and the location of the disturbance have led to considerable disagreement. Switching to integrating process tuning rules when the primary time constant is much larger than the dead time and the realization that most disturbances occur as a process input as a load upset result in a convergence of rules when the objective is minimizing integrated error. Often the difference in factor settings quibbled about is less than the estimation error and variability in the dynamic terms (gains, dead time, and time constants). The claims by authors of the more than 100 tuning methods are often more a result of gamesmanship than reality. Some conclusions are based on tests where the dynamics are changed to make the control less stable. The most common test is to show how the Ziegler-Nichols method causes excessive oscillation for a 25%

increase in process gain. The solution could have been to reduce the PID gain by 25%. The other test is to show excessive overshoot for a set-point change. The solution could be to simply add a set-point lead-lag with the lag time equal to the reset time and the lead time equal to 20% of the lag time or to use a two DOF structure to eliminate the overshoot and still get to the set point quickly. Overshoot is a greater problem for integrating and runaway processes because of the lack of process self-regulation.

The Ziegler-Nichols method and most other tuning methods in use before the 1980s were developed to minimize the peak and integrated error for an unmeasured step disturbance on the process input (load upset). Although this aggressive action is important in certain applications to prevent activating relief systems or initiating a runaway condition, the robustness is generally insufficient and will not deal with practical problems and other objectives. More recent tuning methods, such as lambda tuning, focus on adding robustness, thus minimizing the effect of nonlinearities, interactions, and resonance.

Key CONCEPT

PID tuning is an emotional issue. Many different tuning methods are touted, often without proponents realizing that most methods can account for the inherent trade-off between robustness and performance by simply adjusting a factor and that the tuning should first be done for a load disturbance tested by a PID output change for a PID momentarily in manual and the set-point response obtained by using a set-point lead-lag or DOF PID structure.

Using lambda rather than lambda factors offers many important advantages. Consultants who use lambda tuning for industrial processes do not enter a lambda factor but an actual lambda that includes the total dead time, which is a closed-loop time constant for self-regulating processes and an arrest time for integrating processes. Lambda typically ranges from one to three dead times to meet application requirements. Limits are imposed on the reset and rate times to help maximize load rejection. Finally, processes with a time constant to dead time ratio greater than four are treated as near-integrating processes, and lambda tuning rules for integrating processes are used.

All the equations assume the maximum open-loop self-regulating process gain or maximum open-loop integrating process gain, minimum primary time constant, and maximum dead time and secondary time constant were evaluated at different operating conditions. The worst-case condition is used that results in the smallest gain and rate time and largest reset time. At low production rates, the dead time and open-loop gain are often the largest for composition and temperature control of well-mixed liquid volumes. If these dynamic terms vary, adaptive tuning or scheduling tuning settings is needed.

> **Key CONCEPT** PID tuning should be done for worst-case operating conditions, which for composition, temperature, and *p***H** control are often at the lowest feed rate.

Benefits of Signal Characterization [6]

The obvious benefit of signal characterization is to create a more linear response of the PID PV to the controller output (CO). There are additional benefits in terms of better identification of dynamics from step tests and the moderation of the effect of other nonlinearities seen by the PID controller, such as increases in dead time from resolution and lost motion.

The improvement in dynamic identification is the result of using the transformed value to compensate for the effect of local slope changes on the open-loop process gain (change in the PV percentage divided by the step change in the CO percentage) for different step sizes. For small step changes, the open-loop self-regulating process gain approaches the slope of the plot of PV versus CO but becomes quite different for larger steps spanning various changes in slope.

An example of extraordinary changes in slope is a *p***H** system with a strong acid and strong base. Process output signal characterization is applied to both the PV and the SP, transforming the controlled variable from *p***H** (titration curve ordinate) to reagent demand (titration curve abscissa). The abscissa for signal characterization is typically converted from the ratio of reagent to influent flow to a simplified 0% to 100% reagent demand scale for the controlled variable, where 0% corresponds to the minimum *p***H** and 100% is the maximum *p***H** in the system.

For *p***H** systems, the change in process gain with step size can be an order of magnitude or more, as shown in Figure 8-11a, where the full change in *p***H** is denoted E100 and the 63% change in *p***H** is denoted E63 on the abscissa. The time to the 63% change in *p***H** would usually be used to estimate the time constant. However, the significant change in slope will distort the determination of the time constant unless the nonlinearity is accounted for. Note that the E63 abscissa change is not 63% of the E100 abscissa change as it would be if the steady-state relationship were linear between the *p***H** and reagent ratio. The change in the reagent-flow-to-influent-flow ratio for E63 is extremely small due to the steep slope and is consequently reached much faster than the true time constant determined by the material balance seen in the reagent-flow-to-influent-flow

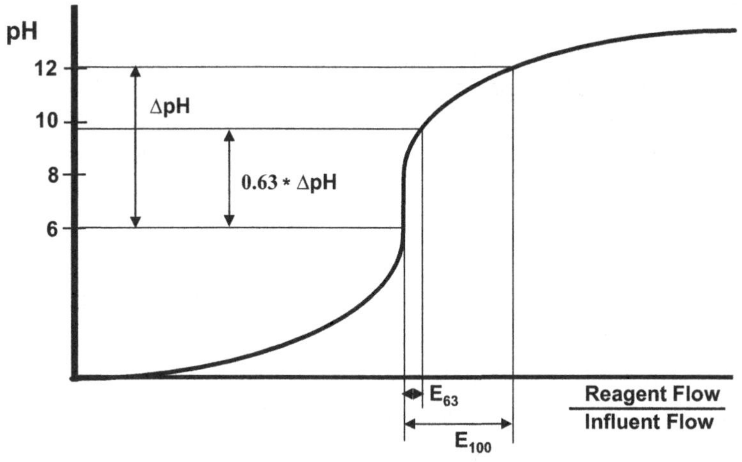

Figure 8-11a. Translating open-loop errors to identify the time constant seen by the *p*H loop.

ratio. Figure 8-11a is based on a strong acid and a strong base, with a conjugate salt that provides considerable moderation of the maximum slope.

Without signal characterization, the observed time constant of *p*H response for a well-mixed vessel and a strong acid and strong base without a conjugate salt could be incorrectly observed to be 0.04 min for a 5 to 12 *p*H change where the estimated actual time constant from the material balance is 20 min. Signal characterization could potentially restore the observed time constant for the controlled variable (percent reagent demand) to about 20 min. Operators are generally more reassured by the changes in the new controlled variable in the neutral region from upsets, resolution, and noise being much smaller due to transformation and from the resulting better control, often eliminating oscillations from linearization. It is recommended that the actual *p*H value be visible to the operator and available for historization and maintenance. Also, a *p*H set point entered by an operator should be automatically converted to the corresponding value of the transformed controlled variable (percentage of reagent demand).

The reduction in dead time by output signal characterization from resolution and lost motion stems from not having to tune the PID with a smaller gain for the steeper portions of the installed flow characteristic for valves and VFDs. A higher PID gain translates to a greater CO rate of change for a given disturbance that reduces the time to get through the resolution and lost motion deadband. The amplitude of a limit cycle from a resolution limitation is proportional to the open-loop process gain that is the product of the valve gain, process gain, and measurement gain. Consequently, the amplitude from a resolution limitation will be reduced in the regions where the process gain was high. This is particularly important for *p*H control involving steep titration curves.

> **Key Concept** Signal characterization improves the linearity and consequently can help improve the identification of the process dynamics and reduce the dead time from less-aggressive tuning settings that increase the amount of time required to get through resolution and sensitivity limitations and lost motion deadband.

In addition, there are operational and maintenance benefits. The reduction in variability boosts engineering and operations' confidence and appreciation of signal characterization. For better operator understanding and maintainability, the characterization is best done in the main control system rather than in field devices and with the original uncharacterized signals clearly visible. The signal characterization must be applied to the *p***H** loop set point in addition to the PV.

> **Key Concept** Signal characterization is appreciated by operators and process engineers when they see tighter and smoother control from reduced oscillations and more gradual changes of the PID's PV, but the signal characterization must be applied to the set point in addition to the PV and operations and maintenance must be able to easily see the *p***H**.

Nomenclature for Tuning Calculations

AR_{-180} = amplitude ratio at –180 degrees phase shift (dimensionless)

a = gain factor in traditional open-loop methods (0.4 to 1.0)

b = reset time factor in traditional open-loop methods (0.5 to 4.0)

c = rate time factor in traditional open-loop methods (0 to 1.0)

K_c = controller gain (dimensionless)

K_i = open-loop integrating process gain (1/s)

K_m = measurement gain (%/m for level)

K_p = process gain (m/kg for level)

K_v = valve or variable speed drive gain (kg/s/%)

K_o = open-loop self-regulating process (steady-state) gain (%/%) (dimensionless)

K_u = controller ultimate gain (dimensionless)

L = Ziegler-Nichols lag graphically estimated as an intersection with the original PV of a tangent to the inflection point of the PV open-loop response (s)

R = Ziegler-Nichols ramp rate sensitivity graphically estimated as the slope of the tangent to the inflection point of the PV open-loop response (% per s per %)

T_i = controller integral time (reset time) (s)
T_d = controller derivative time (rate time) (s)
T_q = quarter amplitude period (s)
T_u = ultimate period (s)
$\%CO$ = operating point controller output (%)
$\%SP$ = operating point the set point (%)
γ_f = gamma factor for internal model control (IMC) closed-loop time constant or arrest time (dimensionless)
γ = gamma for IMC closed-loop time constant or arrest time (s)
λ_f = lambda factor for closed-loop time constant or arrest time (dimensionless)
λ = lambda for closed-loop time constant or arrest time (s)
θ_o = total loop dead time (s)
τ_o = primary time constant (open-loop time constant—the largest time constant in the loop) (s)
τ_s = secondary time constant (second largest time constant in the loop) (s)
ω_n = natural frequency (critical frequency) (radians/s)
ρ = liquid density in surge tank (kg/m^3)

Key CONCEPT

Instead of getting into the emotional issues relating to the merits of tuning methods, we conclude this section with emphasis on the need to accurately identify the open-loop response for operating conditions, switch to integrating tuning rules for lag-dominant processes, and adjust tuning factors realizing the inherent trade-off between robustness and performance and the need to meet peak and integrated error requirements for load upsets to the respective loop without causing upsets, interactions, or resonance in other loops that cannot be mitigated by using ERF or decoupling.

Lambda Tuning for Self-Regulating Processes

The authors of this book advocate using the lambda tuning method for self-regulating processes when the open-loop time constant is less than four times the total loop dead time. For self-regulating processes that are not near-integrating ($\tau_o < 4 \bullet \theta_o$), we have the following series of equations.

The reset time is set equal to the open-loop time constant. The reset time steadily decreases as the ratio of the time constant to dead time decreases from its maximum of four dead times if this tuning method is used when the open-loop time constant is less than four times the total loop dead time. For systems with an open-loop time constant that is much less than the dead time (dead time dominant), the PID action becomes essentially integral-only due to a steady decrease in the controller gain and the reset time. This type of control helps deal with the abrupt near-step changes and noise in a dead time dominant system. Thus, the PID controller takes over the role of providing a gradual smooth response that is missing in the process. The best scenario is to have a large process time constant to fulfill this role, but for plug flow and sheet line processes, this is not possible.

$$T_i = \tau_o \tag{8-11a}$$

A low limit is added to traditional equations to prevent a reset time smaller than one-fourth the dead time for loops where the time constant is very small to provide some gain action.

$$T_i = Max\left[0.4 \bullet \theta_o , \tau_o\right] \tag{8-11b}$$

In industrial applications, the lambda factor multiplication of the open-loop time constant is replaced with lambda, the closed-loop time constant ($\lambda = \lambda_f \bullet \tau_o$), which is the time to reach 63% of a set-point change. Using lambda rather than the lambda factor provides the recognition that lambda is set as a multiple of the dead time to minimize variability in the PID input or output and to provide the gain margin needed to deal with extenuating circumstances (e.g., interaction, inverse response, and resonance).

$$K_c = \frac{T_i}{K_o \bullet (\lambda + \theta_o)} \tag{8-11c}$$

For maximum unmeasured disturbance rejection, a lambda equal to the dead time is used:

$$K_c = 0.5 \bullet \frac{\tau_o}{K_o \bullet \theta_o} \tag{8-11d}$$

A lambda of three to five dead times minimizes the consequences of nonlinearities, inverse response, and resonance.

Usually, a rate setting is not included as part of the lambda tuning for self-regulating processes. If the primary time constant is greater than one-half the dead time, rate action may be beneficial. If a secondary time constant (next largest time constant) can be identified, the rate time is set equal to the secondary time constant. The rate time should be larger than one-half the dead time but not be greater than one-fourth the reset time for an ISA Standard Form.

If the primary time constant is greater than one-half the dead time ($\tau_o > 0.5 \bullet \theta_o$):

$$T_d = Min\ [0.25 \bullet T_i, Max\ (0.5 \bullet \theta_o, \tau_s)] \tag{8-11e}$$

Lambda Tuning for Integrating Processes

The lambda tuning method for integrating processes is advocated by the authors for use on self-regulating processes when the open-loop time constant is greater than four times the total loop dead time (near-integrating processes) and for use on true integrating and runaway processes to provide a faster return to the set point for load upsets.

For PID control, the reset time is twice the arrest time plus the dead time.

$$T_i = 2 \bullet \lambda + \theta_o \tag{8-11f}$$

For PI control, a low limit is added so the reset time is not less than four dead times.

$$T_i = Max\ [4 \bullet \theta_o, 2 \bullet \lambda + \theta_o] \tag{8-11g}$$

$$K_c = \frac{T_i}{K_i \bullet \left(\frac{\lambda_f}{K_i} + \theta_o\right)^2} \tag{8-11h}$$

In industrial applications, the lambda factor multiplication of the inverse of the integrating process gain is replaced with lambda, the closed-loop arrest time ($\lambda = \lambda_f / K_i$), which is the time to stop an excursion for an unmeasured disturbance (time to peak error). Using lambda rather than the lambda factor provides the recognition that lambda is set as a multiple of the dead time to minimize variability in either the PID input or output and to provide the gain margin needed to deal with extenuating circumstances (e.g., interaction, inverse response, and resonance) or is set to provide the maximum absorption of variability.

The numerator uses the unlimited reset time.

$$K_c = \frac{2 \bullet \lambda + \theta_o}{K_i \bullet (\lambda + \theta_o)^2} \tag{8-11i}$$

For maximum unmeasured disturbance rejection, a lambda equal to the dead time is used:

$$K_c = 0.75 \bullet \frac{1}{K_i \bullet \theta_o} \tag{8-11j}$$

A lambda of three to five dead times minimizes the consequences of nonlinearities, inverse response, and resonance.

The rate time should be greater than one-half the dead time but less than one-fourth the reset time.

$$T_d = Min\ \left[0.25 \bullet T_i, Max\ (0.5 \bullet \theta_o, \tau_s)\right] \tag{8-11k}$$

Internal Model Control Tuning for Self-Regulating Processes

The IMC method was developed to provide a controller that is the inverse of the process dynamics. Similar to lambda tuning, IMC tuning uses pole-zero cancellation theory. The IMC tuning rules vary by author, dead time approximation, and vintage. Many authors have offered improvements for different dynamics, particularly lag-dominant processes, to better handle load disturbances.

For self-regulating processes:

$$K_c = \frac{T_i}{K_o \bullet (\gamma_f \bullet \tau_o + 0.5 \bullet \theta_o)} \tag{8-11l}$$

In industrial applications, the gamma factor multiplication of the open-loop time constant is replaced with gamma, the closed time constant ($\gamma = \gamma_f \bullet \tau_o$), for a set-point change.

$$K_c = \frac{T_i}{K_o \bullet (\gamma + 0.5 \bullet \theta_o)} \tag{8-11m}$$

The reset time is set equal to the open-loop time constant plus one-half the total loop dead time. By including the dead time, IMC tuning prevents the PID from using incredibly small reset times and small gains when the time constant is much less than the dead time ($\tau_o \ll \theta_o$). Thus, the IMC tuning behaves more like the modified Ziegler-Nichols tuning for dead time dominance. However, the reset time is even larger than optimum for systems where the time constant is much greater than the dead time

($\tau_o \gg \theta_o$), resulting in an even slower recovery and, consequently, much larger integrated error from unmeasured disturbances. The user can take the same approach recommended for lambda tuning, which is to use a near-integrator tuning when the time constant becomes four times larger than the dead time ($\tau_o > 4 \bullet \theta_o$).

$$T_i = \tau_o + 0.5 \bullet \theta_o \tag{8-11n}$$

For maximum unmeasured disturbance rejection, a gamma equal to the dead time is used:

$$K_c = 0.7 \bullet \frac{\tau_o}{K_o \bullet \theta_o} \tag{8-11o}$$

$$T_i = Max\ (2 \bullet \Delta t_x, \tau_o + 0.5 \bullet \theta_o) \tag{8-11p}$$

The IMC rate term is quite different in being proportional to the product of the open-loop time constant and total loop dead time. The rate time should not be greater than one-fourth the reset time for an ISA Standard Form.

$$T_d = Min\left(0.25 \bullet T_i, \frac{\tau_o \bullet \theta_o}{2 \bullet \tau_o + \theta_o}\right) \tag{8-11q}$$

Internal Model Control Tuning for Integrating Processes

The IMC tuning method is similar to the lambda tuning method for integrating processes except for using one-half of the dead time instead of the whole dead time in the denominator to calculate the controller gain.

In industrial applications, the gamma factor multiplication of the inverse of the integrating process gain is replaced with gamma, the closed-loop arrest time ($\gamma = \gamma_f / K_i$) for an unmeasured disturbance.

$$K_c = \frac{T_i}{K_i \bullet (\gamma + 0.5 \bullet \theta_o)^2} \tag{8-11r}$$

The reset time is twice the arrest time plus the dead time.

$$T_i = 2 \bullet \gamma + \theta_o \tag{8-11s}$$

For maximum unmeasured disturbance rejection, a gamma equal to the dead time is used:

$$K_c = 1.3 \bullet \frac{1}{K_i \bullet \theta_o} \tag{8-11t}$$

$$T_i = 3 \bullet \theta_o \tag{8-11u}$$

If a secondary time constant (next largest time constant) can be identified, derivative action can be used with the rate time set equal to the secondary time constant. The rate time should not be greater than one-fourth the reset time for an ISA Standard Form.

$$T_d = Min\ [0.25 \bullet T_i, \tau_s] \tag{8-11v}$$

Skogestad Internal Model Control Tuning for Self-Regulating Processes

Sigurd Skogestad recognized the problem with incredibly slow and fast integral action in IMC and lambda tuning for processes with extremely large and small primary time constants, respectively, because the reset time was set proportional to the time constant. The Skogestad Internal Model Control (SIMC) rules prevent the reset time from becoming exceptionally large or small and end up with about the same controller gain as lambda tuning for self-regulating processes. Here we will use gamma (γ) employed by IMC tuning instead of the closed-loop time constant (τ_c) used in the paper "Simple Analytic Rules for Model Reduction and PID Controller Tuning" [11] The rules presented here are for PID control of a process with dead time and a large open-loop time constant (τ_o) and a small secondary time constant (τ_s).

$$K_c = \frac{T_i}{K_o \bullet (\gamma + \theta_o)} \tag{8-11w}$$

$$T_i = Min\ [\tau_o, 4 \bullet (\gamma + \theta_o)] \tag{8-11x}$$

$$T_d = Min\ [0.25 \bullet T_i, \tau_s] \tag{8-11y}$$

SIMC Tuning for Integrating Processes

For integrating processes, the SIMC reset time is about 270% larger than the reset time per the lambda and IMC tuning methods for maximum disturbance rejection where gamma is set equal to the dead time. The SIMC controller gain is about 30% smaller for this case. The product of the SIMC gain and reset time is about twice as large as that determined by lambda tuning, giving more margin to prevent slow rolling oscillations.

$$K_c = \frac{1}{K_i \bullet (\gamma + \theta_o)} \quad (8.11z)$$

$$T_i = 4 \bullet (\gamma + \theta_o) \quad (8.11aa)$$

$$T_d = Min\ [0.25 \bullet T_i, \tau_s] \quad (8.11bb)$$

Traditional Open-Loop Tuning

The following equations are for an ISA Standard Form PID controller. The limit on the rate time ensures the rate time is not larger than one-fourth the reset time. The a, b, and c factors are decreased as the ratio of time constant to dead time decreases. The minimum numbers in the nomenclature definition are for a time constant much less than the dead time (dead time dominant). Without the limits on the controller gain and reset time, the controller becomes exceptionally slow for dead time dominant loops.

$$K_c = Max \left[0.2 \bullet \frac{1}{K_o}, a \bullet \left(\frac{\tau_o}{K_o \bullet \theta_o} \right) \right] \quad (8\text{-}11cc)$$

$$T_i = Max\ [0.4 \bullet \theta_o, b \bullet \theta_o] \quad (8\text{-}11dd)$$

$$T_d = Min\ [0.25 \bullet T_i, c \bullet \theta_o] \quad (8\text{-}11ee)$$

Modified Ziegler-Nichols Reaction Curve Tuning

The Ziegler-Nichols reaction curve method, unlike the ultimate oscillation method, is conducted with the controller in manual. A step change is made in the manual controller output ($\Delta\%CO$), and a maximum ramp rate per percent change in the controller output is estimated. The original paper shows the *R* parameter as being graphically estimated as the slope of a tangent line to the inflection point of a self-regulating process open-loop response that goes to completion. The *L* parameter is used to denote a lag that is estimated as the time from the controller output change to the intersection of the tangent with the original PV. The official definition of a *lag* is any phase lag that can be due to a dead time or a time constant. In most publications today, lag time is used interchangeably with time constant, whereas Ziegler-Nichols used lag time as a dead time. The method is modified to provide a smoother than quarter amplitude response and to add some robustness by applying a 0.5 factor to the Ziegler-Nichols gain.

Parameter R is really the integrating process gain that can be measured online by passing a new $\%PV$ through a dead time block whose parameter is the total loop dead time to create an old $\%PV$ that is subtracted from the new $\%PV$ to create a delta $\%PV$. The maximum delta $\%PV$ divided by the dead time (θ_o) and the change in controller output is the maximum ramp rate per percent change in controller output.

$$K_c = 0.5 \bullet \frac{1}{R \bullet L} \tag{8-11ff}$$

$$T_i = 3 \bullet L \tag{8-11gg}$$

The maximum ramp rate per percent change in controller output is the integrating process gain.

$$R = \frac{Max\left(\Delta\%PV / \Delta t\right)}{\Delta\%CO} = K_i \tag{8-11hh}$$

The L parameter is the observed loop dead time.

$$L = \theta_o \tag{8-11ii}$$

If we substitute the equations for the definition of R and L parameters, we end up with the lambda tuning equations for integrating processes when lambda is set equal to the dead time.

$$K_c = 0.5 \bullet \frac{1}{K_i \bullet \theta_o} \tag{8-11jj}$$

$$T_i = 3 \bullet \theta_o \tag{8-11kk}$$

The integrating process gain for a near-integrating process is the open-loop gain divided by the open-loop time constant.

$$K_i = \frac{K_o}{\tau_o} \tag{8-11ll}$$

If we substitute Equation 8-11ll into Equation 1.11kk, we end up with Equation 1.11nn, which is lambda tuning Equation 8-11d for a self-regulating process for a 3:1 time constant to dead time ratio when lambda is set equal to the dead time. The reset time is the same as in Equation 1.11b because the time constant is equal to three times the dead time.

$$K_c = 0.5 \bullet \frac{\tau_o}{K_o \bullet \theta_o} \tag{8-11mm}$$

Modified Ziegler-Nichols Ultimate Oscillation Tuning

The Ziegler-Nichols ultimate oscillation tuning method is conducted with the controller in automatic. If the loop is lined out, the controller is momentarily put in manual, a step change is made in the controller output, and the controller is immediately returned to automatic. With the reset time at a maximum (more than 100 times greater than the dead time) and the rate time set to zero to give essentially a proportional-only controller, the controller gain is increased until there are equal amplitude oscillations. The controller gain that caused these oscillations is the ultimate gain, and the period of the oscillations is the ultimate period. Generally, this technique is too exciting in that the loop is on the borderline of instability where the oscillations could rapidly grow in magnitude. Consequently, the manual quarter amplitude oscillation method or the relay auto tuner is preferred to get the ultimate gain and period. Here we look at how the Ziegler-Nichols ultimate oscillation method gives about the same results as the much more practical Ziegler-Nichols reaction curve method.

Starting with the fundamental relationship that ultimate gain is the inverse of the product of the open-loop gain and amplitude ratio at –180 degrees phase shift:

$$K_u = \frac{1}{K_0 \bullet AR_{-180}} \tag{8-11nn}$$

For self-regulating, single, time constant processes, the amplitude ratio at –180 degrees is:

$$AR_{180} = \frac{1}{\sqrt{1+(\tau_o \bullet \omega_n)^2}} \tag{8-11oo}$$

$$K_u = \frac{\sqrt{1+(\tau_o \bullet \omega_n)^2}}{K_o} \tag{8-11pp}$$

Using the natural frequency relationship to the ultimate period $\omega_n = \dfrac{2 \bullet \pi}{T_u}$

$$K_u = \frac{\sqrt{1+(\tau_o \bullet 2 \bullet \pi / T_u)^2}}{K_o} \tag{8-11qq}$$

For a loop dominated by a large time constant ($\tau_o \gg \theta_o \Leftrightarrow T_u \ll \tau_o$), the ultimate gain equation simplifies to:

$$K_u = \frac{2 \bullet \pi \bullet \tau_o}{K_o \bullet T_u} \tag{8-11rr}$$

For an ultimate period being about four dead times ($T_u \cong 4 \bullet \theta_o$):

$$K_u = \frac{2 \bullet \pi \bullet \tau_o}{K_o \bullet 4 \bullet \theta_o} \tag{8-11ss}$$

If we use the Ziegler-Nichols ultimate oscillation equations for a PID controller, we end up with the Ziegler-Nichols reaction curve method tuning, except the reset factor is 2 instead of 3. The rate time for the ISA Standard Form is one-fourth the reset time. We can convert back and forth between a lag-dominant and near-integrator calculation for the controller gain using Equation 8-11ll to convert the dead time to time constant ratio to an integrating process gain. The method is modified to provide a smoother than quarter amplitude response and to add some robustness by applying a 0.5 factor to the Ziegler-Nichols gain, as shown in Equation 8-11tt.

$$K_c = 0.5 \bullet 0.6 \bullet K_u = 0.5 \bullet 0.6 \bullet \frac{2 \bullet \pi \bullet \tau_o}{K_o \bullet 4 \bullet \theta_o} = 0.5 \bullet \frac{\tau_o}{K_o \bullet \theta_o} \tag{8-11tt}$$

$$T_i = 0.5 \bullet T_u = 0.5 \bullet 4 \bullet \theta_o = 2 \bullet \theta_o \tag{8-11uu}$$

$$T_d = 0.125 \bullet T_u = 0.125 \bullet 4 \bullet \theta_o = 0.5 \bullet \theta_o \tag{8-11vv}$$

Even if the Ziegler-Nichols ultimate oscillation method is not used, knowing the ultimate gain offers knowledge of how close the loop is to instability. The gain margin is the ratio of the ultimate gain to the current PID gain. This gain margin is used extensively to deal with changes in open-loop gain as well as dead time and time constants.

8.12 External Reset Feedback [6]

General

ERF is an inherent capability from the positive feedback implementation of integral action where a simple filter is used instead of an integrator. The controller output or manipulated variable is the input to the filter, whose time constant is the reset time. The filter output is added to the contribution of the proportional mode. Laplace transform analysis shows it is equivalent to the integral action obtained with an integrator.

History

Pneumatic controllers utilized a bellows and restrictor equivalent to a filter to implement integral action through positive feedback. A simpler mechanical arrangement of the integral and derivative bellows favored implementing the Series Form of the PID algorithm. When electronic controllers appeared, most companies switched to using an integrator for integral action, which has carried over to digital implementations of today. A few realized the value of ERF in preventing the integral mode output from changing faster than the manipulated variable can respond and stayed with the positive feedback implementation; others switched back. One must be careful that the PID algorithm truly uses the positive feedback implementation when ERF is touted.

Key CONCEPT Nearly all suppliers were unaware of the loss of the potential power of ERF when they moved from pneumatic to electronic and digital controllers replacing the bellows and restrictor filter in the positive feedback implementation of the integral mode with the newly available signal integrator.

Standard Form with Structure Options

Most of the literature uses the Series Form to show a simple positive feedback implementation of integral action and consequential ERF. Today, ERF is available in Standard Form with two DOF structure options by using the set-point weight factors beta (β) for the proportional mode and gamma (γ) for the derivative mode and a bias when integral action is not used. The algorithm is consequentially more complex, as shown in the block diagram in Figure 8-12a.

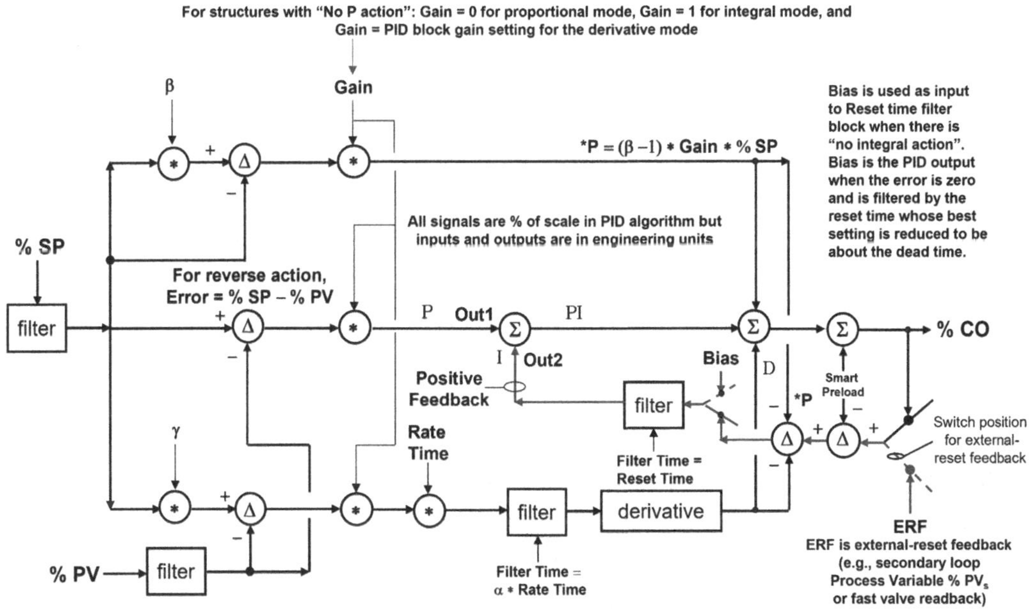

Figure 8-12a. ERF block diagram with structures for the Standard Form.

Feedback Source and Connection

ERF is activated in the controller by switching to use ERF from a signal representing the manipulated variable instead from the controller output. When valve position readback is used, it is important that readback updates are fast and precise. The feedback signal should have the same units as the PID output and ideally be linear with the PID output. If signal characterization or split-range is used, the manipulated variable signal is reversed back to its corresponding originating value for the feedback to correspond with the PID output that is the signal characterizer or split-range block input. Some blocks have this signal reversal capability built-in because it is needed for output initialization even if ERF is not used. If there are problems with a bump on download, the PID can be put in manual momentarily. ERF prevents the contribution of the integral mode from changing faster than the manipulated variable can respond but does not slow down the contribution of the proportional or derivative modes. Stopping the integral mode is enough to suppress oscillations. The feedback of the manipulated variable should be accurate and at least as fast as the execution rate of the PID. Wireless and HART auxiliary variable updates are generally not fast enough. The actuator position readback may not represent the closure member position, particularly for rotary valves, and flow measurement signals may have an insufficient signal-to-noise ratio and inadequate rangeability for accurate feedback.

> Key CONCEPT
>
> There is a general lack of recognition of the ease of ERF use, which involves simply turning it on in the PID controller and connecting a fast and accurate update of the actual manipulated variable (e.g., valve stroke, VFD speed, or secondary loop PV).

Slow Secondary Loop Response Time

By turning on ERF in the primary PID with the secondary loop PV as the feedback, oscillations are prevented from occurring when the secondary loop is too slow, violating the cascade rule. If the primary CO scale does not match the secondary PV scale, the ERF is converted to the corresponding value on the CO scale. Oscillations without ERF may only occur for large changes in PID output, leading to confusion and a false sense of not needing ERF or a faster secondary loop. The cascade rule may be obeyed in the original design but may be violated when sensors are fouled, greatly increasing their time constant. For *p***H** electrodes, the time constant can go from 3 s to 300 s due to an almost imperceptible coating. This can also occur from dehydration or aging of the glass measurement electrode that accelerates as the process temperature increases above 25°C or the concentration of strong acids or bases increases. It is important to improve the design, installation, and tuning of the secondary loop so it is as fast as possible and can reject secondary loop disturbances.

Secondary Loop Failure

If the secondary loop PV stops responding, possibly caused by the sensor (e.g., broken *p*H electrode) or a stuck valve, the PID output will freeze at the last PV response when ERF is used. Some control systems have the capability to bypass the secondary loop when the secondary PID is unavailable (e.g., with a bad PV). In this case, the primary loop writes directly to the secondary PID output, bypassing the secondary loop. The ERF source must be switched to what the secondary loop output is manipulating, and the primary loop tuning settings must be adjusted to match the associated new dynamics. Signal characterization of the secondary controller output can help the primary loop deal with the nonlinearities of the installed flow characteristic of the final control element. When the bad PV is caused by operating below the rangeability limit of the secondary loop (a common problem for flow loops), an inferential measurement of flow from the installed flow characteristic can be used. A smooth transition between inferential and actual flow measurement is possible by initializing the inferential measurement.

Slow Valve or VFD Response Time

Poor positioner sensitivity can greatly increase valve response times to 50 s or more for small changes in PID output (e.g., <0.25%). Large actuators can greatly increase stroking time and, therefore, response time for large changes in PID output (e.g., >25%). Slow rate limits added to the VFD drive setup can cause a slow VFD response time. Turning on ERF with a fast, accurate readback of the actual valve closure member position or VFD speed will prevent the PID output from integral action changing faster than the valve or VFD can respond and stop action for a stuck valve or frozen response. ERF can also deal with the set-point rate limits imposed on the VFD to reduce overdrive and loop interaction.

Valve or VFD Resolution

Using ERF with fast readback of actual valve stroke or VFD speed can stop limit cycles from valve resolution limitation due to friction and VFD resolution limitation from an insufficient number of bits in the I/O signal input card.

Valve or VFD Deadband

Using ERF with fast readback of actual valve stroke or VFD speed can stop oscillations from valve deadband due to lost motion in the valve actuator or linkages or VFD deadband in the drive setup that occurs when there are two or more integrators in the system.

Valve or VFD Failure

If the valve (e.g., stuck valve) or VFD stops responding, the PID output will freeze at the last PV response when ERF is used. When the valve or VFD starts to respond, the PID bumplessly resumes integral action.

Directional Move Suppression

Set-point rate limits for advanced control can be different for up and down directions and can be programmatically set to address operating conditions and objectives. Although slowing manipulated variable action generally is undesirable from the viewpoint of reducing a respective loop's performance, there are practical issues related to the need to slow down the propagation of disruptive flow changes and the prioritization of actions from a broader view extending to other loops and other objectives. Set-point rate limits can be used for optimum moderation of flow changes without the need for retuning by simply tuning on ERF. The set-point rate limits can be different for up and down directions providing directional move suppression. They are set faster or not even used in the direction for a required correction where a fast response is more important, particularly to prevent environmental violations, abnormal operation (e.g., compressor surge), or SIS activation. The set-point rate limit is slower for movement in the direction of gradual optimization or a direction that would adversely affect more important loops. The most disturbing disturbances are fast, which is why control loops optimally manipulating a flow cause less variability than a sequence, on-off control, or an operator changing a flow. A control loop is usually less disruptive than a manual or stepped change in flow during normal operation because the PID output change from one execution to another can be much smaller through PID tuning, 2DOF structure, and directional move suppression. Also, feedback control continually corrects for the unknowns and changes in the action required. For start-up, shutdown, and abnormal operation, procedure automation (state-based control) can be used to provide intelligent actions, but with the goal of being on feedback control when possible. Additional logic may be required to manage the PID output when its mode is switched to manual and the PID output does not match the rate-limited set point of the final control element or the secondary loop PID.

Optimization

Turning on ERF can simplify tuning a valve position controller which is often used to optimize a process. The typical strategy of using integral-only control with a slow reset time is sluggish and in danger of running out of the valve. Adaptive tuning can be used, but extremely low gains around the set point can violate the low gain limit for near-integrating, integrating, and runaway processes.

Set-point rate limits placed on a valve position controller output offer directional move suppression enabling a slow, gradual optimization and a fast getaway for abnormal operation to prevent running out of the valve (e.g., they prevent the valve position of the process loop being optimized from approaching the wide open position where the upper installed flow characteristic is the flattest or closed position and where resolution is poorest due to seating and sealing friction). The valve position controller can be tuned for the fastest response when ERF is used.

To maximize the reactor production rate, ERF helps facilitate using minimal set-point filters on the multiple reactant, secondary loop PID set points that are set to match the slowest secondary loop response. Set-point filters do not interfere with the secondary loop's ability to reject secondary disturbances. They do reduce the primary loop's ability to reject primary disturbances, but the production rate changes are often desired to be slow. The set-point filters help maintain a moment-to-moment stoichiometric balance of reactant feed which is better than the approach of detuning all the feed controllers to match the response of the slowest feed controller and thereby causing poorer secondary loop disturbance rejection in the faster loops. There are other solutions, such as using the totalization of the stoichiometric error as a controlled variable to correct a ratio of the reactants, but these are after the fact.

RCRA pH Limits

Set-point rate limits can be placed on a secondary PID, valve, or VFD to provide a gradual approach to the RCRA *p***H** limit that saves on reagent use and is a fast getaway from the RCRA limit for a large upset. The *p***H** controller can be tuned for the fastest response when ERF is used.

Maximizing Proportional Action

Set-point rate limits can be placed on a secondary PID, valve, or VFD to prevent excessive manipulated variable changes from a large PID gain setting without upsetting other loops or the operator when ERF is used. The secondary set-point rate limits would be set fast enough to not affect the ability to react to most disturbances. Also, operator training, decoupling, and a two DOF structure or filter on the primary PID set point may reduce the concern for a high gain, eliminating the need for a secondary rate limit. This enables a smaller peak error and integrated error that is inversely proportional to the PID gain and a faster set-point response that can be particularly important for near integrating, true integrating, and runaway processes and override control. The PID can be tuned for the fastest manipulated variable response when ERF is used.

Dead Time Compensation

Dead time compensation can be achieved by simply inserting a dead time block in the ERF path for the total loop dead time in the process or automation system that is accurately identified and updated. A small filter is also inserted on the PID output to smooth out oscillations from a dead time mismatch. The ERF block delays—by the block's dead time setting—further integral mode reaction to any change in the PID output. Unlike a Smith predictor, there is no need to identify and set a process gain or time constant or to restore the PV to the operator. A larger PID gain and an order of magnitude smaller reset time may be possible for an accurate dead time setting. Performance does not improve until the tuning is made more aggressive.

Intelligent Split-range Control

A rate limit block can be added to the PID output and set-point rate limits can be set to be slower when the output is near the split-range point. The approach is toward the split-range point, and the PID is tuned for the fastest response when ERF is used. This eliminates unnecessary split-range point crossings that tend to cause oscillations because of the discontinuity and the nonlinearity from the installed flow characteristic, the resolution being worse near the closed position, and changes in the manipulated variable. Preventing oscillations at the split-range improves process efficiency by stopping the cross-neutralization of reagents for *p*H and unnecessary heating and cooling for temperature control. In mammalian cell bioreactors, using ERF prevents the unnecessary use of sodium bicarbonate in *p*H control, helping to avoid increases in osmotic pressure and cell lysis. The ERF feedback can simply be the split-range block output.

Override Control

Without additional logic, using ERF will prevent the unselected PID from ramping off due to integral action and enables a better understanding of how proportional action will determine the PV at which the PID output will be selected.

> **Key Concept:** There is a general lack of understanding of the many ERF opportunities to prevent oscillations from slow valves, slow VFDs, slow secondary loops, split-range point discontinuities, resolution limitations, and lost motion deadband; to improve ratio control and override control, prevent RCRA *p*H violations, and enable optimization and dead time compensation; and to ride out valve and measurement failures.

8.13 System Selection

Because a *p*H loop is more sensitive than any other control system to the performance of its components and usually acceptable deficiencies can cause a *p*H loop not only to fail but also to fail miserably, let us quickly review the main best practices for the measurement, valve, equipment, and reagent.

The electrodes should provide fast and reproducible measurements. This is best achieved, where possible, by using a spherical bulb and a flowing reference junction.

The electrodes should be located at a representative point in the process with a minimum transportation delay and a velocity sufficient to keep the electrodes clean and responsive. When the electrode life expectancy is sufficient, select the middle value of three measurements (electrodes and estimators) to improve the measurement reliability and signal-to-noise ratio. If the exposure time must be limited, use automatically retractable probes.

The control valve resolution should be less than half of the allowable reagent error for the control band and meet the rangeability requirement for minimum and maximum reagent demand. Usually, this is best achieved by using a properly sized sliding stem (globe) valve with flow friction packing and a digital positioner.

The equipment should minimize the magnitude of the process dead time from a plug flow volume. This can be realized by maximizing the back mixing in a vessel or minimizing the length of an in-line system and avoiding putting loops on horizontal tanks and sumps.

The reagent delivery system should minimize the transportation delay, flatten the titration curve, and provide a high turbulent reagent flow free of solids or gas. Practically, this is achieved by prediluting the reagent, selecting a weak acid or base with a dissociation constant close to the set point, keeping reagent systems full of reagent, and premixing the reagent into a high feed or recirculation stream close to the point of entry into the vessel. Subsequent to accurately identifying the titration curve, signal characterization and linear reagent demand control can be used to further increase performance by greater linearity.

> **Key CONCEPT**
>
> To make the most out of the extraordinary rangeability and sensitivity of *p***H** control, the dead time from equipment, piping, and the automation system must be minimized; the precision of reagent valves and VFDs must be maximized; linearity must be maximized by using weak acids and weak bases and signal characterization; and *p***H** electrode reliability and repeatability must be maximized.

For control system selection, the main questions pertain to the number of stages, type and size of equipment, location of the set points, and the most appropriate advanced control technology. The question of the number of stages and type of equipment mostly applies to neutralization systems. For *p***H** control of unit operations for production such as fermentation, reaction, and crystallization, the process equipment is predetermined.

The traditional rule has been that one stage of neutralization is required for every 2 *p*H units the untreated influent stream *p*H is away from the set point. For example, to go from an influent stream at 1 *p*H to a set point of 7 *p*H requires three stages of neutralization, shown in Figure 8-13a. Typically not stated is the critical requirement that the reagent injection not be done via a dip tube but instead in a recirculation line to prevent a huge reagent injection (detailed in Chapter 7). A lower-cost alternative for a neutralization stage is to control the recirculation line *p*H (shown in Figure 8-13b) using an existing volume, eliminating the appreciable cost of designing, purchasing, and installing an agitated, well-mixed vessel.

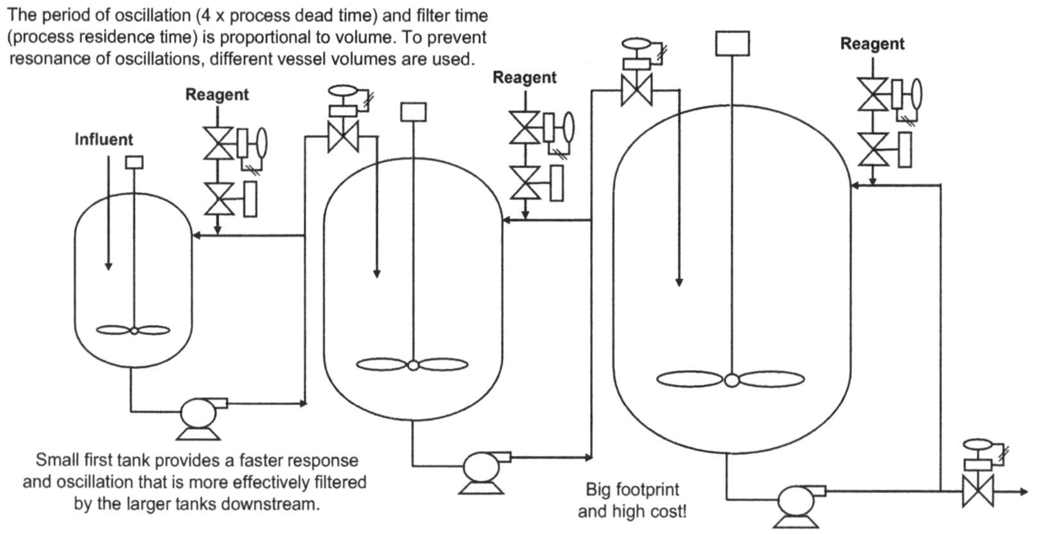

Figure 8-13a. Three-stage vessel neutralization system for difficult *p*H applications where the influent *p*H is 6 or more *p*H units from the control band.

A more exact rule would add another stage of neutralization if the control valve resolution was estimated to be more than half the allowable reagent error for the control band. This means that control valves with excellent resolution, minimal lost motion, and advanced control strategies that enable manipulating a trim and a coarse valve on the same stage could eliminate a stage of neutralization.

Key CONCEPT

The traditional rule is that one stage of neutralization is needed for every 2 *p*H units the influent is away from the final set point. Three stages are typically required for a final set point of 7 *p*H and influent significantly below 3 *p*H or above 11 *p*H, and the largest well-mixed volume usually is the last to provide the greatest filtering and best dead time to time constant ratio.

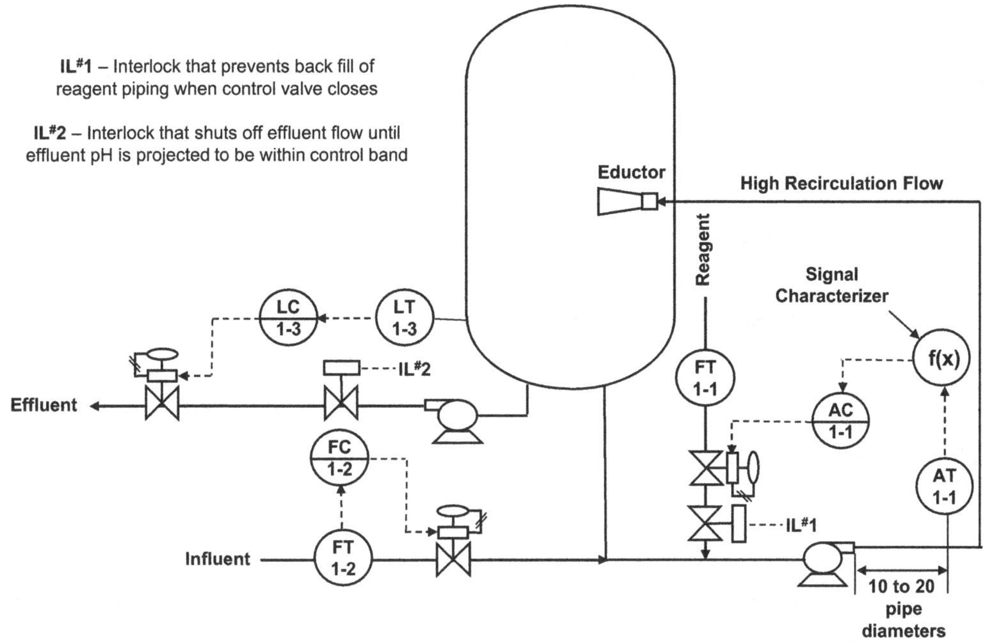

Figure 8-13b. Using recirculation flow *p*H control on an existing unagitated volume, eliminating the need to install a well-mixed vessel.

The traditional rule states that the volume of each vessel differs by a factor of 5 or more to reach the goal of the oscillation period for each vessel to differ by a factor of 5 or more. Not realized is that the oscillation period depends on dead time and tuning. The ultimate period ranges from about three to four dead times as the ratio of the open-loop time constant increases from 1 to 20. The actual period for a moderately aggressive tuning is about six dead times; therefore, to get a 5:1 change in oscillation period by using well-mixed tanks in series is challenging. The dead time from mixing is proportional to the volume divided by the sum of the agitator pumping rate and the recirculation line flow rate. Because the tightness of the control and the filtering action of short-term disturbances is proportional to the size of the process time constant, which is the residence time for a well-mixed vessel, the largest vessel is often used for the last stage. The first stage should have the fastest response, which corresponds to the least dead time, to rapidly start the correction for disturbances. An in-line *p***H** control system may serve this purpose. However, the in-line *p***H** set point may need to be optimized slowly to help compensate for the nonlinearity in the titration curve and the oscillations above and below the set point not being equal, resulting in an offset in the average *p***H** seen by the next stage.

> **Key CONCEPT** The oscillation period and, thus, the total loop dead time of each stage must differ by more than a factor of 5 to prevent resonance.

Key CONCEPT An in-line system (e.g., static mixer) with an optimal set point can replace a well-mixed vessel as a stage of neutralization if there is a downstream volume to attenuate oscillations.

If the influent rapidly changes back and forth between being acid and basic, a larger volume may average out the reagent demand, reducing reagent use. This is best done by a large volume upstream, which does not have *p*H control and does not need to be particularly well-mixed.

Key CONCEPT Any large upstream volume before the stages of *p*H control can help average out changes in influent reagent demand, particularly for rapid changes in acidic and basic influent.

Precise reagent control valves are essential in all *p*H stages. The coordinated use of small control valves for fine adjustment and large reagent control valves for coarse adjustment (discussed in Sections 7.6 and 8.8), combined with Coriolis mass flowmeters for flow ratio control and accurate signal characterization to give linear reagent demand control, can possibly eliminate one stage in a three-stage system. This possibility, along with the opportunity to use a fast in-line system instead of a vessel, can translate to $400,000 or more savings in project cost by eliminating the need for a well-mixed vessel and the associated piping, mechanical design, and installation.

Key CONCEPT Exceptionally precise control valves combined with feedforward or ratio control and accurate signal characterization can possibly eliminate one stage of required neutralization.

The set point should be located on the flattest part of the titration curve that keeps the *p*H within the control band. The oscillations will often be fewer, and the whole control problem will be reduced if the *p*H is shifted away from the neutral point toward the edge of the control band.

Using a dynamic process model in a digital twin is essential to identifying and detailing the *p*H control system with the best performance and lowest project cost. The process model should use a simplified charge balance with carbon dioxide and conjugate salt concentrations adjusted to match the laboratory titration curve in the observed range of operating conditions. The process model should use first-principle material and energy balances to get the effect of operating conditions on process composition and temperature. All sources of dead times and lags must be in the model—such as mixing delays, transportation delays, sensor lags, reagent valve and VFD resolution, lost motion, and 86% response time—and all automation system filters and update rates. The actual control system configuration must be used. The digital twin, historically called a *virtual plant* (described in Chapter 9, "Digital Twin"), is the best means of accomplishing the required fidelity and functionality of the model and control system and, consequently, the best plant design. Modeling in a digital twin is more critical for *p*H than for any other type of control loop because of the incredible nonlinearity, rangeability, and precision of *p*H systems. For examples of how the digital twin has been used to improve the *p*H control system and plant design and to promote and sustain innovation, see References 7, 8, 9, and 10.

> **Key CONCEPT** A dynamic model in a digital twin (described in Chapter 9) that accurately replicates the titration curve; includes all the equipment, piping, and automation system dynamics; and uses the actual control system configuration is essential to identify, develop, and test the best control strategies and tuning needed to minimize project cost and maximize process performance.

The least expensive equipment with the least total dead time is the in-line *p*H system. Even though it does not have the good dead time to time constant ratio of the well-mixed vessel, the integrated error and the amount of off-spec material are much less. The system is also easier to tune once it is realized that strong acid-base systems require a *p*H filter time of 0.2 min. Figure 8-13c shows that although the in-line system has a high peak error for a step upset, it recovers much faster than a well-mixed vessel does. The figure also shows how disastrous a horizontal vessel is for *p*H control. There is not enough room on the trend to show the ridiculously slow and severe oscillations from a sump.

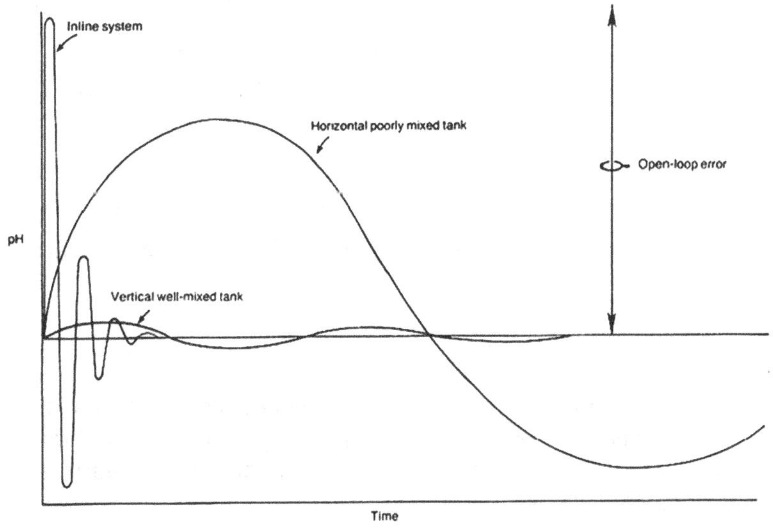

Figure 8-13c. Performance of an in-line system, well-mixed vessel, and horizontal tank.

Key CONCEPT

Exceptionally precise control valves combined with feedforward or ratio control and accurate signal characterization can possibly eliminate one stage of required neutralization.

The in-line system must have a volume downstream sufficient to attenuate the oscillations. Equation 5-3j can be used to estimate the volume size for the worst-case oscillations. The in-line system can take many forms. It can be a static mixer, centrifugal pump, or dynamic in-line mixer that has high-speed rotating blades. It can be located on a feed or recirculation stream and can be set up to identify the titration curve online. An in-line system should be used for the first stage rather than the final stage.

To reiterate the guidelines for multiple stages, the period of oscillations must be different by a factor of 5 or more to eliminate resonance. This means the total dead time from each stage and the volume for each well-mixed vessel should be different by more than a factor of 5, shown in Figure 8-13a. The smallest volume should be first, so this stage that bears the brunt of the assault creates the least amount of off-spec product. The biggest control valve should be on the first stage and the smallest on the last stage. The set point for the first stage should take the *p*H the furthest distance on the titration curve.

If magnetic flow or Coriolis meters providing higher accuracy, rangeability, and reliability are installed to measure the reagent and influent flows, cascade control of *p*H control to flow control, and flow feedforward or ratio control, online titration curve identification should be utilized to facilitate start-up and recovery from the flat portions of the titration curve and improve overall system performance and knowledge. If the titration curve is relatively fixed or can be calculated or automatically identified online, linear reagent demand control should be used to suppress oscillations and noise and to restore process time constants to their true value, and *p*H estimators should be developed to improve the reliability and speed of response of the *p*H measurement. If the titration curve changes shape, a temporary ramp to an in-line *p*H control system can be used to identify changes in the slope, and a model predictive controller can correct the dissociation constant and influent concentration in an online charge balance model [7].

Key CONCEPT

Magnetic flow and Coriolis meters can provide online identification and verification of the titration curve for different operating conditions and the best feedforward and ratio control, procedural automation, and linearization by signal characterization.

If there are benefits from maximizing waste or less-expensive reagent flows or simultaneously controlling a coarse and trim valve, consider using MPC. Where possible, the controlled variables should be linear reagent demand by accurate signal characterization. Finally, if the best set-point changes with process conditions, explore real-time optimization as an opportunity to home in on the set point for the greatest process efficiency.

> **Key CONCEPT** Consider MPC optimize linear reagent demand control and simultaneously position precise coarse and fine reagent valves.

8.14 Best Practices

1. Minimize dead time in the piping, equipment, valves or VFDs, measurement, and controller, realizing the total loop dead time is the sum of the pure dead times from mixing and transportation delays and digital components, equivalent dead times from sensor lags and signal filters, valve or VFD response time and lost motion, and resolution or sensitivity limitations.

2. Use digital twin and first-principle models with the charge balance adjusted to match lab titration curves for a range of operating conditions and the dynamics of equipment, piping, and automation system components included to identify the design that minimizes project and operating costs and prevents an unacceptable number or persistence of deviations outside the desired control region.

3. Consider an in-line system in the feed or the recirculation line to any existing tank to replace the need for a much more expensive well-mixed vessel.

4. Consider the opportunity for a combination of precise reagent valves, accurate signal characterization, and ratio control using Coriolis flowmeters to reduce the number of stages for neutralization by one.

5. Consider using procedure automation to deal with operational problems and for start-up and batch *p*H control. Include a future value block (detailed in Chapter 9) to predict a future *p*H by employing signal characterization to minimize disruption from failures and reduce start-up and batch time with minimal overshoot.

6. If possible, move the set point to the flatter part of the titration curve to reduce oscillations and consequential excessive use of reagent, particularly for the first stage.

7. Use an open-loop backup where the PID is momentarily put in output tracking with the future value block (detailed in Chapter 9) employed to provide a rapid, proactive, predictive correction that ensures preventing an RCRA *p***H** violation or SIS activation with a smooth transition back to feedback control.
8. Use ERF to stop oscillations from a slow secondary loop, slow valves or VFDs, stuck valves, or lost motion.
9. Use ERF and intelligent directional velocity limits to prevent upsets to other loops from high PID gain settings and unnecessary split-range point crossings and to provide gradual optimization with a fast getaway for abnormal conditions.
10. Use the digital twin model to test the configuration and to train operators and process, mechanical, maintenance, and automation professionals on continual performance and improvement and the ability to proactively respond to plant problems.
11. Use accurate signal characterization to enable the PID to remove most of the process nonlinearity in terms of change in *p***H** for a change in the ratio of reagent flow to feed flow and enable users and the PID to see the beneficial effect of the actual process time constant for vessels.
12. Tune for a feed disturbance with good robustness (e.g., lambda three to five times dead time) at low and high feed rates, realizing that the hidden gain factor in the open-loop process gain is inversely proportional to feed flow, then use set-point lead-lag or a two DOF structure to achieve the desired set-point response.

References

[1] Gregory K., McMillan, *Tuning and Control Loop Performance*, 4th ed. (New York: Momentum Press, 2015).

[2] Gregory K. McMillan, *Process Control Case Histories* (New York: Momentum Press, 2010).

[3] Gregory K. McMillan, Mark S. Sowell, and Peter W. Wojzsnis, "Adaptive Control—Shifting Your Process into High Gear," *Chemical Processing* (September 2004).

[4] Gregory K. McMillan, *Good Tuning: A Pocket Guide*, 4th ed. (Research Triangle Park, NC: ISA [International Society of Automation], 2015).

[5] Gregory K. McMillan and Robert A. Cameron, *Models Unleashed: Model Predictive Control and Virtual Plant Applications* (Research Triangle Park, NC: ISA [The Instrumentation, Systems, and Automation Society], 2003).

[6] ISA TR5.9 PID, *Algorithms and Performance* (Research Triangle Park, NC: ISA [International Society of Automation], 2023).

[7] Gregory K. McMillan and Mark S. Sowell, "Virtual Control of Real pH," *Control* (November 2007).

[8] Gregory K. McMillan, Roger D. Reedy, and John P. Moulis, "Virtual Plant Provides Real Insights," *Chemical Processing* (September 2008).

[9] Gregory K. McMillan and Martin E. Berutti, "Virtual Plant Virtuosity," *Control* (August 2017).

[10] Gregory K. McMillan, "Simulation Enhances Career and System Performance," *Control* (August 2020).

[11] Sigurd Skogestadt, "Simple analytic rules for model reduction and PID controller tuning," *Journal of Process Control 13,* (April 2003).

9 Digital Twin

9.1 Introduction [1]

The digital twin is a representation of the physical plant assets (e.g., process equipment, instrumentation, controls) and the processes that take place within them (e.g., chemical reactions, separation processes, heat transfer). The actual control system configuration, operator interface, and alarms are used and exchanged between the plant and digital twin. This eliminates the traditional time-intensive tasks of emulation and translation inherently prone to differences between plant and digital twin automation systems.

Although much of the emphasis in this book is on neutralization systems, good *p***H** control plays an important role in many unit operations that determine product quality in continuous and batch processes, particularly in the food, pharmaceutical, and specialty chemical industries. Most notable is the need for *p***H** control in biological and chemical reactors where prolonged deviations of just 0.1 *p***H** can be detrimental. In modern-day bioreactors for producing newly patented high-value-added biologics, *p***H** deviations of just 0.05 *p***H** can cause a decline in cell growth and product formation rates. There is also a great need in these bioreactors to prevent unnecessary split-range point crossings that cause excessive addition of the base reagent sodium bicarbonate, which greatly increases the accumulation of sodium ions. High sodium ion concentrations lead to high osmotic pressure in the cells that can trigger cell death from lysis (rupture of the cell membrane). There is also the complication of making sure that adding the acid reagent carbonic acid by carbon dioxide sparge does not interact with the air and oxygen sparge for dissolved oxygen control and that bubbles from the sparge do not reach the electrodes. Because of the critical importance of this unit operation for nearly all new pharmaceuticals, part of this chapter provides guidance on using digital twin models to provide better bioreactor *p***H** control which can save a batch that takes weeks to complete and is potentially worth $10 million or more.

The requirements for managing change are stricter than ever in the process industry. Validation requirements for pharmaceuticals involving extensive and intensive design, qualification, and verification documentation pose an even larger motivation to minimize change. The sheer complexity and effort involved limits the time for creativity and discourages doing anything that changes the process defined by research and development (R&D). The pharmaceutical industry's Process Analytical Technology (PAT) initiative seeks to address these concerns to promote innovation and continuous improvement.

Often the process is set based on laboratory benchtop and pilot plant runs. Even here, the opportunities are narrow because batch cycle times are extremely long, especially for new biologics (e.g., 10 days for mammalian cell bioreactors), and the time to market must be minimized to make the most out of patent protection for high-value-added products.

The key to addressing this reality is the digital twin. The actual configuration, including the operator interface and alarm systems, is the same as used in the lab, pilot plant, or production plant. The signals can use actual input/output (I/O) channels for communication by simulated input/outputs (SIOs) or a virtual input/output module (VIM). If necessary, an actual distributed control system (DCS) can be connected with a first-principle model of the process running with communications to the DCS via Open Platform Communications (OPC). Recent breakthroughs enable model parameters to be set so the batch profiles in the digital twin match those in the lab and plant.

Increasingly, R&D is realizing the value of the increased capability and flexibility of a small DCS to run batches automatically, greatly reducing cycle time and nonintentional variability. Transiting from the lab to pilot plant to the production plant control system plus digital twin is seamless and fast, greatly reducing the validation burden.

The key to the scientific method and innovation is experimentation. Presently, experimentation is largely limited to the early phases of R&D. It is increasingly recognized that the digital twin enables rapid and free experimentation with speedup factors greater than 500, so a 10-day batch can be completed and analyzed in 30 min.

Dynamic simulators became the key to the success of digital systems starting more than 30 years ago in the first migrations from analog control. These simulators simplify checking out and developing increasingly complicated displays, alarms, and control systems. Moreover, a working simulator created in lockstep with control development affords an efficient transition to operator training and provides future opportunities for experimentation to improve process efficiency. Large process manufacturers have realized the benefit of offline dynamic simulations to improve processes and control systems. However, these offline simulations required writing code to functionally duplicate controls, severely limiting the scope and reality of the plant being investigated.

Today, first-principle dynamic simulations for gaining process knowledge and improving product quality, process capacity, and efficiency can become one with the simulations for operator training and system checkout in the digital twin. Advancements in packaged commercial simulation software enable users to focus on the heart of the simulation rather than writing and rewriting code for the models.

Key CONCEPT First-principle dynamic simulation has become more accessible than ever and offers deep insight into process behavior and exciting avenues for building the understanding of operators and control engineers.

Problems with traditional integrated control and safety systems (ICSS) testing and commissioning are all too familiar. Plant operators commonly face delays in start-up and normal operation due to ICSS configuration and graphics issues, lack of trained operators, and last-minute changes required in control strategies and operating procedures. A digital twin solution with project milestones aligned to those of plant commissioning and start-up, especially in the case of greenfield projects, enables one to leverage the tremendous benefits of dynamic simulation during commissioning and start-up and across the plant lifecycle. It is important to adequately plan system maintenance considerations at the beginning of a digital twin project. The maintenance focuses on equalizing the offline control system and plant model with any significant changes in the real plant.

The digital twin is a relatively new concept that is easily confused with existing simulation methods for process design, configuration checkout, and operating training systems. Users may not realize that most of the existing batch process simulations are offline and noninteractive and that most real-time dynamic process simulations were originally designed for continuous processes. These real-time process simulations can develop severe numerical errors or even fail under the extreme conditions of batch operations and require interfaces for communicating I/O, controlling inventory, and coordinating with the control system in speeding up, slowing down, pausing, or resuming. The control system engineer is probably most familiar with tieback simulations because these have been predominantly used for configuration checkout and operating training systems. The process response in these tiebacks is mimicked by using trial and error to adjust ramp rates. Ramps are triggered by manually including flow path logic, including opening and closing valves or turning pumps on or off.

A properly maintained digital twin provides an offline, safe, and reliable system to support a variety of phases of the plant lifecycle (including conducting an ICSS factory acceptance test—FAT, enabling virtual commissioning, reviewing operating procedures, and training operators well ahead of plant checkout and start-up).

Key CONCEPT A true digital twin imports control modules and exact copies of operator interface files, providing a seamless experience for trainees and engineers.

Section 9.2 of this chapter discusses the key features that distinguish a digital twin from existing simulation systems and itemizes the important advantages of the digital twin approach. Section 9.3 provides a spectrum of uses for digital twins. Section 9.4 concludes with the basic requirements and issues concerning implementing a digital twin, plus an opportunity for nonintrusive adaptation of key model parameters.

Section 9.5 gives an overview of the data required. Section 9.6 introduces the concept of *p***H** modeling with the charge balance. The remaining sections dive into the details of modeling automation system components, metrics, and batch profile and endpoint control.

Appendix M has key examples of how dynamic models are set up for in-line and vessel *p***H** control systems in various combinations and arrangements for batch and continuous processes. The relative merits of an in-line system on the feed or in a recirculation line of a vessel, particularly to provide one stage of neutralization, can be readily prototyped and studied. The models include control strategies and blocks for modeling transportation and mixing delays, resolution limitations, backlash deadband, sensor lags, signal filters, analyzer cycle times, wireless update rates, tuning, performance metrics, and batch profile slope and endpoint prediction. There are links in Appendix M for users to download models and run them in Mimic Simulation Software.

Besides verifying actual titration curves, first-principle models can be improved by including all the sources of dead time and automation system dynamics. Software used to identify the open-loop dynamic response can be applied to both the actual plant and *p*H model to help the digital twin model match the dynamics in the plant. The model can generally be improved by adjusting the key parameters that help associated manipulated variables in the digital twin match those in the plant when loops are in automatic with the same tuning settings and set points. Even if model-predictive control (MPC) is not used for *p*H control, it can be employed, as noted in Section 9.4, to provide automated improvements in key model parameters.

9.2 Key Features [1]

The first key feature that distinguishes a digital twin from process simulators is its ability to use the configuration, historian, displays, and advanced control toolset of the real plant without translation, emulation, special interfaces, or custom modifications. The configuration database from the real plant can be exported and then imported and downloaded into a personal computer or a control system computer as if it were an actual hardware controller. Also, files for operator graphics, process history charts, and data history from the real plant can be copied to the digital twin so the user has the entire control system of the real plant on a computer, as shown in Figure 9-2a (the intent is to convey knowledge paths and not the actual process being simulated).

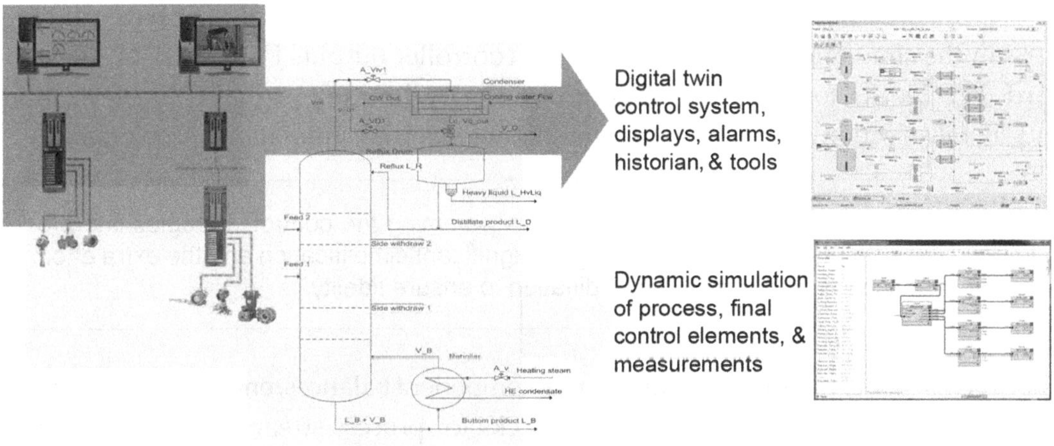

Figure 9-2a. Digital twin with actual DCS and dynamic simulation of process and instrumentation.

> **Key CONCEPT** Using the actual configuration, database, displays, historian, and advanced control toolset without any translation, emulation, special interfaces, or custom modification enables an inherent replication of the real control system.

Most dynamic, high-fidelity process simulation software offers the ability to build a basic control strategy or sequence inside the simulation environment. However, simulation developers tend to have a process rather than a control background and focus. It is unrealistic to expect the process and batch control capability offered in simulation software to be in the same realm as the control capability of DCS software, which is the culmination of 100 engineering years (i.e., 5 years by 20 engineers) or more of an effort by process control experts. The overall control functionality in process simulators is primitive compared to the capabilities offered in the modern DCS. The DCS offers capabilities such as Sequential Function Charts and basic function blocks, the batch manager, and advanced control tools like multivariable MPC. These are almost nonexistent in simulators. The effort to duplicate even a simplified version of a control system in a dynamic simulation is great. At best, the user ends up with two control systems (one actual and one simulated) with no assurance that they will match well and no method for automatically managing changes between them.

Consequently, most simulation software now offers a standard or custom OPC interface. However, for the simulations with material balances to be able to run independently of the DCS for development and testing, the simulations must still have internal pressure and level loops set up to prevent volumes from running dry, overflowing, overpressurizing, or developing flow reversals from pressure gradient reversals, which can lead to fatal numerical errors. Tables must be mapped that transfer control from these internal simulation loops to the DCS loops and initialize the proper controlled variables, set points, and manipulated variables. DCS loops that do not have counterparts in the simulation must still have their controller outputs initialized. Using standard DCS blocks for split-ranged control, velocity limiting, signal characterization, and signal selection makes properly initializing external simulations problematic.

Key CONCEPT

The digital twin eliminates custom programming the control strategies into process simulations and eliminates the significant simplification and the extra effort, verification, revision, and coordination to ensure fidelity.

The process model has the material balance, component balances, energy balance, charge balance, and kinetics. The user can drag blocks for process streams, mass flowmeters, pumps, fans, transmitters, control valves, and actuators into the model and then wire them to the bioreactor model. The process information is automatically conveyed from block to block through the "wires" that function as process and signal paths. The flow in each process path is automatically computed, as described in Chapter 6, "Control Valves," from pressures set for streams entering or exiting the model, bioreactor pressure, and from intervening pump or fan speeds, valve positions, and pressure drops from flow resistance. It is unnecessary to duplicate level and pressure loops for inventory control and sequences for batch operation or to add interface tables for communicating I/O.

Key CONCEPT The graphical configuration of process models linked to the control modules eliminates the need to duplicate inventory and batch control or to add interface tables.

Implementing the process simulation as process models in the digital twin offers advantages beyond eliminating the previously noted issues of duplicating loops and configuring interface tables. Foremost is the digital twin's inherent ability to run the process models at the same real-time execution time multiplier as the control modules. This ensures that the user can set a common real-time multiplier and that the simulation and control system will slow down or speed up in unison. The user can set a real-time execution multiplier of 1/30 to 30 for all modules in the digital twin. When external high-fidelity simulation packages are used, the actual real-time factor may depend on processor loading. This is because of the complexity of the calculations associated with the objectives of the process or equipment design. Often during the times of greatest interest, during disturbances or failures, these simulations slow down because the integration step size has decreased for numerical stability and control system and operator activity have increased. Even if the hook exists between the simulation and DCS real-time multipliers, the DCS is always playing catch-up.

Key CONCEPT The digital twin addresses the difficulty of ensuring that the control system is running at the same real-time multiplier as a separate process simulation during the times of greatest interest.

The process model speedup can be much faster than the real-time multiplier for the control system. The process model speedup is the product of speedup factors for kinetics and the material and energy balances. If the kinetic speedup factor is 10 and the material and energy balance factor is 50, the process model speedup is 500. If the control system speedup is 10, the response times of the instrumentation could be reduced by a factor of 50 to keep the proportional-integral-derivative (PID) controller tuning about the same. However, the process flow ranges of final control elements and flow measurements must be increased by a factor equal to the kinetic speedup.

The process models in a digital twin can be designed to show the proper behavior for failures, start-ups, shutdowns, and batch sequences. This means that the process models can handle zero flows (closed valves and turned-off pumps or fans), empty volumes, nonequilibrium conditions, and imperfect mixing. Sophisticated external (non-digital twin) dynamic process simulations are prone to fatal numerical errors from ill-conditioned matrices used for the simultaneous solution of stream conditions by the pressure-flow solver. These simulations may not only slow down, but they may also shut down during extreme conditions. Consequently, batch process simulation often requires specialized software that runs offline as a single program execution. Functionally, the run conditions are set at the time of execution. These batch process simulations may appear to the user or be cited by the supplier as interactive, but a change in conditions or parameters requires that the entire program and the batch process be re-executed from start to finish. The ability to include the interaction between the process and the batch and process control strategies is severely limited or impossible. Because these programs are not running in real time or a multiple of real time, it is impossible to pause, restore, or play back parts of a batch.

> **Key CONCEPT** It is important to verify whether a dynamic simulation can handle the zero flows, empty volumes, and nonequilibrium conditions associated with batch operation.

> **Key CONCEPT** Most external (non-digital twin) batch process simulation software offered to date is not truly interactive with the user or the control system.

Process simulations developed by process engineers can predict self-regulating process gains relatively accurately because of the sophistication of the physical properties, thermodynamics, interactions, and equations of state needed for continuous process and equipment design. Process engineers tend to think steady state. Batch process design is noticeably absent in books and courses on chemical engineering. Consequently, these engineers learn and think in terms of a steady state and equilibrium that is consistent with continuous processes.

> **Key CONCEPT** Most dynamic, real-time process simulations for chemical processes were developed by process engineers with steady-state process and equipment design in mind.

Process models designed by control engineers can be set up to do a better job than process simulations in modeling system dynamics by including the effects of transportation delays, mixing delays, mass transfer rates, kinetics, analyzer cycle times, sensor lags, noise, resolution, and deadband. The result is a closer match to the dead time, time constant, or integrating process gain and a fidelity that is more in tune with the spectrum of uses for control systems.

> **Key CONCEPT** The process simulation dead time is often less than the actual dead time because transportation delays, mixing delays, mass transfer rates, kinetics, analyzer cycle time, sensor lags, noise, resolution, and deadband are omitted.

The digital twin can simultaneously stop and start the execution and restore and replay simulated conditions for all the control systems and the simulation models, as seen in Figure 9-2b. Now emerging is the ability to replay actual plant data history files at high speeds to adapt and test the process models without having a connection to the actual plant.

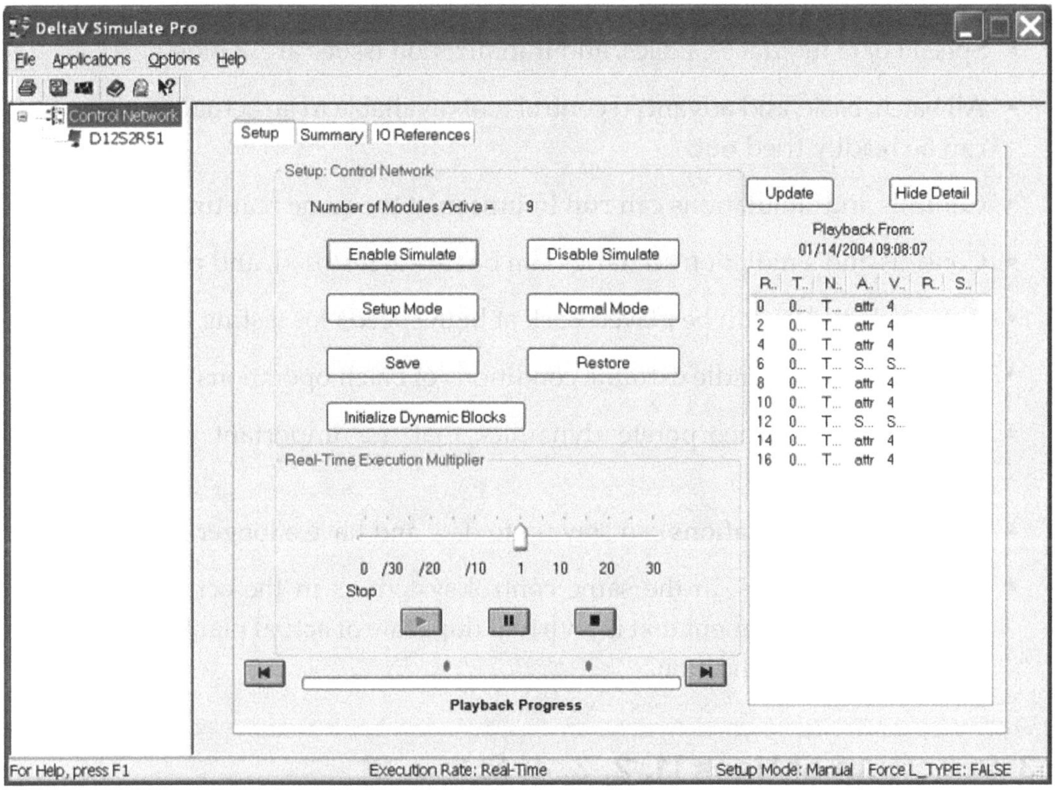

Figure 9-2b. Screen for controlling digital twin speed, modes, capture, and playback.

Most simulations used for control system checkout and operator training become obsolete after start-up. The investment is lost. The digital twin offers a better chance of keeping the control system up to date by simply importing the most current configuration and enabling the simulation to better match process changes using nonintrusive automatic adaptation.

External (non–digital twin) simulations used for control system checkout and operator training become obsolete after start-up because they cannot be automatically updated and adapted as the process and automation system changes.

Finally, in a digital twin, everything is done in the same configuration environment that is used for the actual control system. The focus can be more on the application than learning the inevitable undocumented features and tips and techniques associated with any new simulation software and interface. The advantages offered by a digital twin are summarized as follows:

- Control systems and graphics do not need to be duplicated, emulated, or translated.
- Special data interfaces, tables, and initialization issues are avoided.
- All batch, basic, and advanced control tools available in an actual control system can be readily tried out.
- Controls and simulations can run in unison at the same real-time multiplier.
- Controls and simulation scenarios can be saved, restored, and played back.
- Actual plant data can be played back at high speeds for testing and adaptation.
- Simulations can handle extreme conditions of batch operations and failures.
- Simulations can incorporate dynamics that are important for tuning and performance.
- Controls and simulations can stay up to date and have a longer lifecycle.
- Engineers can work in the same control system as in the actual plant and a simulation environment that is a virtual duplicate of actual plant enabling them to focus on the application.

9.3 Spectrum of Uses [1, 2, 3, 4, 5, 6, 7, 8]

The most familiar use of a digital twin is for testing and training. To checkout batch sequences and train operators, it is important to be able to repetitively and rapidly simulate batch phases. The ability to stop, start, save, restore, and replay scenarios and record operator actions is critical. For first-pass testing and familiarization of sequences and graphics, an automated tieback simulation may be sufficient. To test and learn about the interaction and performance of control strategies and the process, the higher fidelity dynamics offered by process models is important. It opens the door to upgrading the process and control skills of technology, maintenance, and configuration engineers who support operations. A process simulation with high dynamic fidelity is also important for testing process and control system interaction and performance.

Not well recognized is that the dynamic models often used for training operators as part of an automation project have a much wider utility that today is more important than ever. There is a great opportunity to use the digital twin to maximize the synergy between the operators, process control engineers, and control systems. To start on this path, process control engineers must be given the time to learn and use a digital twin and set up online metrics for process capacity and efficiency. The digital twin offers flexible and fast exploring => discovering => prototyping => testing => justifying => deploying => testing => training => commissioning => maintaining => troubleshooting => auditing => continuous improvement showing the "before" and "after" benefits of solutions from online metrics. Figure 9-3a outlines the major steps in continuous improvement and maximizing innovation.

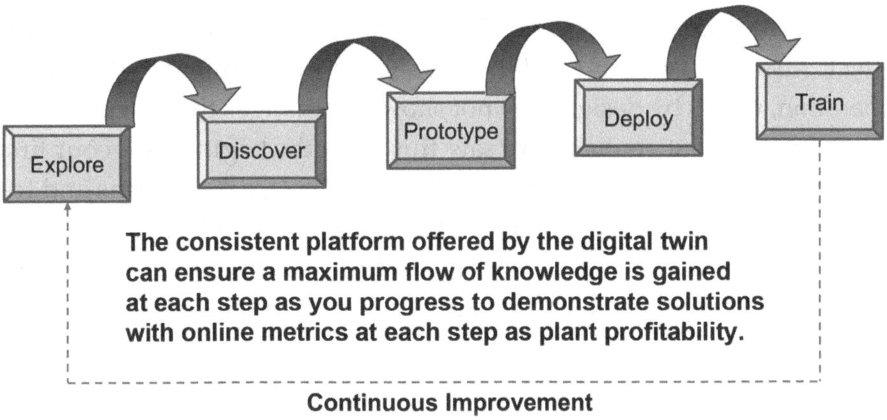

Figure 9-3a. Continuous improvement can become an inherent part of the digital twin, maximizing the synergy of operational, process, and control system knowledge.

The capability of dynamic models to improve system performance has greatly increased, even though the use has focused mostly on training operators as an automation project nears completion. The digital twin should detail the tasks needed to address difficult situations based on the best operator practices and process knowledge and eliminate the need for special operator actions through state-based control. Advanced process control (APC) and MPC can respond to disturbances and address constraints intelligently, continually, and automatically with great repeatability. Compare this with what operators can do in terms of constant attention, deep knowledge, and timely predictive corrections considering dead time, multivariable situations, and uncertainty in human behavior. Some operators may do well, but not all operators. Then, of course, an operator can have a bad day. Automation enables continuous improvement and recognition of abnormal conditions by a much more consistent operation. A better understanding by the operator of control system functionality and process performance from online metrics greatly reduces disruptions by an operator unnecessarily taking a control system out of its highest mode and/or making changes in flows. Furthermore, procedural automation can eliminate manual operations during start-up when the risk is the greatest compared to steady-state operation.

While we have singled out operators and process control engineers, the need for knowledge to attain the best performance extends to maintenance technicians, process engineers, mechanical engineers, and information technology specialists. Think of what can be realized if we are all on the same page understanding the process and operational opportunities and the value of the best measurements, valves, controllers, and software.

Here is a summary of the many possible uses of a digital twin.

- **Cause-and-effect relationships** – Data analytics can point to many possibilities, but the relationships identified are correlations, not necessarily cause and effect. Also, the data sets used in data analytics are limited to the range of plant operation, which by design may not show the changes that are possible. Process control is all about changes. We saw this from the beginning in our first course on control theory, where all the variables for Laplace transforms and frequency response were deviation variables. We must see the effect of changes in process inputs on the changes in process outputs. A digital twin can make all the changes you cannot make in a real plant. You can be creative, freeing your mind to make changes and learning the most from mistakes. We can do this all in the safe environment of the digital twin. The resulting data can verify correlations; provide richer data sets for data analytics; identify the dynamic compensation needed in data analytics for the predictions by projection to latent structures, also known as *partial least squares* (PLS); and provide a much leaner, faster, safer, and more focused design of experiments for the actual plant. Root cause analysis can be greatly improved.
- **Interactions and resonance** – The degree of interaction and resonance depends on the dynamics, tuning, and oscillation periods. The digital twin can show all this and much more using power spectrum analyzers to track down the culprit if the process and automation system dead times are modeled.
- **Valve and sensor response** – In most loops, the largest source of dead time and oscillations can be traced back to the valve and sensor responses. In particular, most of the valves designed for high capacity and tight shutoff not only have a poor response time but will also cause continuous limit cycles from stiction and backlash. Showing the effect of slow or noisy sensors and on-off valves posing as throttling valves can provide the justification for getting the best measurements and valves. You can show the effect of sensor and valve failures and the value of redundancy and a smooth recovery.

- **Process safety stewardship** – The digital twin should be used for extensive and continual testing of all the layers of protection in addition to the detailed design and implementation of the safety instrumented system (SIS). The interplay and performance of layers should be thoroughly evaluated and documented for all the important scenarios. Measurement and valve dynamics are often not thoroughly understood, and the consequences of failures should be analyzed in these scenarios.

- **Control system and SIS knowledge** – One of the most difficult aspects of control system and SIS design is recognizing the implications of measurement and final control element repeatability, rangeability, reliability, resolution, and response time. The digital twin enables you to explore and determine the value of the best instrumentation, PID control, and MPC, including the solutions and parameters needed. You must see how well the systems play together and help the operator in abnormal conditions. You can try all kinds of what-if scenarios to see how the system and operator perform.

- **Validation and regulatory support** – The validation and regulatory compliance of automation systems in pharmaceutical production requires a large expenditure of time and expertise in automation projects. The digital twin offers the performance verification needed for confirmation and documentation.

- **Code checkout** – Checking out the code and control sequences before the plant start-up enables the control engineer to validate that the code was written as required and designed. It enables debugging mistakes made during the coding offers validating code with consequential improvements seen in the better coordination of each sequence. The more time the control engineer spends checking out the code and exercising the sequences, the fewer issues he or she will face during the plant start-up and commissioning.

- **Process and equipment knowledge** – Most of our deep understanding of the implications of process and mechanical designs has come from running dynamic models. This was particularly true for the specialties of *p***H** and compressor control. For *p***H** control, the extraordinary sensitivity and rangeability and the consequential incredible nonlinearity of a system of strong acids and bases lead to strict design requirements in terms of minimizing mixing and injection dead time and seeking better set points, and the possibility of using weak acids or weak bases and conjugate salts to help reduce the steepness and nonlinearity. Today, we have the ability to handle conjugate salts as well as a wide spectrum of weak acids and weak bases by using an expanded charge balance.

- **Process equipment degradation –** The consequences of fouling heat transfer surfaces, thermowells, and electrodes; loss of catalyst activity; and plugging filters can be studied, and more optimum times for maintenance and cleaning can be scheduled. Better installation practices and measurements redundancy can be justified. For *p*H, three electrodes with middle signal selection provide the best protection against reaction to noise, failures of any type, and fouling.

- **Start-ups, transitions, shutdowns, and batch operation –** The design and value of procedural automation and state-based control for continuous processes are best achieved by trying out all conceivable scenarios in a digital twin. The digital twin is also the key to optimizing batch profiles, end points, and cycle times. If an operator claims his actions cannot be automated, there is an even greater motivation for a digital twin to identify and test the best actions. The automated, repeatable best actions reveal the other sources of changes by eliminating the variability from operator response.

- **Optimum operating points –** Finally, you can identify and document the benefits of better set points and achieve these set points more reliably and extensively by feedforward control, override control, batch profile control, full throttle batch control, and MPC.

Before the configuration even starts in the front end of a project, the process models can be used to evaluate control strategies and advanced control tools. In the past, this was done with offline dynamic simulations. Having ready access to an industrial tool set for basic and advanced control and simulations that are adapted to benchtop or pilot-plant runs offers rapid prototyping opportunities. This can lead to control definitions that have better detail and potential performance. Benchtop or pilot-plant systems with a mini version of the industrial DCS are now available that greatly facilitate developing and scaling up the control system [6]. Benchtop systems and pilot plants that have all the functionality of the main manufacturing systems are not yet prevalent because the development groups of these types of companies traditionally do not have the expertise (and, more importantly, the interest) to configure, maintain, and engineer these systems. The digital twin enables synergy between scientists and control engineers to make the incredible capability of DCS part of the process R&D.

The digital twin can be demonstrated with the industrial batch, basic, and advanced control systems used in the benchtop or pilot-plant system. The process models can be adapted by connecting to a benchtop or pilot-plant system or by built-in, high-speed playback of process data from experimental runs. An opportune time to take advantage of the PAT initiative is during the research and process development phase. The value of benchtop and pilot-plant systems that have industrial DCSs is significant in terms of project schedule, cost, and effectiveness because the DCS enables process control to be designed into the plant at an early stage and puts the biochemist, process technology, and configuration engineer on the same page [6]. However, the best return on investment (ROI) for PAT is realized by eventually implementing advanced process analysis and control in large-scale manufacturing processes.

In a digital twin, innovative strategies such as effectively switching the controller output can be trialed and tuned. Advanced control tools such as adaptive control, autotuning, fuzzy logic, MPC, neural networks (NNs), principal component analysis (PCA), and PLS can be demonstrated, adjusted, and evaluated "faster than real time."

> **Key CONCEPT** A benchtop or pilot-plant bioreactor that has an industrial control system offers significant opportunities for reducing configuration costs, improving communication, developing a digital twin, and prototyping advanced control, but the greatest ROI is realized by implementing PAT in large-scale manufacturing processes.

As a general rule of thumb, five changes are needed for each process input to develop an experimental model of a process output. For MPC, this corresponds to a minimum of five step changes in each process input, at least one of which is held long enough for the process output to reach a steady condition. For NN, PCA, and PLS, it means a minimum of five batches of process input in which the respective process input differed from the normal value. Although actual plant operation is obviously the best source of data, the long batch cycle time and the desire to minimize disruptions from introducing perturbations severely restrict the amount of useful plant data available for the development of MPC, NN, PCA, and PLS. More changes in inputs are typically needed for identifying empirical models. For example, developing a PCA with four inputs for detecting an abnormal batch would require at least 20 batches with varying inputs and at least five with normal inputs. If the batch cycle time is about two weeks, it would take about a year of plant production to have enough data. If you consider that you cannot deliberately optimize the spectrum of variability in the inputs, then it may take several years of production runs to have enough data.

In contrast, perturbations (e.g., traditional open-loop step changes to manipulated variables) can be automated and introduced to a digital twin running faster than real time. In two weeks, there is enough data to identify models for MPC, train NNs, develop latent variables and discriminant analysis for PCA, and predict economic variables via PLS. The predictive ability of MPC, NNs, PCA, and PLS can then be verified and evaluated using the high-speed playback of previous plant batches.

Conventional PCA assumes that all process inputs other than the ones used for the PCA are fixed. A digital twin running in real time that is synchronized with the actual plant can be used to predict the effect of variations in other process inputs by using the model-based PCA algorithms.

The digital twin can also be used to help explore more optimal operating conditions and investigate what-if scenarios. These scenarios are important for identifying the cause of an abnormal batch. Today, PCA for batch fault detection only identifies a batch as abnormal. Logic must be added to diagnose the fault. Fuzzy logic rule sets have been used in conjunction with PCA to provide real-time predictive fault analysis. A digital twin can help develop these rule sets offline faster than in real time by creating scenarios and evaluating the rule sets through the high-speed playback of previous batches. An online digital twin synchronized with the real plant can be sped up and run to batch completion to predict and analyze abnormal situations based on current batch conditions. A digital twin can also be used to help create the predictive capability of the PLS "Y space" of economic variables from the PCA "X space" of process variables.

Key CONCEPT A digital twin can eliminate process testing and provides years of process data for advanced control, fault analysis, and performance prediction within a few weeks.

Additionally, a digital twin can provide inferential real-time measurements of important process outputs such as nutrient, biomass, and product concentration. The built-in material balances can be used with kinetic models to predict concentrations. These predictions are then delayed so the values are synchronized with online or lab analysis. The concentration is shifted by a bias that is a fraction of the difference between the inferential and actual measurement, similar to what is done in the feedback correction of an NN used for property estimation.

Key CONCEPT A digital twin can provide faster, more reliable, and smoother concentration measurements than online analyzers.

MPC can use the rate of change of these concentrations to optimize batch profiles. A digital twin can also be run in some systems "faster than real time" to batch completion to provide an online prediction of key performance indicators (KPIs), such as batch cycle time and yield, based on current batch conditions and inferential measurements. Presently, management's understanding of the value of process control is decreasing, and the emphasis on profitability is increasing. It cannot be taken for granted that practitioners will be given the time and funding to make process control improvements. The retirement of most of the process control experts has exasperated the situation. We are at a critical point in terms of the decreasing rate of innovation and increasing tendency to copy jobs in migration projects. The best way to turn this around is by using KPIs with actual dollars correlated to process control improvements that can be revealed and tested by dynamic first-principle models in a digital twin and made available in real time to operators and maintenance, process, mechanical, and automation professionals.

Figure 9-3b shows a digital twin's functional value, highlighting the bidirectional flows of the control system and process/equipment knowledge of process control and analysis tools, including online performance metrics for greater analysis and justification of improvements. The two-way knowledge flow is the key to improving the process/equipment and the control system in addition to the dynamic model and data analytics. As the fidelity of the dynamic model increases, opportunities arise for these tools to get results from the digital twin that can be used in the actual plant. The dynamic model can be run faster than in real time with the tuning corrected by applying the speedup factors. New control functionality can be developed and included in the dynamic model for evaluation. If online metrics show significant improvements in

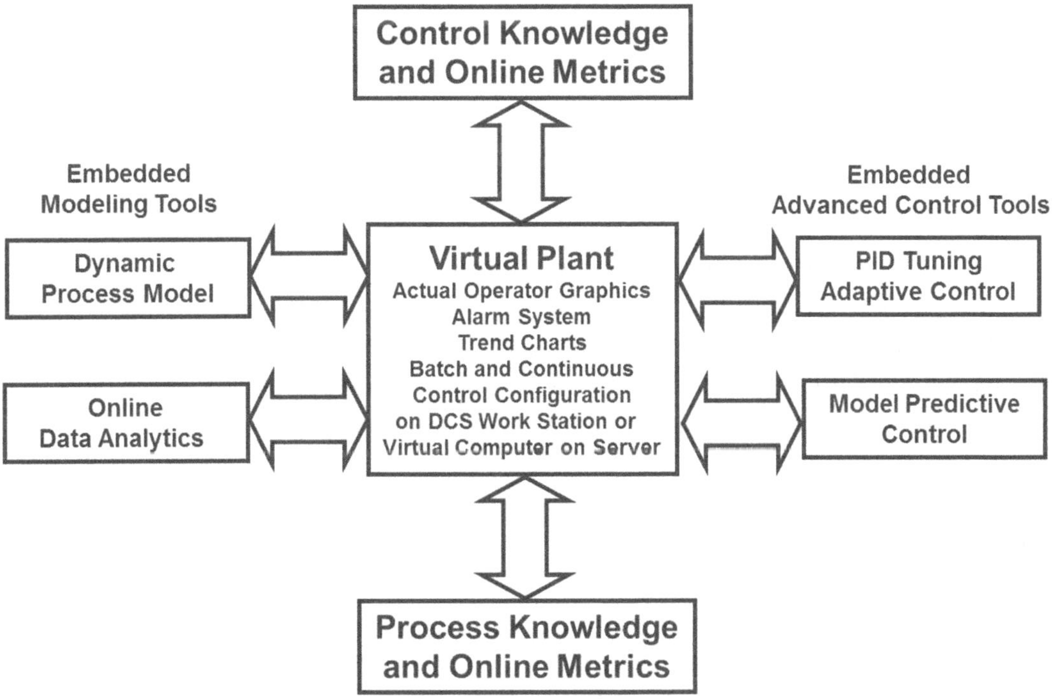

Figure 9-3b. The two-way flow of knowledge in the digital twin between tools, models, and the actual control system is the source of the increasing synergy of knowledge between the process, control system, engineers, technicians, and data scientists.

control and process performance, the functionality prototyped can be added as new blocks or as improvements to existing blocks in the DCS.

The digital twin can speed up the benefits gained by offering users the ability to use process and end point monitoring and control, continuous improvement, and knowledge management tools in an integrated and accelerated manner.

9.4 Implementation [1]

Composite blocks for equipment, piping, pumps, streams, final elements, and transmitters are inserted into the digital twin process model and wired up. External input and output parameter blocks are added and connected to the final control element (e.g., valve, VFD) inputs and transmitter outputs, respectively. The control system configuration is imported, and the analog input blocks in the virtual control system are switched to the simulation mode. The paths of the external inputs for the final control elements are chosen via a browser to be the outputs of the analog output blocks of the virtual control system.

Similarly, the paths of the external output parameters for the transmitters are selected to be the simulation inputs of the analog input blocks of the virtual control system. Analog outputs and analog inputs of loops that are not modeled can be simulated by simple tiebacks in which the analog output is passed through signal characterizer, dead time, and filter blocks and then connected to the simulated input of the analog input block.

The concentrations, pressures, and temperatures of each inlet and outlet stream; the flow rate and pressure rise of each pump as a function of speed; the flow coefficient of each valve as a function of position; the agitator pumping rate as a function of speed; and the initial inventory, concentrations, temperature, and pressure of the bioreactor are then set. Control valve deadband and resolution are set in the *actuator* blocks, and measurement delays and lags are set in the *transmitter* blocks.

The batch sequence, basic process control system, and simulation are run in an offline mode, and the simulation parameters are manually adjusted to match the virtual with the real plant batch profiles of uncontrolled process outputs and manipulated process inputs.

When representative digital twin batch profiles are obtained, the processor loading and profile alteration are checked as the real-time execution multiplier is increased, and the maximum practical speed is noted. Key model parameters are chosen as the manipulated variables, and associated process measurements are chosen as the controlled variables of the MPC for adapting the digital twin. An automated test sequence is then run for the MPC at the highest possible speed offline, and the MPC models are identified and visually checked as reasonable in terms of the direction and relative magnitude of the effect.

The digital twin is then connected to the actual plant in a read-only nonintrusive setup. The digital twin modes, set points, and batch phases come from the actual plant. The set points (targets) of the MPC's controlled variables for adaptation are externally referenced to the manipulated variables by the actual plant's key PID controllers whose process measurements are affected by the adapted model parameters shown in the detailed signal paths outlined by Figure 9-4a. Because the relationship between model parameters and key process variables is generally nonlinear, an NN may be used to enhance the future trajectories of an MPC. Alternatively, an NN could be trained to adjust model parameters on a one-by-one basis. Because NNs are good at interpolation but not extrapolation, it is important to verify whether the inputs are within the training set range of values. Even with a simple, traditional, linear MPC model, MPC offers a multivariable solution for adapting dynamic models.

The MPC targets and controlled variables are the actual plant and digital twin manipulated variables (e.g., flows), respectively, in Figure 9-4a. That figure shows how to

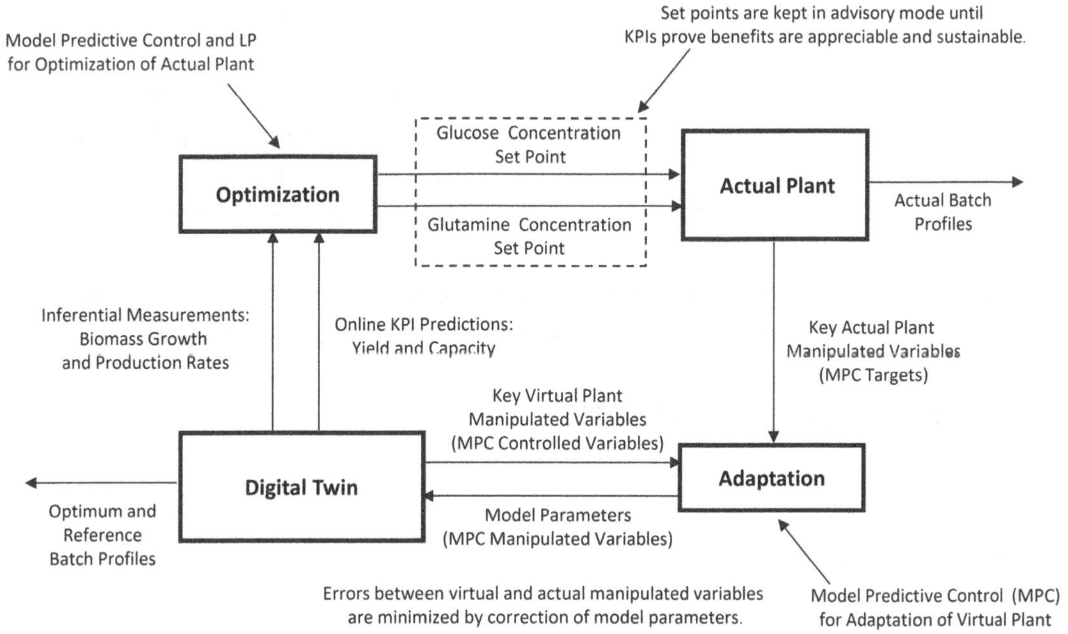

Figure 9-4a. Nonintrusive automated adaptation of model parameters to match manipulated variables with potential future optimization based on improvement in KPIs.

adapt a bioreactor model to provide more accurate inferential measurements of biomass growth and production rates that are slopes of the batch profiles for cell and product concentration. The MPC manipulated variables are model parameters (e.g., kinetic parameters) that affect these profiles.

The MPC can be identified and tested without the digital twin connected to the actual plant. The MPC development is much faster than that of an actual plant MPC. This independence enables extensive adaptation and the ability to gain associated knowledge of process relationships without disrupting the plant.

The adaptation is done without affecting the actual plant because the plant's manipulated variables are being read by but not changed by the digital twin. It is critical that the digital twin has the same set points and tuning settings as the actual plant and that the digital twin is started with controller outputs initialized to match the actual plant.

The optimized set points from MPC with inferential measurements of growth and production rate are done in an advisory mode that does not affect the actual plant. Not shown in Figure 9-4a is that an MPC is run in automatic mode in another digital twin that is a duplicate of the adapted digital twin to generate and study the optimized set points. The set points are only eventually used to automatically optimize the plant if they prove more accurate, beneficial, and reliable per KPIs than an MPC with inferential measurements computed from online and at-line analyzers.

After innovative control, advanced diagnostics, and MPC applications have been prototyped and documented, they can be implemented on benchtop or pilot-plant systems. If the digital twin provides inferential measurements, it is run in real time and synchronized with these systems. After verification, validation, and accounting for scale-up factors, these applications and the digital twin can be installed for plant trials. Initially, these systems run only in the monitoring mode. When process inputs are finally manipulated, they are initially restricted to a very narrow range.

9.5 Data-Driven Dynamics [1]

Key CONCEPT The quality of the model depends directly on the quality of the data used in the model configuration.

Using good data as input leads to a better model. After selecting unit operations expanded to include those that might affect the one with *p*H control, simulation engineers must collect the data that govern the physics of the problem. In a perfect world, key operation parameters are historized and are directly relevant for simulation. Although that is not always the case, these parameters show only the response. To understand the generation of that response, the system's physical characteristics must be considered. The model configurations and tuning parameters follow plant design documents such as piping and instrumentation diagrams, health and material balance diagrams, and equipment and device process data sheets. Without this data, models may be based on plant operating data such as snapshots of control system graphics and historians. This is often the case if there is a significant deviation in plant operation from the design or non-steady-state operation seen in batch processes. We will first consider the physical parameters for vessels, valves, and pumps.

Most of the dynamics of a process system are the interplay of lag and dead time. In the case of a vessel, these are generally functions of agitation and—more importantly—physical dimensions: height, radius, and dimensions of the bottom and top of the vessel. Many simulators do not natively consider residence time distributions or imperfect mixing. However, the flow rates from the tanks are impacted by the hydrostatic head, which directly correlates to vessel dimensions.

Key CONCEPT A critical facet of system dynamics that often is overlooked resides within the dynamics of the valves in the system.

Valve dynamics depend chiefly on resolution, backlash, and slew rate. Resolution dictates the coarseness of the valve response. The backlash accounts for the lost motion deadband of the actuator. The slew rate governs how fast a control valve can respond. (More details on modeling this response are given in Section 9.10.) Frequently, these effects are inaccurately labeled *hysteresis*. Hysteresis fails to accurately encompass the dynamics of the valve. A frictionless valve that can instantly respond and reach the exact actual position is an ideal, never a reality. Innumerable valves with their own dynamic characteristics play together to exert a profound impact on the response of the field. Accounting for these characteristics is critical for showing the proper system response.

Pumps are the prime movers of the digital plant. The most critical piece of information for simulating a pump's response is the pump curve. The pump curve's intersection with the resistance curve of the piping shows the operation point. The actual pump curve will provide the response at a variety of operating conditions and respond appropriately. The pump curve's response is inextricably tied to the piping specification and liquid properties. A perfect pump curve paired with an inaccurate capture of the resistance will provide skewed results. However, without a curve, a single operation point can be used. It is important to remember that for a system of flow equations to be tractable, the curves must be monotonic (ever increasing or decreasing) in the first quadrant of the coordinate system.

9.6 Titration Curve Modeling [1]

To prototype *p*H control strategies using the digital twin, modeling the electrolytes involved in the running system provides the basis for *p*H simulation. The topic of electrolyte modeling is multifaceted and complex and could easily fill an entire book. However, the basic concepts needed to prepare a simulation are straightforward: electroneutrality, dissociation equilibrium for acids, and species balances.

Key CONCEPT Electroneutrality provides a simple kernel for computing ***p*H** after identifying which species are present in the system.

The overall expression for electroneutrality is a simple one. This expression assumes that the sum of the concentrations of species multiplied by their charges is zero, as shown in Equation 9-6a:

$$0 = \sum_{i=1}^{n} \pm \varepsilon_i \bullet [X_i^{\pm \varepsilon_i}] \tag{9-6a}$$

where i is an index over the species (X_i) in the system of charge ε_i.

The next concept required to model *p*H is the concept of the dissociation equilibrium constant. The equilibrium constants for aqueous components are generally given by pKa values. The pKa values express the ratio of products to reactants of the dissociation reaction, such as for an acid as exemplified in Equation 9-6b:

$$H_x A^{\theta} \leftrightarrow H^{+} + H_{x-1} A^{\theta-1} \tag{9-6b}$$

The equilibrium expression is then given by Equation 9-6c:

$$K_{a1} = \frac{[H^{+1}] \bullet [H_{x-1} A^{\theta-1}]}{[H_x A^{\theta}]} \tag{9-6c}$$

Note that θ is the value of the net charge on the species on the left-hand side of the equilibrium expression.

The pKa is simply the base 10 logarithm of K_{a1} as defined in Equation 9-6d:

$$pK_{a1} = -\log(K_{a1}) \tag{9-6d}$$

The auto-ionization of the solvent also must be considered in any system, as shown for water in Equation 9-6e:

$$pK_w = [H_3O^+] \bullet [OH^{-1}] \tag{9-6e}$$

A species balance simply accounts for the amount of material added to the system. Equation 9-6f shows the result of adding a monovalent acid (HB) to a constant volume:

$$X_a = [HB] + [B^-] \tag{9-6f}$$

where X_a is the initial concentration of the acid.

These concepts combined enable formulating a simple charge balance that enables reasonable simulation of *p***H** for systems of low ionic strength and where salts are assumed to dissociate completely. Regardless of the ionic strength or solubility of the salts in the system, the first important step in such simulation is identifying the species involved. The simulation developer should list all salts, acids, and bases used within the system. For each salt involved, the developer must also list the conjugate acid–base pair (e.g., sodium carbonate corresponds to carbonic acid). After identifying the species involved, the pKa for each species (and at each dissociation step) must also be identified. For immensely strong acids and bases, the first dissociative step is often so great as to be unmeasurable. Note that for any aerobic system under study, carbon dioxide will contribute to the charge balance as carbonic acid. See Chapter 2, "The Chemistry," for details on how the charge balance is set up for acids and bases, each with one to three dissociation constants; how ionic strength affects the *p***H** measurement; and typical values of dissociation constants at 25°C. For an in-depth derivation of the charge balance that includes the effect of conjugate salts plus an example, see Appendix H, "Charge Balance to Model pH." A first-principle *p***H** model can be created based on the knowledge of reagents added plus carbon dioxide and by-products produced. When using a commercial dynamic simulator, the user may configure the components and the dissociation data and enable the simulator to compute the *p***H** value automatically.

> **Key Concept**
>
> A simple charge balance can provide an accurate and fast simulation of the *p***H** in process systems if all the acids, bases, salts, and solvents concentrations are right for all the different process conditions, including the effect of temperature and ionic strength on dissociation constants as detailed in Chapter 3, "Titration Curves," and Appendix H.

The *p*H calculation is complicated by the effects of ionic species activity coefficients in higher ionic strength solutions (typically above 0.1M). Rigorous calculations can be incorporated into the *p*H algorithm to consider ionic species activity values. However, the return on the additional computation involved should be evaluated thoroughly. In most cases, ionic species activities cannot be determined precisely with the available correlations. Also, most dissociation constants are known at 1% accuracy, and industrial *p*H measurement is rarely known at much higher precision.

Whenever the user has collected the data for simulation, it is critical to ensure the modeling is accurate by obtaining the titration curve for the system under study.

Repeating this experiment in the digital twin enables the component and assumption set to be verified. In most bioreactors, some form of buffering is added to aid in controlling *p*H.

> **Key CONCEPT** Any system will have some buffering effects that depend on the species present in the system. Even when no buffer is added, carbonic acid and carbonate from simple exposure to air will provide noticeable buffering.

In the titration curve in Figure 9-6a, the first additions of base to the acidic system show relatively small amounts of change. This stems from the buffering provided by adding sodium acetate. It is conceptually useful to consider Le Châtelier's principle under the Brønsted–Lowry acid–base theory. Le Châtelier's principle states that a change on one side of an equilibrium expression (e.g., a change in concentration) results in a

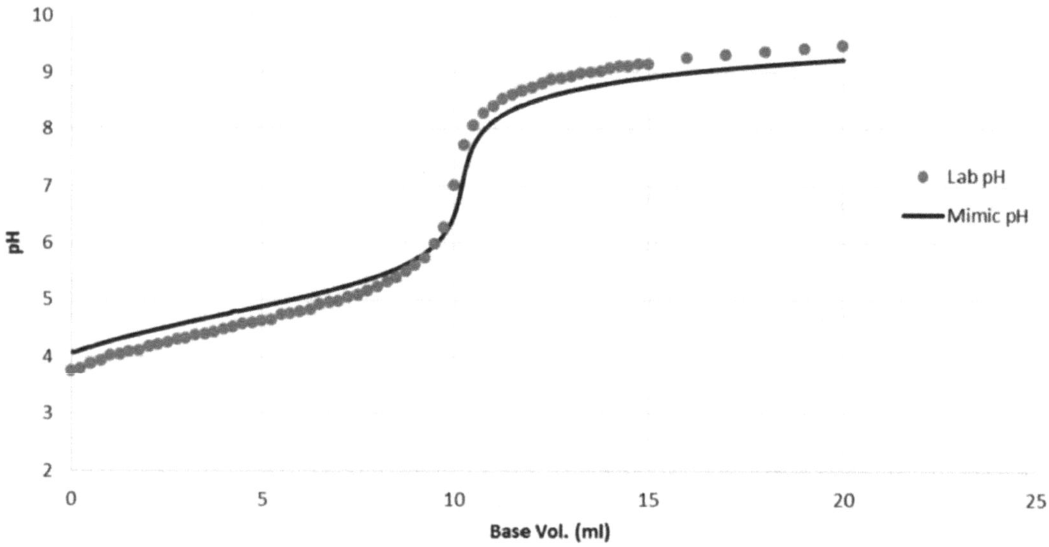

Figure 9-6a. Comparison of *p*H values between simulated and lab titration.

correction mitigating that change. Consider the dissociation of acetic acid as shown in Equation 9-6g:

$$H_2O + HCH_3CO_2 \Leftrightarrow H_3O^+ + CH_3CO_2^- \tag{9-6g}$$

By adding a base to the system (e.g., NH_4OH), more hydroxide ions enter the system. These ions can then neutralize some of the hydronium ions in the solution. The net effect of this neutralization is that there are fewer hydronium ions in the system (on the right side of Equation 9-6g). Le Châtelier's principle then mitigates this effect by causing the reaction to proceed further from left to right. In Figure 9-6a, around the 10 mL mark, there is no longer enough undissociated acetic acid to mitigate the base addition. Thus, the buffer is broken and a rapid change in *p***H** results. This precipitous change in *p***H** poses unique challenges in *p***H** control. When the simulation includes these components, it enables realistic testing for control performance and builds the operator's understanding of the system's response.

9.7 Instrumentation Modeling [1]

The "dynamic" aspects of modeling are particularly important for bioprocesses because a steady state is rare, as most unit operations are batch. Even perfusion processes start out as a batch and remain continuous by recycling and purification only until contaminants build up, requiring a restart as a batch. By computing the slope of the cell growth rate and the product formation rate, you can achieve a pseudo steady state and an insightful and useful process variable for batch profile control, but this requires a good dynamic model and a computation (e.g., future value block) that has a good signal-to-noise ratio.

Most do not realize that industrial PID algorithms work in percent rather than engineering units, most of the total dead time comes from instrumentation, and most disturbances are on an input to the process instead of the process output. This missing understanding has profound implications.

The gain and dynamics seen by the controller are determined by much more than just the process gain or dynamics. The controller sees an open-loop gain as the change in the percentage of the process variable (controller input) divided by the percentage of change in the manipulated variable (controller output), and this gain and the total loop dead time, rather than just process dead time, determine tuning and control-loop performance for process disturbances.

> **Key CONCEPT** Tuning and control-loop performance depends on instrumentation dynamics, sizing, and calibration in addition to the process gains and dynamics.

The open-loop gain is a product of the manipulated variable gain (e.g., control valve or variable speed pump flow span divided by 100% for a linear installed flow characteristic), a hidden factor that is the inverse of the feed flow, a process gain that is the change in the process variable for a change in the ratio of manipulated flow to feed flow, and a measurement gain that is 100% divided by the measurement span as detailed in Chapter 8, "Control System." Assuming the piping pressure drops and inherent valve or pump flow characteristics are properly modeled, a digital twin automatically provides the proper open-loop gain because it uses the actual control system with its input and output scales.

> **Key CONCEPT** A digital twin gives the controller scales that—combined with valve or pump inherent flow characteristic with a dynamic piping pressure drop—provide a proper model of the open-loop gain needed for tuning and control-loop performance.

The total dead time is the sum of the dead times that are pure delays and small lags that are effectively the dead times in manipulated variable response (e.g., valve position or pump speed), manipulated flow transportation (e.g., reagent injection delay), process transportation to sensors (e.g., sample delay), sensor response, transmitter response (e.g., damping and update rate), and controller response (e.g., filter time, scan time, and execution rate). The process model is often missing transportation delays of manipulated flows by valves, pumps, and process flows to sensors. There are also dead times missing that are the resolution, sensitivity, or deadband divided by the rate of change of the corresponding signal.

> **Key CONCEPT** Models of instrumentation response and transportation delays of associated manipulated and process flows provide the total dead time needed for tuning and control-loop performance.

A control loop cannot do anything to correct an unmeasured load disturbance until it starts to see the disturbance and correction arrive at the same point in the loop. The total loop dead time for the disturbance to propagate through the process, measurement, controller, valve or variable frequency drive (VFD), and piping and return to its point of entry determines what a controller can do. In a first-order plus dead time approximation, the time constants in series as you traverse the loop that create an equivalent dead time are added to the pure dead times from transportation delays; digital device update and execution rates; analyzer cycle time; and instrumentation sensitivity, resolution, and deadband.

The best controller gain for load disturbance rejection is inversely proportional to the open-loop gain and total loop dead time. The minimum peak and integrated error are proportional to the dead time and dead time squared, respectively. Blocks that model each of the major sources of dead time should be included in the model to achieve the dynamic fidelity needed.

Key CONCEPT The minimum peak error and integrated error for a load disturbance are proportional to the total dead time and dead time squared, respectively.

Instrumentation characteristics (i.e., response time, sensitivity, resolution, and deadband) depend on the operating point. These characteristics are generally worse for low flows. In low-flow situations, valves experience higher stiction during operation near the seat and low velocities at sensors (especially for less-expensive actuators, positioners, and valves). Many positioners designed in the 1970 to 1990 time frame have a response time that increases by two orders of magnitude (e.g., from 0.5 to 50 s) for step sizes less than 0.25%. Even more recently designed positioners have this problem, which remains unrecognized due to the traditional step test of 5% or more. The less-expensive positioners (e.g., spool) have a particularly slow response time. Thus, a variable filter block and a dead time block that enable the time constant and dead time to be bumplessly updated online should be used. A control valve designed for throttling (e.g., diaphragm actuator, digital valve positioner, and sliding stem globe body) has a sensitivity resolution approaching 0.1% and a backlash deadband of 0.2% above 20% open. A control valve designed for tight shutoff and on-off control with a piston actuator and spool positioner has a resolution of 1% to 10% and a backlash deadband of 2% to 20%. The lower price tag, lower leakage, and greater capacity of the on-off valve posing as a throttle valve and the lack of requirements for sensitivity, resolution, or response time on valve specifications make this a common problem.

The slew rate limit of a throttling valve can be kept at a rate of 50% per second or faster (100% stroke time less than 2 s) by using a volume booster on the positioner output. Omitting the volume booster for large valves can slow the slew rate by two orders of magnitude. These are important factors to include when modeling valves.

The time constant of a new *p***H** electrode in a buffer solution for a decrease in *p***H** is more than twice what it is for an increase in *p***H**. The time constant is two to four times larger for a strong acid and strong base solution. The time constant increases by an order of magnitude for a decrease in *p***H** of 0.5 ft/s or less velocity at the electrode. Almost imperceptible coatings cause a dramatic increase in the electrode time constant. For example, a slime coating of 1 mm thickness can increase the electrode's

98% response time from 10 s to 7 min. Premature aging of the glass due to prolonged exposure to process temperatures greater than 50°C can increase the 98% response time to 160 min. Middle signal selection of 3 *p*H measurements will inherently ignore the slowest electrode, including a single failure of any type, as well as reduce the effect of noise and provide much more intelligent diagnostics and a calibration schedule. An 86% response time is a better metric because a 98% response time takes more than twice as long and is much more susceptible to errors in response time from noise and disturbances.

A temperature measurement time constant is mostly governed by the thermowell design and the fit of the sensor, but it is also adversely affected by a low fluid velocity. Air acts as a great insulator. If the sensor tip is not touching the bottom of the thermowell, the increase in time constant can be orders of magnitude. An annular clearance of 0.055 in (very loose fit) instead of 0.005 in (extremely tight fit) can increase the largest time constant for even a high-velocity liquid stream from 4 to 228 s. The time constant for 1/4-inch sheathed bare elements is much less than the largest thermowell time constant.

> **Key CONCEPT** The time constants for electrodes and thermowells can vary from a few seconds to several minutes depending on the process operating conditions and the sensors' design, condition, and installation.

A block used to model a discrete device or analyzer should have sensitivity and/or resolution input parameters, a cycle time or update time, and an analysis time or latency. The resulting dead time is the sensitivity and/or resolution divided by the process variable rate of change plus the dead time from periodic updates that is the sum of one-half the cycle or update time and the total analysis time or latency. The analysis time is typically the cycle time for analyzers, and the latency is typically near 0 for digital devices not overloaded, so the dead time is 1.5 times the cycle time and 0.5 times the update time, respectively. The resolution of older thermocouple and resistance temperature detector input cards (e.g., those from the 1980s) caused a noticeable problem with dead time and a poor signal-to-noise ratio. The analyzer sensitivity is often a concern.

> **Key CONCEPT** The dead time from periodic updates, analyzer sensitivity, and many less-expensive actuator and positioner designs is a significant problem best recognized and corrected by using models with instrumentation dynamics.

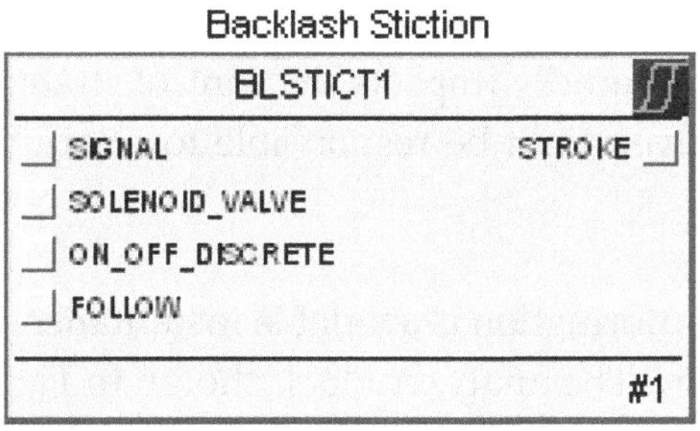

Figure 9-7a. Backlash stiction modeling block.

A backlash stiction block, as shown in Figure 9-7a, enables the nonideality of valves to be encapsulated simply. The signal represents the input from the control system. It is a value ranging from 0% to 100%. The backlash represents the valve's lost motion. The resolution represents the minimum change in the requested position the valve can manifest. A prestroke delay enables modeling the pure delay between when a signal is requested from the control system and when it reaches the actuator. In addition, two stroke lags enable modeling the change as a second-order process. If additional restrictions on the rate of change are required, a slewing rate model directly limits the rate of change that can pass into the backlash stiction object. To summarize, the valve signal passes through a dead time block, two lag blocks, a slew rate block, and the backlash stiction block before becoming the stroke. The stroke represents the actual opening of the valve after accounting for all the nonidealities explained. If a positioner's sensitivity drastically deteriorates for small signal changes, particularly in the opposite direction, the dead time parameter can be increased accordingly. For example, for a positioner that exhibited a dead time of only 0.5 s for a 2% or larger signal change, the dead time increased to 50 s for a 0.25% signal change in the same direction and increased to 100 s for a 0.25% signal change in the opposite direction.

A slew rate can be applied via the block shown in Figure 9-7b to the signal resulting from the stroke or used to intercept the signal to the block. A slew rate provides a maximum rate of change per model execution in the upward (increasing slew rate) and downward (decreasing slew rate) directions. Combining the slew rate and backlash stiction models enables fully modeling valve nonidealities. Determining these parameters may require step tests if the measurements are unavailable from the manufacturer.

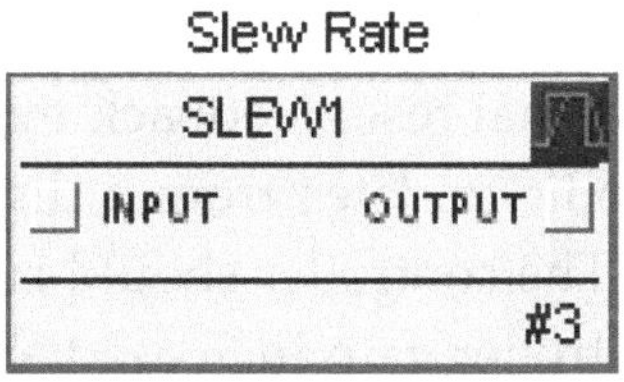

Figure 9-7b. Slew rate modeling block.

Valves play an important role in capturing the dynamics critical for modeling bioprocesses. A control valve profoundly impacts the control strategy in all processes, especially where a control valve might be responsible for modulating sensitive variables like *p*H.

In a simulation, detailed information is available instantaneously. This is a luxury that real processes do not enjoy. The analyzer block shown in Figure 9-7c allows for simulating an offline analyzer. The simulated value is transmitted through the measurement input. The sample time average represents the expected time between samples taken by an analyzer or a technician. The analyze time average represents the expected time from a sample being taken to the result being delivered to the control system. In addition, a deviation for both the sample time and the analyze time can be added to incorporate the nonuniformity inherent in a real system. Analysis noise provides a random fluctuation in the measurement. Resolution dictates the minimum amount of change in the composition that results in a reported change in the measurement. Once the process is complete, the sample with noise injected is returned as the new value. The new flag becomes true for one execution to signal other parts of the model when a fresh sample analysis is complete.

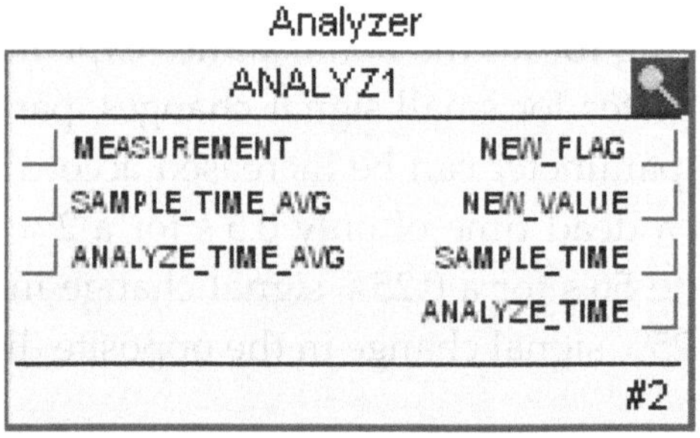

Figure 9-7c. Analyzer modeling block.

Using an enhanced PID block (e.g., PIDPlus), as shown in Figure 9-7d, with external-reset feedback (e.g., dynamic reset limit) that suspends integral action between periodic updates and uses the time between updates for the derivative calculation can enable much more aggressive tuning independent of changes in time between updates when this time is larger than the 63% response time of the process. This capability is achieved by enabling the external-reset feedback mode and connecting the actual process variable result of periodic updates from a digital device or analyzer as the external-reset feedback signal. The controller should automatically become more stable, and in many cases, the controller gain increased and reset time decreased. If the

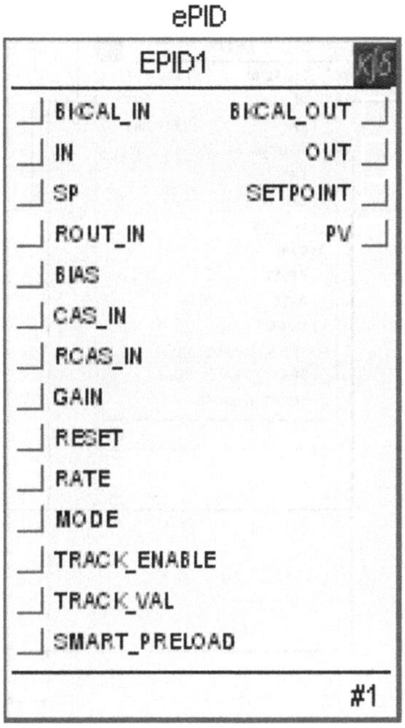

Figure 9-7d. Example of a PID block.

time between period updates is more than the 63% response time, increases in the time between updates do not require retuning, which means that incredibly long analyzer cycle times, possibly from offline analysis in labs, do not require retuning or cause oscillations.

The PID block's in and out can be connected to a Rough N Ready Tuner, as exemplified in Figure 9-7e. A PID can be placed in remote output mode to enable the Rough N Ready Tuner to perform a series of step tests. This allows for rapidly identifying initial tuning settings. It is important to keep the scales and control actions synchronized in both blocks. The Rough N Ready Tuner only works whenever the system has settled. A set of initial tuning settings for PID control can assist when prototyping configurations in the digital twin. Table 9-7 lists possible initial settings (shown as the first number outside the range of settings in parentheses) for various applications. The initial settings and ranges of settings shown are not meant to be indiscriminately used in the actual system deployed. They can be used as a check for unusual settings to trigger an investigation into whether a tuning mistake has been made or there are some unusual application conditions. There are exceptions to every rule, especially when it comes to the dynamics of industrial processes, which are usually severely nonlinear and a function of raw materials, production rate, cycle time, by-products, and operating/equipment conditions.

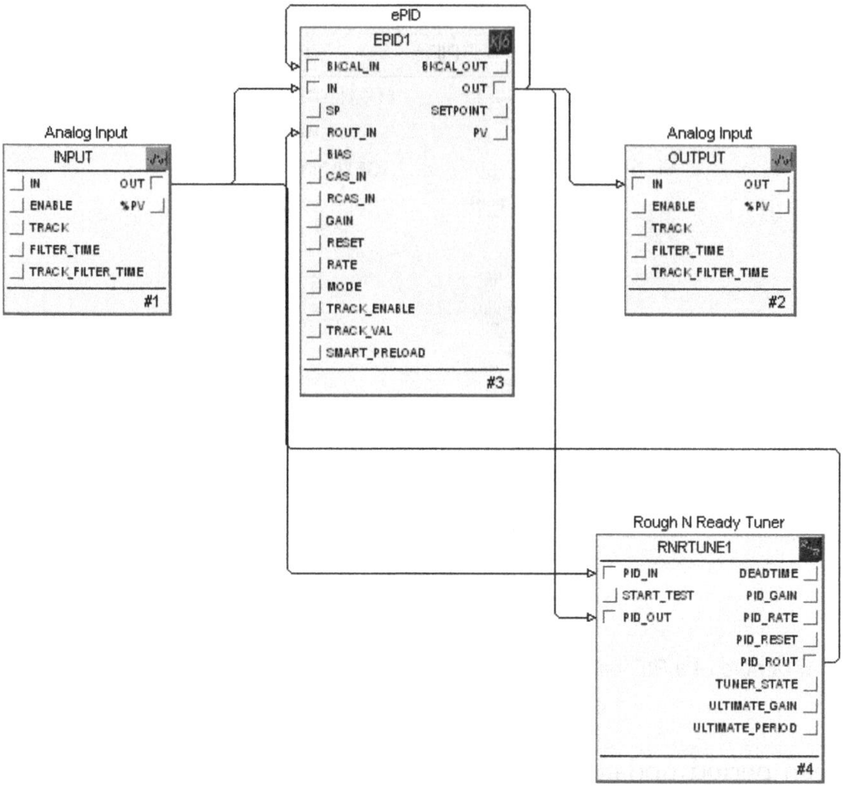

Figure 9-7e. Example of a Rough N Ready Tuner in a simulated loop.

Table 9-7. Typical PID execution rates and tuning settings for common loops.

Application Type	PID Execution Rate (s)	PID Gain	Reset Time (s)	Rate Time (s)
Liquid Flow	0.5 (0.2–1)	0.3 (0.2–0.8)	6 (1–12)	0 (0–2)
Liquid Pressure	0.1 (0.02–0.5)	0.5 (0.2–1.0)	3 (1–6)	0 (0–2)
Tight Liquid Level*	5 (1.0–30)	10 (5–25)	600 (120–6000)	0 (0–60)
Gas Pressure (psig)*	0.2 (0.1–0.5)	5.0 (0.5–20)	300 (60–600)	3 (0–30)
Reactor *p*H*	1 (1.0–5)	1 (0.01–50)	120 (60–600)	30 (6–30)
Neutralizer *p*H	2 (1.0–5)	0.1 (0.001–5)	300 (60–600)	60 (6–120)
In-line *p*H	1 (0.2–2)	0.2 (0.01–0.4)	30 (15–60)	0 (0–3)
Liquid Reactor Temperature	5 (2.0–15)	5.0 (1.0–50)	60 (30–300)	12 (12–60)
Gas Reactor Temperature	1 (0.5 –2)	1.0 (0.5–5)	15 (5–30)	3 (1–6)
In-line Liquid Temperature	2 (1.0–5)	0.5 (0.2–2)	30 (15–60)	6 (3–15)
Distillation Column Temperature	10 (2.0–30)	1.0 (0.2–10)	1200 (600–12,000)	120 (60–1200)

9.8 Speedup [1]

Bioprocess batch operations can have incredibly long batch cycle times (e.g., 10 days). The speedup factors shown in Figure 9-8a can be used to complete a batch in hours rather than days. The media speedup factor can be as fast as 20 times real time; the kinetics speedup factor can be 10 or more times. The total speedup for media response is the product of the two factors, which means a speedup of 200 or more times real time is possible. The flow ranges of valves, variable speed drives (VSDs), and flow transmitters should be increased by the kinetic speedup factor. If the capacity and flow span are adjusted correctly, the flow controller tuning should not change much. However, the primary process time constant and integrating process gain will be faster for composition, dissolved oxygen (DO), *p***H**, and temperature control. The instrumentation dead time should be decreased by the total speedup factor to keep the same ratio of the total dead time to the actual primary process time constant or the equivalent primary process time constant that is the inverse of an integrating process gain. For composition, DO, *p***H**, and temperature control, the PID gain and reset time must be decreased based on autotuner results. Many batch processes have too small a reset time, even without speedup, as seen by an overshoot and delayed correction of PID output until after the process variable crosses the set point. Also, some sources of dead time cannot be reduced enough to match speedup. Consequently, the PID gain must be decreased and the decrease in reset time moderated. A reset time smaller than four times the maximum dead time or an oscillation period greater than six dead times is indicative of a reset time that is too small. A faltering (temporary settling of PID output just before reaching the set point) on the approach to the set point or oscillations with periods about four times the dead time are indicative of the need to decrease the PID gain. An autotuner should be used. The Rough N Ready autotuner can be used to quickly identify the controller settings at different batch times.

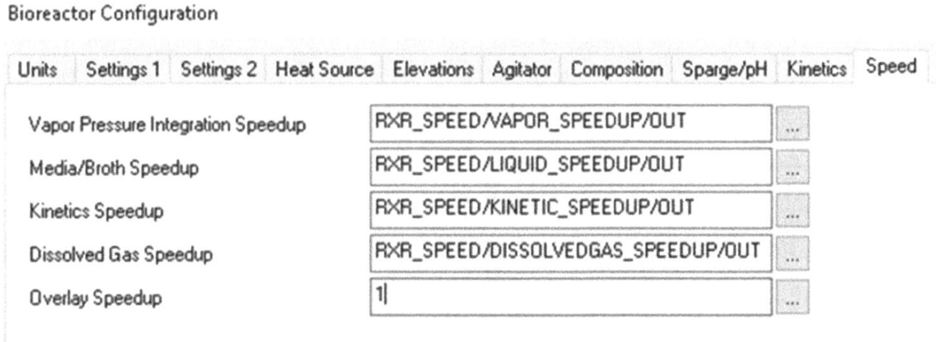

Figure 9-8a. Configuration of individual speedup parameters.

> **Key CONCEPT**
> The batch cycle time can be speeded up by the product of kinetic and model-media speedup factors, but the flow capacities of valves and VSDs and the flow measurement spans must be increased in proportion to the kinetic speedup factor. The instrumentation dead times should be decreased by the total speedup factor. Flow controller tuning should not need to be adjusted if capacities and spans are adjusted properly. However, for composition, DO, pH, and temperature control, the PID gain should be decreased, and the reset time should possibly be decreased per the autotuner results.

9.9 Performance Monitoring [1]

Performance metrics can be computed automatically by a block shown in Figure 9-9a for process input (load) disturbances and set-point changes to assess and improve the performance of control strategies, instrumentation, and tuning.

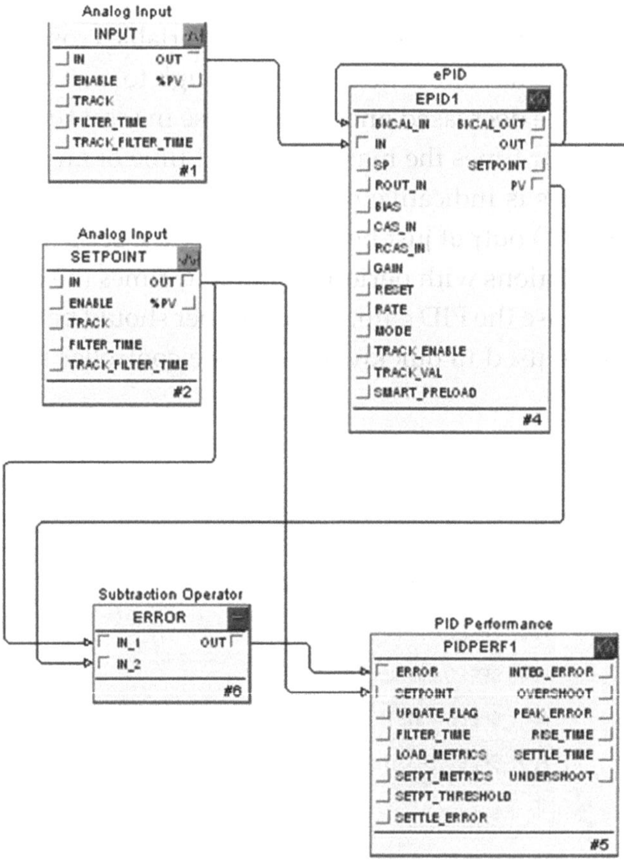

Figure 9-9a. Example of a PID performance metrics setup.

If load metrics are enabled, this block continually computes peak and integrated absolute errors. The load metrics are reset to zero if this metric is disabled. It is up to the user to periodically disable these metrics or apply forgetting logic. If the set-point metrics are enabled and a set-point change larger than the threshold setting occurs, the rise time (time for the process variable to first reach within the settling error of the set point) is computed with subsequent metrics. The maximum error in the same direction of the set-point change after the first set point crossing is the overshoot. The undershoot is the maximum error in the opposite direction after the second set point crossing. The settling time is the time after overshoot for the filtered error to stay within the settling error. Note that an error filter is used to prevent reversals in the sign of the direction of the error change. The set-point metrics are reset to zero if there is another set-point change or the set-point metrics are disabled. Thus, there is no need for a period reset or forgetting factor for set-point metrics. The peak error is not useful for set-point metrics because it will simply be the set-point change plus the overshoot if it is larger than the previous peak error value.

9.10 Generating and Fitting Profiles [1]

See Section 8.10 for more on performance metrics, such as rise time; for key diagnostics, such as the excessive time the PID is at the output limit or not in best mode; and for the distinguishing characteristics of oscillations. The conditions that trigger these metrics and diagnostics can be tested in the digital twin, leading to a better understanding and avoidance of conditions that cause problems.

The future value block shown in Figure 9-10a provides a continuously updated rate of change with a good signal-to-noise ratio for each model execution. This block enables computing the slope of profiles for any type of batch operation (e.g., the cell growth rate and product formation rate). The slope can be used as an MPC- or PID-controlled variable, changing a batch process effectively into a continuous process. The future value block uses the rate of change to predict a value that is one dead time or more into the future to enable preemptive evaluation and adjustments to improve batch operation, typically at the midpoint and end point of the batch cycle. The results of a future value block can be used to improve the set-point response and operator interface of most loops by showing where a key process variable will be 1 to 1.5 dead times into the future.

Key CONCEPT

A future value block can continually update batch profile slopes that can be used for model fitting and as controlled variables, enabling a batch process to have a continuous response.

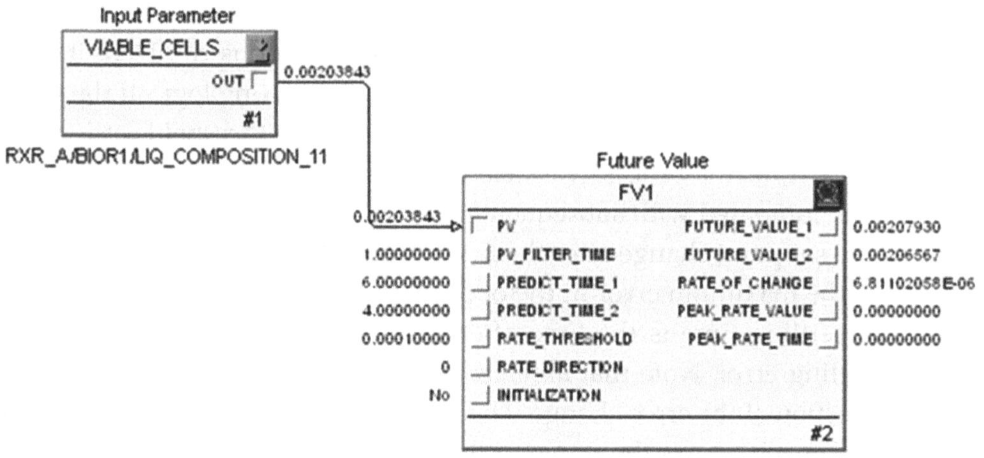

Figure 9-10a. Example of future value block estimating cell compositions.

> **Key CONCEPT** A future value block can provide predictions one or more dead times into the future of key process variables for greater operational and process understanding and better set-point response.

The future value block should be tuned for the dead time of the system for the prediction time. With a profile in place, the predicted value can be compared to an expected value to determine when a batch might be failing or should be proactively corrected. Observing a full run of values and change rates enables fitting a profile.

> **Key CONCEPT** To fit a profile, the agitation, concentrations, temperature, liquid volumes, feed and reagent flow rates, and *p*H must be known.

9.11 Best Practices [1]

1. Implement a digital twin that requires no changes to the DCS alarms, graphics, or logic.

2. Use a digital twin with the actual set points, control capability, historian, and operator interface.

3. Use first-principle process models with energy and material balances for all processes and charge balances for *p*H with representative physical properties and components.

4. For *p***H** models, add a small but significant concentration of carbonic acid from exposure to atmospheric CO_2 and conjugate salts so the titration curve generated matches lab results, particularly regarding the *p***H** range and titration curve slope in the control region.

5. Use an actual download of an automation system (not an imitation by emulation) that enables free exchange between the digital twin and the actual system of

   - the configuration, alarms, and APC tools (e.g., adaptive control, MPC, and data analytics) by exporting and importing modules and models and

   - the historian and graphics by copying and pasting files.

6. Achieve at least medium fidelity by adjusting model parameters so operating points sufficiently match process variables and flows on accurate process flow diagrams.

7. Use a variable dead time block to simulate mixing and injection delays.

8. Ensure the process model can show the dynamics in start-up and transitions.

9. Have process engineers evaluate process operating conditions and responses.

10. Include automation system dynamics such as the deadband and resolution limits in control valves (e.g., backlash stiction) and VFDs (e.g., I/O cards, speed sensor, setup), sensor transportation delays, sensing element lags (e.g., thermowells), transmitter lags (e.g., damping), wireless update rates, and analyzer cycle and analysis times.

11. Develop, test, and document scenarios, including transitions, start-up, and abnormal conditions (e.g., mechanical and automation system failures).

12. Develop and use online metrics for process efficiency and capacity.

13. Analyze and improve alarm management and the human-machine interface using the Center for Operator Performance and International Society of Automation standards for guidance.

14. Detail best operator responses during start-ups and to abnormal situations (e.g., equipment, valve, or sensor failures) for procedure automation (state-based control).

15. The digital twin can provide methods and practical considerations in online metrics computations, be adapted to improve fidelity by manually and automatically adjusting model parameters to match up virtual and actual flows, and be used to discover and realize process control improvement opportunities.

16. Ensure the operator training system (OTS) is continually updated and available for refreshing and enhancing operator performance.

17. Use OTS to educate technicians and engineers in maintenance and in automation, mechanical, and process technical support to improve the synergy with operations. Use a multidisciplinary team that includes plant operations. If the models will be deployed online, engage the operators early in the cycle. Start with your business objectives. What problem should be addressed, and how will you measure success in accomplishing these objectives?

18. Use a high-fidelity digital twin to develop online process metrics.

19. Synchronize process inputs with process outputs for online metrics by using dead time and filter blocks for synchronization time (e.g., total dead time and open-loop time constant).

20. Use running averages of raw materials, energy, and resulting product for metrics.

21. Use different periods for running averages to show metric short- and long-term effects.

22. Improve the metric signal-to-noise ratio by a filter time of less than one-fifth of the synchronization time.

23. The shortest process evaluation period should be greater than twice the synchronization time. Additional metrics to alert operations can have shorter periods.

24. A period equal to the shift time shows the operators' performance, a period equal to the batch cycle time shows the batch operations performance, a period equal to a month helps predict effects on accounting cost sheets, and a period as short as six total loop dead times can alert the operator to consequences of actions taken (e.g., changing set points or modes) provided the signal-to-noise ratio is sufficient.

25. Involve accounting, plant management, and marketing in developing and using online metrics for real-time accounting.

26. Use a digital twin for faster and more effective metrics and process control improvement (PCI) implementation.

27. Use a digital twin to reduce plant testing required for metrics and PCI.

28. Develop an MPC model to adapt the digital twin without connecting to the actual plant.

29. Use actual plant set points and controller tuning settings adjusted for model speedup in the digital twin.

30. To adapt the digital twin, adjust the model parameters to match the manipulated process inputs by the MPC, PID loops, and noncontrolled process outputs.

31. To reduce the amount of identification needed in the actual plant for online process metrics synchronization, add injection and mixing delays and automation system dynamics to the digital twin to develop metrics.

32. Use a digital twin to test new strategies and process metrics when a good model is available.

33. Use historized data to tune a model by matching model operating points, manipulated flows, and batch slopes with the convenient cardinal equations (use the set points and alarm settings for growth and product formation with the convenient cardinal equations). Model other dynamics using Michaelis-Menten equations by using low and high alarm settings for the limitation and inhibition parameters. Fine-tune the model by subsequently adjusting parameters including yields and by-product formation.

34. Account for the nonidealities of valves and sensors, including slow response time, deadband, and resolution limits when building models and tuning loops.

35. Take advantage of the future value algorithm capabilities to compute rates of change (e.g., batch profile slopes) and to predict key process parameters (e.g., batch end points and set-point responses) one to one and one-half dead times into the future.

36. Include the buffering effect of salts, weak acids, weak bases, and dissolved carbon dioxide along with electrode lags, and reagent injection and mixing delays whenever modeling and controlling *p***H**.

37. Use speedups to quickly test new strategies and model parameters, realizing that model dead time must be reduced proportionally for existing tuning settings to be reasonable.

38. Model wireless update rates and analyzer sample transportation delays and cycle times and use an enhanced PID block (e.g., PIDPlus) with external reset (e.g., dynamic reset limit) to accommodate wireless updates and analyzer cycle times. Turning on external reset can stop oscillations from deadband, stiction, and slow secondary valves, VFDs, and secondary loops by connecting the manipulated stroke, speed, or flow as external-reset feedback.

39. Use Rough N Ready tuning settings as a first step to check PID tuning and response capability.

40. Set up the control system and metrics to understand possible opportunities and loop performance.

41. Build all stakeholders' understanding by using the digital twin as a laboratory for exploration, discovery, education, innovation, and excellence using "before" and "after" online metrics.

References

[1] Gregory K. McMillan, Christopher Stuart, Rehman Fazeem, Zachary Sample, and Timothy Schieffer, *New Directions in Bioprocess Modeling and Control*, 2nd ed. (Research Triangle Park, NC: ISA [International Society of Automation], 2021).

[2] Gregory K. McMillan and Mark S. Sowell, "Virtual Control of Real pH," *Control* (November 2007).

[3] Gregory K. McMillan, Roger D. Reedy, and John P Moulis, "Virtual Plant Provides Real Insights," *Chemical Processing* (October 2008).

[4] Gregory K. McMillan and Martin E. Berutti, "Virtual Plant Virtuosity," *Control* (August 2017).

[5] Gregory K. McMillan and P. Hunter Vegas, *Process/Industrial Instrumentation and Controls Handbook*, 6th ed. (New York: McGraw-Hill Education, 2019).

[6] Christopher Stuart, Rehman Fazeem, and Gregory K. McMillan, "Improve pH Control," *Chemical Processing* (March 2017).

[7] Gregory K. McMillan and Christopher Stuart, "Simulation Breeds Innovation," *Control* (November 2019).

[8] Gregory K. McMillan, "Simulation Enhances Career and System Performance," *Control* (August 2020).

Appendix A Automation System Performance Top 10 Concepts

Automation system performance depends not only on tuning but also on the performance of each component in each control loop. System performance can be maximized by attention to design and implementation details that affect the ability of the controller, final control element(s), equipment, piping, sensors, and transmitters to do their jobs. Here is a summary of the top 10 concepts to guide the automation engineer in this endeavor [1]. We finish with a perspective on the relative importance of accuracy and precision.

A.1 Delay

"Without dead time, I would be out of a job."

- Fundamentals
  - A more descriptive name would be *total loop dead time*. The total loop dead time is the amount of time required for the start of a change that has occurred

in the loop to travel around the complete control loop and end up at the point of origin. For example, an unmeasured disturbance cannot be corrected until the change is seen in the controlled variable and the correction arrives in the process at the same point as the disturbance.

  - Process dead time offers a continuous train of values, whereas digital devices and analyzers offer noncontinuous data values at discrete intervals. The resulting delays add a phase shift and increase the ultimate period (decrease the natural frequency) like process dead time.

- Goals
  - Minimize delay (the loop cannot do anything until it sees and enacts change).
- Sources
  - Pure delay from process dead times and discontinuous updates
    - Piping, duct, plug flow reactor, conveyor, extruder, spin-line, and sheet transportation delays (pure process delays set by mechanical design; remaining pure delays set by automation system design)
    - Digital device scan, update, reporting, and execution times
    - Analyzer sample processing, analysis cycle time, and multiplex time
    - Sensor threshold sensitivity
    - Signal resolution
    - Valve stiction
    - Valve backlash (lost motion)
    - Variable speed drive deadband
    - Wireless trigger level
    - Wireless update rate
  - Equivalent delay from lags
    - Mixing, column trays, dip tube size and location, heat transfer surfaces, and volumes in series (process lags set by mechanical design; remaining lags set by automation system design)
    - Thermowell lag

- Electrodes
- Transmitter damping
- Signal filters

A.2 Speed

"Speed kills"—high-speed processes and disturbances with low-speed control systems can kill performance.

- Fundamentals
  - The rate of change in four dead time intervals is most important. By the end of four dead times, the control loop should have completed most of its correction. Thus, characterizing self-regulating processes with a large time constant to dead time ratio as near-integrating processes is consistent with performance objectives.
- Goals
  - Make control systems faster and make processes and disturbances slower.
- Sources
  - Control system
    - Proportional-integral-derivative (PID) tuning settings (gain, reset, and rate)
    - Slewing rate of control valves and velocity limits of variable speed drives
  - Disturbances
    - Steps: batch operations, on-off control, manual actions, safety instrumented systems (SISs), start-ups, and shutdowns
    - Oscillations: limit cycles, interactions, and excessively fast PID tuning
    - Ramps: reset action in the PID controller
  - Process
    - Degree of mixing in volumes due to agitation, boiling, mass transfer, diffusion, and migration

A.3 Gains

"All is lost if nothing is gained."

- Fundamentals
    - Gain is the change in output for a change in input to any part of the control system. Thus there is a gain for the PID controller, valve, disturbance, process, and measurement. The open-loop gain that is the product of the valve, process, and measurement gain is needed for controller tuning. Knowing the disturbance gain (e.g., the change in manipulated flow per change in disturbance) is important for sizing valves and feedforward control.
- Goals
    - Maximize control system and process gains.
        - Maximize valve sensitivity.
        - Maximize sensor sensitivity.
        - Maximize transmitter gain (minimize calibration span).
        - Maximize controller gain.
        - Maximize process sensitivity (e.g., find the best distillation column tray).
    - Minimize disturbance gain (size and impact).
        - Minimize the size of upstream disturbances by automation (e.g., elimination of oscillations and transfer or absorption of variability by control).
            - Decrease limit cycles.
            - Decrease interactions.
            - Replace operator actions and on-off control with throttling control.
            - Increase absorption of flow variability by surge tank level tuning.
            - Use redundancy to minimize failures (e.g., middle signal selection).
        - Minimize the size of process disturbances by enabling process self-regulation (e.g., cancel or moderate disturbances).
            - Get internal reflux control by tight distillate level control.
            - Use weak acid and base reagents to buffer *p*H systems.

- Sources
  - Abnormal operation and failures (e.g., pump, measurement, and valve failure)
  - PID controller gain
  - Inferential measurement location (e.g., the sensitivity of temperature change as an inference of composition change in the distillation column)
  - Sensor type (e.g., the threshold sensitivity of radar versus differential pressure [d/p] level measurement and resistance temperature detector [RTD] versus thermocouple)
  - Control valve or variable speed drive installed characteristic (gain is slope that includes the effect of inherent characteristic and system pressures)
  - Measurement calibration (gain is 100%/span)
  - Process design
  - Backlash and deadband (limit cycle amplitude)
  - Stiction and resolution limits (limit cycle amplitude)
  - Interaction (fighting loops oscillation amplitudes)
  - Operator actions
  - SISs
  - Sequences
  - Variations in stream composition, density, temperature, and physical properties, particularly for recycle streams, heat integration, and stressed utility systems

A.4 Resonance

"Don't make things worse than they already are."

- Fundamentals
  - An oscillation period that is close to the ultimate period can be amplified by feedback control, as seen in Figure A-4a.
- Goals
  - Make the disturbance oscillation period slower or the control loop faster.
  - For interacting loops with similar priorities, make the fast loop faster.

- For interacting loops with different priorities, make the least important loop slower.

- Sources
  - Control loops in series with similar dead times (e.g., multiple-stage *p***H** control)
  - Control loops in series with similar tuning and valve stiction and backlash
  - Control loops that affect each other with similar dynamics
  - Control loops in series with improperly set priorities and levels of importance
  - Day-to-night ambient changes to slow loops (e.g., column temperature control)

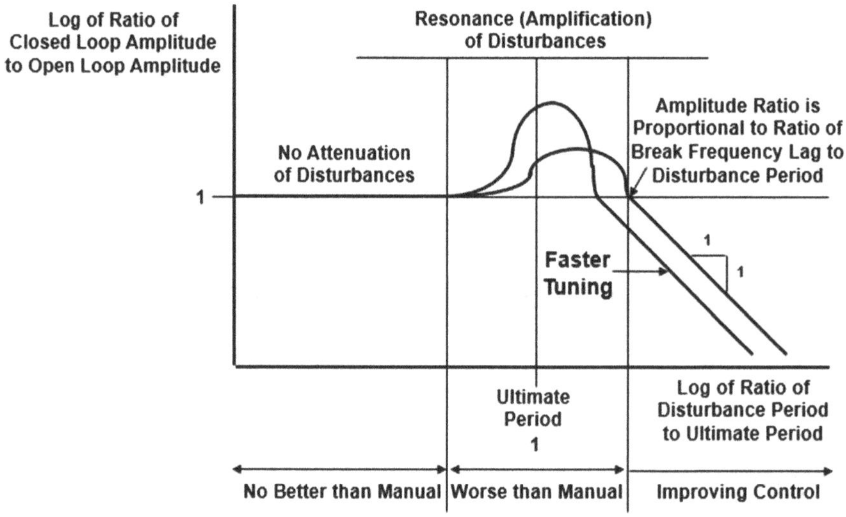

Figure A-4a. An oscillation with a period near the ultimate period (frequency near the natural frequency) is amplified by the PID controller due to *resonance*.

A.5 Process Attenuation

"If you had a blend tank big enough, you would not need control."

- Fundamentals
  - Attenuation increases as the volume of the blend tank increases and the ultimate period of the control loop decreases (the filter time constant is back-mixed volume residence time).

- Goal
  - Maximize attenuation by increasing volume and mixing and making loops faster.

- Sources
  - Mixed volume size and degree of back mixing (a lower feed rate increases attenuation by increasing residence time for continuous processes and makes the rate of change more gradual for batch processes)

$$A_f = \frac{A_o}{\sqrt{1+(\tau_f \bullet \omega)^2}} \quad \text{(A-5a)}$$

The attenuated (filtered) amplitude (A_f) can be estimated from the original amplitude (A_o) and an attenuation (filter) time constant (τ_f) that is much greater than the oscillation period (T_o) using the relationship between oscillation frequency and period ($\omega = \frac{2 \bullet \pi}{T_o}$):

$$A_f = \frac{A_o \bullet T_o}{2 \bullet \pi \bullet \tau_f} \quad \text{(A-5b)}$$

Equation A-5b is also useful for estimating the filtering effect of the thermowell lag, electrode lag, transmitter damping, and signal filtering on process oscillation. In addition, the equation can be used to estimate the original process amplitude given the measurement amplitude when the measurement lag is causing an attenuated version of the real process variability.

A.6 Threshold Sensitivity and Resolution

"You cannot control what you cannot see."

- Fundamentals
  - The minimum possible change measured or manipulated is set by sensitivity and resolution. Once the threshold sensitivity limit is passed, the full change is seen or used, but the resolution limit will quantize the change (stair step where the step size is the resolution limit) as shown in Figures A-6a and A-6b. Both will cause a limit cycle if there is an integrator in the process or control system.
- Goal
  - Improve threshold sensitivity and resolution.
- Sources
  - The minimum change measured and communicated (e.g., sensor threshold and wireless update trigger level) and the quantized change of analog-to-digital (A/D) and digital-to-analog (D/A) converters
  - The minimum change manipulated (e.g., valve stick-slip and speed resolution)

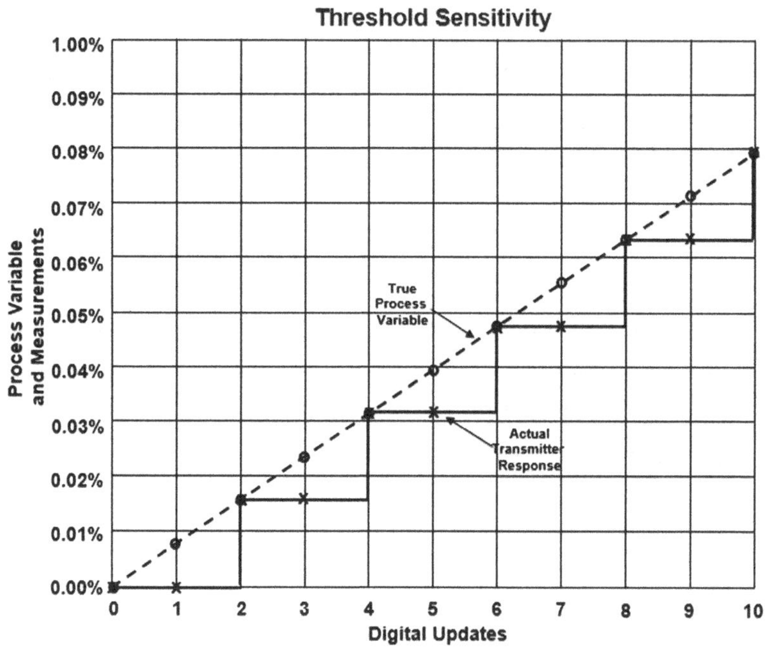

Figure A-6a. When the change in the process variable exceeds the *threshold sensitivity limit*, the full change in the process variable is seen.

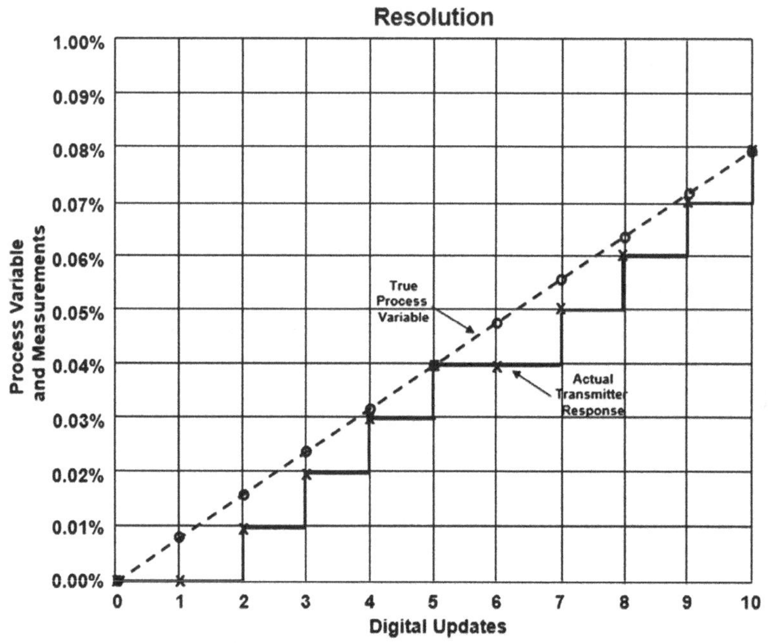

Figure A-6b. When the change in the process variable exceeds the *resolution limit*, the change in the process variable is seen as a step the size of the resolution limit.

A.7 Backlash (Lost Motion), Configuration Deadband, and Hysteresis

"No problem, man—if you never change direction."

- Fundamentals
  - Hysteresis is the bow in a response curve between full-scale traverses in both directions seen in Figure A-7b. It is usually much smaller and less disruptive than backlash deadband seen in Figure A-7a.

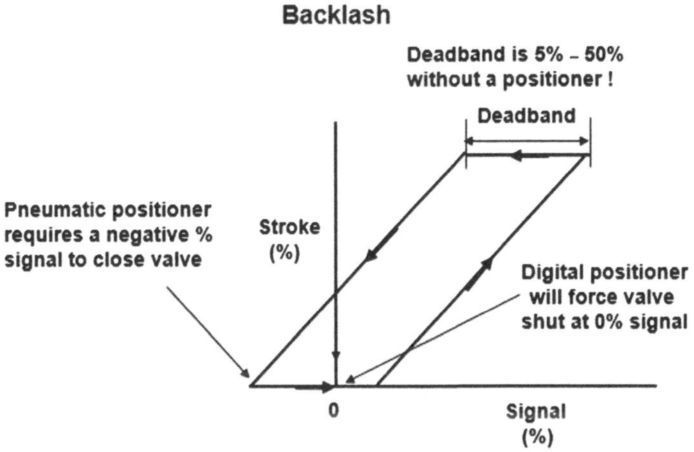

Figure A-7a. When the controller output changes direction and the change exceeds the *deadband*, the full change in the valve stroke or drive speed is seen.

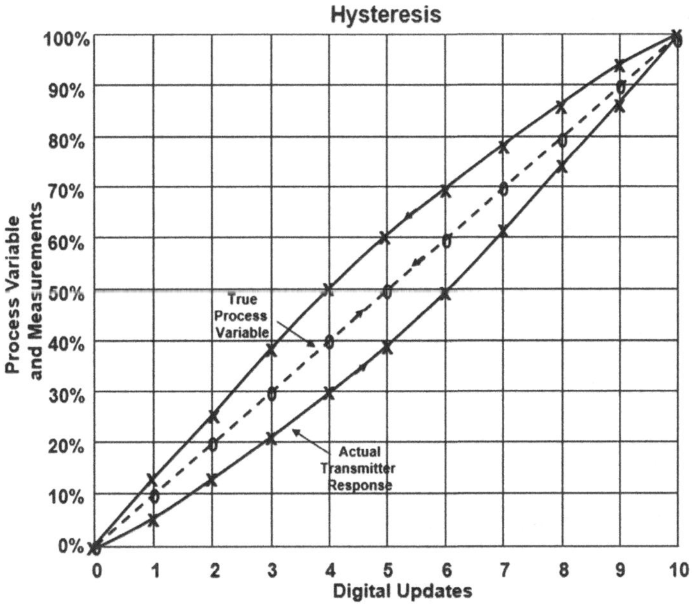

Figure A-7b. When the controller output changes direction, the path is bowed by the *hysteresis*, but the change is immediately and fully seen.

  - Backlash deadband is the minimum change measured or manipulated once the direction changes. Once past the backlash (lost motion) deadband limit, you achieve full change.
  - Backlash (lost motion) deadband will cause a limit cycle if there are two or more integrators in the process or control system.
  - Deadband can be a configuration parameter in a PID controller or a variable frequency drive (VFD).
- Goals
  - Minimize hysteresis in measurements (eliminate mechanical components).
  - Minimize backlash (lost motion) deadband in final control elements.
  - Minimize configuration deadband settings in the PID controller and VFD.
- Sources
  - Pneumatic instrument flappers, links, and levers (hopefully, these are long gone)
  - Rotary valve and damper links, connections, and shaft windup
  - VFD configuration parameters to eliminate hunting and chasing noise

A.8 Repeatability and Noise

"The best thing you can do is not react to deception."

- Fundamentals
  - Noise is extraneous fluctuations in measured or manipulated variables.
  - Repeatability is the difference in readings for the same true value in the same direction seen in Figure A-8a.
  - Repeatability is often confused with noise.
- Goals
  - Minimize the size and frequency of noise.
  - Do not transfer noise to the process.

- Sources
  - Noise
    - Bubbles
    - Concentration and temperature nonuniformity from imperfect mixing
    - Electromagnetic interference (EMI)
    - Ground loops
    - Interferences (e.g., sodium ions on a *p***H** electrode)
    - Velocity profile nonuniformity
    - Velocity impact on pressure sensors
  - Repeatability
    - Threshold sensitivity, resolution, and extraneous effects

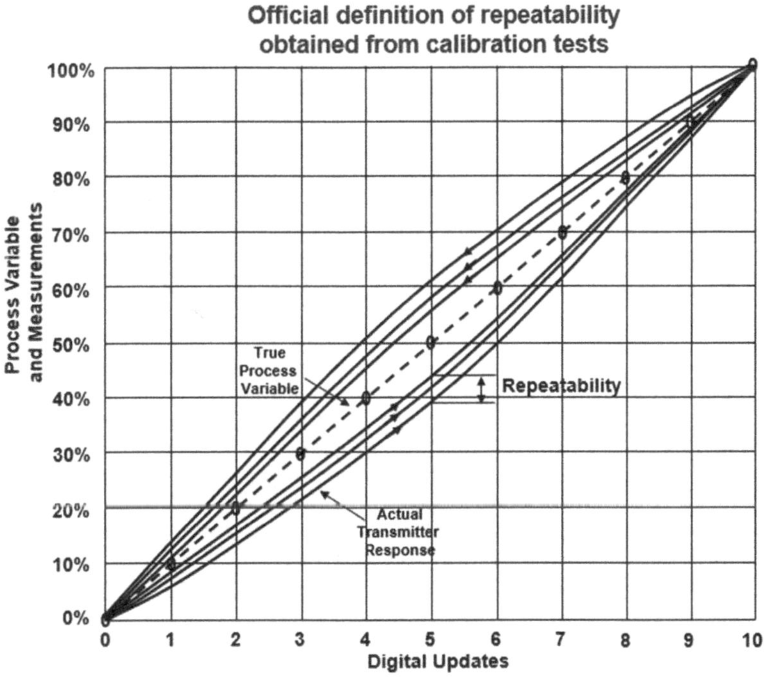

Figure A-8a. The maximum separation of the traverses of the scale range in the same direction is the *repeatability*.

A.9 Offset and Drift

"There is always an offset and drift; it is a matter of their size and consequence."

- Fundamentals
  - Offset is the deviation of the peak in the distribution of actual values from the true value seen in Figure A-7a.
  - Drift shows up as a slowly changing offset seen in Figure A-9b.
- Goals
  - Minimize offset and nonlinearity (see Concept A.10) by using smart transmitters, sensor matching, and smart-tuned digital positioners with accurate internal closure member feedback.
- Sources
  - Manufacturing tolerances, degradation, decalibration, and installation effects (process and ambient conditions and installation methods and location)

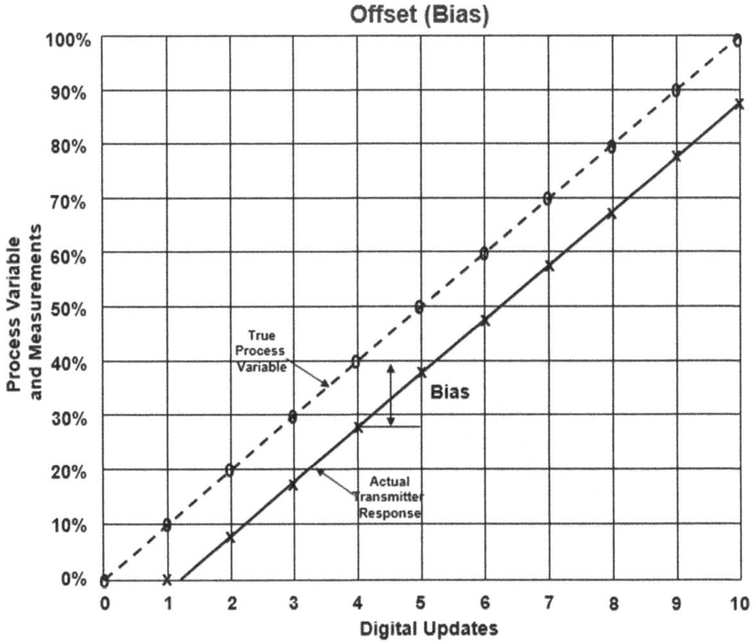

Figure A-9a. An *offset* (bias) is a constant difference between the process variable and the measurement in a complete traverse of the scale range.

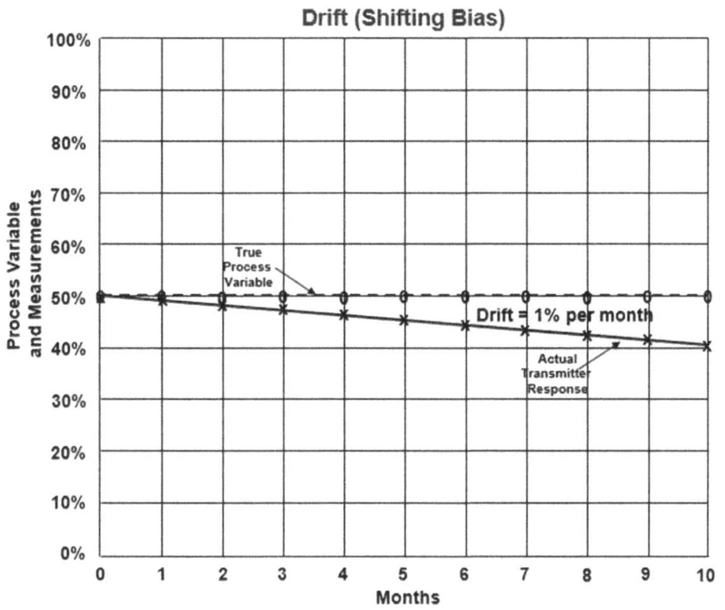

Figure A-9b. Drift is a shifting difference between the process variable and the measurement with time.

A.10 Nonlinearity

"Not a problem if the process is constant, but if the process is constant, you do not need a control system."

- Fundamentals
  - Although usually associated with a process gain that is not constant, as seen in Figure A-10a, in a broader view, a nonlinear system result occurs if a gain, time constant, or delay changes anywhere in the loop. In addition, all process control systems are nonlinear to some degree.
- Goals
  - Minimize nonlinearity with process and equipment design (e.g., reagents and heat transfer coefficients), smart transmitters and sensor matching, valve selection, signal characterization, and adaptive control.
- Sources
  - Control valve and variable-speed drive installed characteristics (flat at high flows)
  - Process transportation delays (inversely proportional to flow)

- Digital and analyzer delays (the loop delay depends on when change arrives in the discontinuous data value update interval)
- Inferred measurement (the conductivity or temperature versus composition plot is a curve)
- Logarithmic relationship between sensor output and process concentration (glass *p*H electrode and zirconium oxide oxygen probe)
- Process time constants (proportional to volume and density)
- Shape of equipment volume (particularly cross-sectional area for level control)

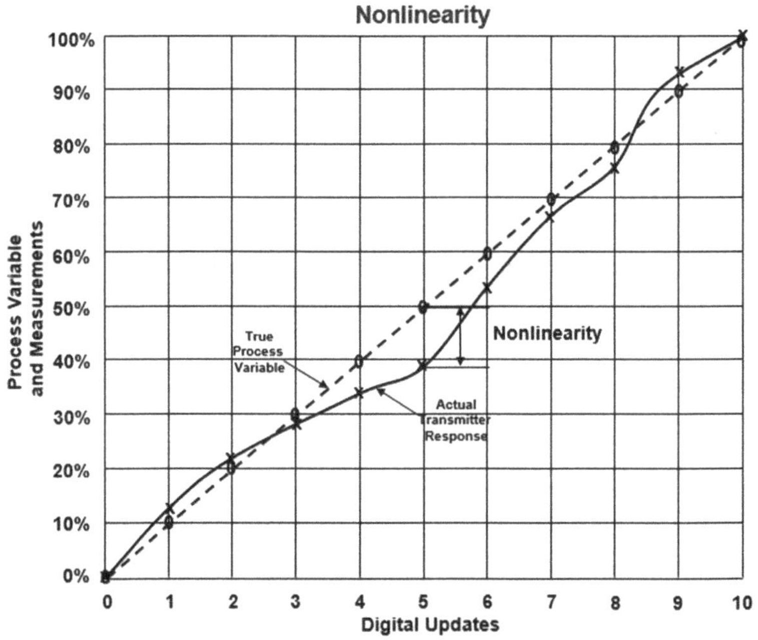

Figure A-10a. A difference between the process variable slope and the measurement slope for a traverse of the scale range is a gain *nonlinearity.*

A.11 Which Is More Important, Accuracy or Precision?

The accuracy of a measurement is determined as the difference between the true value and the measured value. The precision is how much the measured value changes for repeated measurements. In a plot of measurement frequency versus measured value for the same true process variable value (Figure A-11a), a mean measurement value closer to the true process value indicates better accuracy. A tighter distribution of measurement values (small standard deviation) indicates better precision.

Inaccuracy can be corrected by changes in the set point, either manually or by upper-level loops, so the effect is less on performance and more on process calculations and analysis. In contrast, a lack of precision adds an uncertainty for which there are few remedies for the resulting poor loop performance. For example, control valve stroke inaccuracy will introduce errors in system pressure drop and installed flow characteristic calculations, but the effect on loop performance will be minimal because the process control loop throttling the valve will adjust the signal to the valve as needed. Similarly, secondary flow measurement inaccuracy will create errors in material balance calculations, but the effect on process performance will be minimal because the primary controller in a cascade control system will eliminate the bias by adjusting the secondary controller set point.

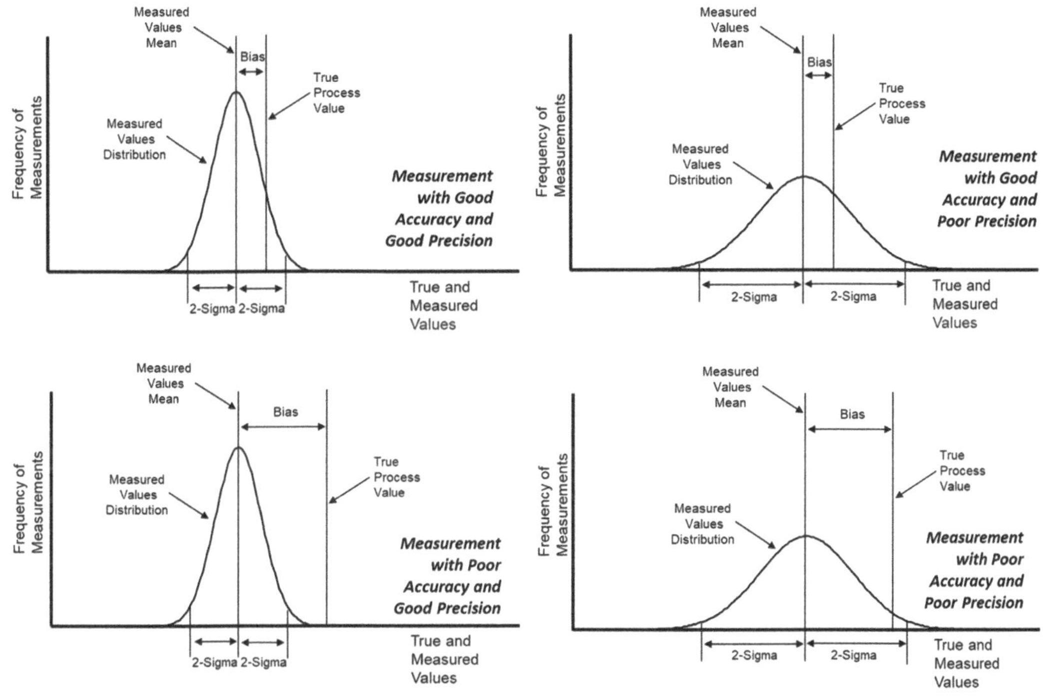

Figure A-11a. A set point can correct *accuracy*, but *precision* is prevalent uncertainty.

Consequently, improving factors that affect precision, such as threshold sensitivity, repeatability, and resolution—and minimizing backlash, deadband, and noise—are more important than reducing factors that affect accuracy, such as offset, drift, and nonlinearity. Besides the precision of measurements and final control elements, minimizing dead time and resonance, slowing down disturbances and making them smaller, optimizing process gains, and maximizing process attenuation are important for loop performance. What is left are the control strategies, controller tuning, and possibly adding feedforward or ratio control.

Reference

[1] Gregory K. McMillan, Christopher Stuart, Rehman Fazeem, Zachary Sample, and Timothy Schieffer, *New Directions in Bioprocess Modeling and Control*, 2nd ed. (Research Triangle Park, NC: ISA [International Society of Automation], 2020).

Appendix B
Questions and Answers

Questions

Chapter 1

1.1 For the strong acid and base system, what is the order of magnitude decrease in acid flow required to go from 1 *p*H to 6 *p*H?

1.2 For the strong acid and base system, why would the control error caused by reagent control valve stick-slip be less at 1 *p*H than at 6 *p*H?

1.3 For the strong acid and base system, what is the order of magnitude change in *p*H process gain if the *p*H increases from 1 *p*H to 6 *p*H?

1.4 For the strong acid and base system at 25°C, at what *p*H is the *p*H process gain the largest?

Chapter 2

2.1 If an influent stream has a specific gravity of 1.1 and a normality of 2.75 due to the presence of sodium hydroxide, what is the weight percentage of sodium hydroxide in the influent?

2.2 If the reagent contains 20% by weight hydrochloric acid, what flow in kilograms per minute of reagent is required to neutralize 100 kg/min of the influent described in Question 2.1 for an effluent specific gravity of 1.08?

2.3 At 7 *p*H, does the acid concentration equal the base concentration in normality units for a strong acid and base system?

2.4 At 7 *p*H, does the acid concentration equal the base concentration in normality units for a weak acid and base system?

2.5 If the temperature of a solution is 50°C, at what *p*H does the hydrogen concentration equal the hydroxyl concentration in normality units?

Chapter 3

3.1 Rank the titration curves in Figures 3-1a through 3-1d from least to most difficult for control at a 6 *p*H set point.

3.2 What location of *p*H set point would be the best for controlling a weak acid with a single dissociation constant titrated with a strong base?

3.3 What problems could occur with a laboratory titration that has sodium hydroxide in the sample?

3.4 Will the interval-halving search method ever fail to converge on the *p*H that corresponds to a zero excess charge?

3.5 Why are laboratory titration curves more accurate than computer-generated curves for mixtures of many different ions at high concentrations?

3.6 Why can't the pipeline method of titration curve generation be used for lime reagent?

Chapter 4

4.1 If the hydrogen activity is 0.0000001 normality for the internal fill of a measurement electrode without any application problems, at what *p*H is the isopotential point?

4.2 What is the slope of the millivolt versus *p*H curve if the solution temperature is 100°C and there are no application problems?

4.3 What are the potential causes of an erratic *p*H reading?

4.4 What are the potential causes of a slow *p*H reading?

4.5 What is the potential cause of a shortened span only at the high end of the *p*H scale?

4.6 What is the potential cause of a constant 5 *p*H reading?

4.7 What are the potential causes of a constant 7 *p***H** reading?

4.8 What are the potential causes of a constant negative *p***H** reading?

4.9 What is the sample transportation delay due to installing the electrodes in a 2-inch schedule 40 pipeline 50 ft from the effluent discharge nozzle for an effluent flow of 100 gal/min?

4.10 List the methods employed to reduce electrode coating problems.

Chapter 5

5.1 What is the equipment dead time and time constant for a vertical tank if the impeller diameter is 2 m, the inside tank diameter is 4 m, the agitator speed is 100 rpm, the working volume is 200 m^3, the density is 1000 kg/m^3, and the total input flow is 20,000 kg/min?

5.2 Which type of vessel provides the most effective filtering action for upstream disturbances for a recirculation flow that is 4 times the total of the feed and reagent flow?

5.3 Which type of equipment has the smallest dead time?

5.4 Which type of equipment has the largest dead time?

Chapter 6

6.1 What is the maximum allowable stick-slip for each titration curve shown in Figures 6-2a and 6-2b?

6.2 If the capacity of a control valve is doubled due to the lack of a trim size in the model number preferred for spare parts, what happens to the *p***H** error caused by reagent error for a titration curve slope equal to 10 *p***H** per percent reagent valve position?

6.3 From the control room, how could you detect a leaking acid reagent control valve for a split-ranged acid and base reagent system if there were no reagent flow measurements?

6.4 What is the transition point between split-ranged air fail closed acid and base reagent valves if the acid valve normality is 10, the base valve normality is 2.5, the capacity of each valve is 10 kg/min, and the split-range gap is 1%?

6.5 What is the dead time from a control valve with a deadband of 5% and a vessel *p***H** controller output signal changing at the rate of 0.1% per second?

Chapter 7

7.1 List the advantages of a properly installed caustic reagent dilution.

7.2 Why would a test solution with a high buffer capacity be a desirable *p***H** meter calibration?

7.3 What is the residence time required for a 99% conversion in a single vessel for the dry lime dissolution time shown in Figure 7-4b?

7.4 If the reagent injection length below the vessel surface cannot be decreased, what other options exist to reduce the reagent delivery time delay from back-filling the reagent pipe with vessel fluid?

Chapter 8

8.1 What happens to the time to the first peak for a given disturbance to a *p***H** feedback loop if the electrode becomes coated?

8.2 What advantages does a linear reagent demand controller have over a notch gain controller?

8.3 Why would a *p***H** feedforward controller be added?

8.4 If the influent *p***H** is below 0 *p***H**, should *p***H** feedforward be used?

8.5 If the spray water to the top of an absorber column needs to be *p***H** controlled and the *p***H** is measured at the discharge of the circulation pump, should the reagent be added to the spray as it enters the column, to the sump, or to the suction of the circulation pump?

8.6 Why should a linear reagent demand controller be used for adaptive control?

Answers

Chapter 1

1.1 The hydrogen ion concentration must decrease by 10^5. Because the hydrogen ion concentration is proportional to the acid flow for a strong acid and base system, the acid flow must also decrease by the same factor.

1.2 The slope of the titration curve, and therefore the *p***H** process gain, is less at 1 *p***H** than at 6 *p***H**, so changes in the net acid or base concentration along the abscissa due to control valve stick-slip translate into smaller *p***H** changes. The limit cycle amplitude from the oscillation in reagent concentration from valve stick-slip is smaller due to the smaller process gain.

1.3 The $p\mathbf{H}$ process gain increases by a factor of 10^5 because the difference between 1 $p\mathbf{H}$ and 6 $p\mathbf{H}$ is 5 $p\mathbf{H}$, and the $p\mathbf{H}$ gain increases by a power of 10 for each $p\mathbf{H}$ unit increase toward 7 $p\mathbf{H}$ for a strong acid and strong base system.

1.4 The titration curve slope and, therefore, the $p\mathbf{H}$ process gain are the greatest at 7 $p\mathbf{H}$ for a strong acid and strong base system at 25°C.

Chapter 2

2.1 The influent density is 1100 g/L for a 1.1 specific gravity. The molecular weight is 40.01 per Table 2-3a. The valence or charge of sodium hydroxide is 1 per Table 2-2a. Equation 2-1e is solved for the weight fraction of sodium hydroxide in the influent and multiplied by 100%.

$$X_{ii} = \frac{N \bullet N}{z \bullet d} = \frac{40.01 \bullet 2.75}{1 \bullet 1100} = 0.1$$

weight percent = 0.1 • 100% = 10%

2.2 The hydrochloric acid concentration equals the sodium hydroxide concentration in the effluent for complete neutralization because they are both fully ionized. The effluent density is 1080 g/L for a 1.08 specific gravity. The molecular weight of hydrochloric acid is 36.46, and the molecular weight of sodium hydroxide is 40.01 per Table 2-3a. The valence or charge of hydrochloric acid and sodium hydroxide is 1 per Table 2-1a. The equation from Question 2.1 is ratioed for the weight fraction of hydrochloric acid to sodium hydroxide in the effluent. Equation 2-1i is used to calculate the required reagent flow in kg/min.

$$X_{re} = \frac{M_r \bullet N_r}{z_r \bullet d}$$

$$X_{ie} = \frac{M_i \bullet N_i}{z_i \bullet d}$$

Because $N_r = N_i$ and $z_r = z_i$:

$$X_{re}/X_{ie} = M_r/M_i = 36.46/40.01 = 0.9125$$

$$F_r = \frac{x_{re} \bullet x_{ii}}{x_{ie} \bullet x_{rr}} \bullet F_i = 0.9125 \bullet \frac{0.10}{0.20} \bullet 100 = 45.6\ kg/min$$

2.3 At 7 $p\mathbf{H}$ and for a pK_w equal to 14, the concentration of hydrogen ions equals the concentration of hydroxyl ions. The concentration of acid ions must then equal the concentration of base ions. Because a strong acid or base is completely ionized, the concentration of the acid or base must be equal if the solution temperature is 25°C.

2.4 Because a weak acid or base is partially ionized, the concentration of the acid equals the concentration of the base only if the acid dissociation constants are equal to the base dissociation constants and the solution temperature is 25°C.

2.5 At 50°C, the pK_w is 13.26 per Table 2-3b. Equation 2-31 shows that for equal hydrogen and hydroxyl ion concentrations, the hydrogen ion concentration is equal to the square root of the ionic product, or the *p*H is equal to half of the pK_w.

$$p\mathrm{H} = pK_w / 2 = 13.26/2 = 6.63$$

Chapter 3

3.1 The systems ranked from least to greatest slope and therefore largest reagent sensitivity and *p*H process gain at 6 *p*H are:

- Weak acid and weak base
- Weak acid and strong base
- Strong acid and weak base
- Strong acid and strong base

3.2 The best *p*H set point would be at a *p*H equal to the pK_a because this is the midpoint of the flat portion of the titration curve.

3.3 The caustic solution would absorb carbon dioxide, which would add a buffering effect that may not be in the process; the number of data points may be insufficient near the equivalence point due to the steep slope; and the error at the high *p*H end of the scale may be different than the field due to a sodium ion error.

3.4 The interval-halving search method will always converge because there is only one zero crossing of the charge balance if the *p*H is within the original search interval selected and if the arithmetic precision of the computer is sufficient for the *p*H system.

3.5 The laboratory titration curve would be more accurate due to the uncertainty about the activity coefficients from the high ion interaction.

3.6 The lime would not be completely dissolved with its ions in the solution.

Chapter 4

4.1 The internal fill is at 6 *p***H** per Equation 2-2b, so 6 instead of 7 *p***H** should be used in Equation 4-1i, and the isopotential point is shifted from 7 to 6 *p***H**.

4.2 The slope is the absolute temperature term in Equation 4-1i multiplied by the deviation of the *p***H** from the isopotential point.

$$\text{slope} = 0.1984 \bullet (T + 273.16) = 0.1984 \bullet (100 + 273.16) = 74.04 \text{ mV per } p\mathbf{H}$$

4.3 An erratic reading (noise) can be caused by measurement bulb abrasion, dehydration, etching, pure water, or gas bubbles.

4.4 A slow response can be caused by measurement bulb abrasion, bulb dehydration, bulb etching, bulb coating, non-aqueous solutions, or low sample velocity.

4.5 The alkalinity error causes the shortened span or bending over of the millivolt versus the *p***H** line at the high end of the scale.

4.6 A constant *p***H** reading is most likely due to a broken measurement electrode.

4.7 The measurement electrode terminal could be shorted to ground or to the reference electrode terminal, the *p***H** meter switch could be in the standby position, or the circuit could be open due to a broken electrode wire or a complete nonconductive coating of either electrode.

4.8 Because no known electrode failure could cause such an extreme millivolt, the source must be either a concentrated strong acid or an open circuit in the transmitter output.

4.9 The transportation delay is calculated via Equation 4-7b, where volume is the product of area and length after the cross-sectional area is calculated as 0.0233 ft and the flow is converted from 100 gpm to 13.37 ft^3/min.

$$\tau_{dc} = \frac{A \bullet L}{Q_s} = \frac{0233 \bullet 50}{13.57} = 0.086 \text{ min} = 5\ s$$

4.10 The preferred methods of reducing electrode coating problems are jet washing and flushing using a high sample velocity.

Chapter 5

5.1 Using Equations 5-3a through 5-3c:

$$N_q = \frac{0.4}{(D_i/D_t)^{0.55}} = \frac{0.4}{(2/4)^{0.55}} = 0.586$$

$$F_a = \rho \bullet N_q \bullet N_s \bullet D_i^3 = 1000 \bullet 0.586 \bullet 100 \bullet 2^3 = 468{,}800 \text{ kg/min}$$

$$\tau_{dp} = \frac{\rho \bullet V}{F_i + F_r + F_c + F_a} = \frac{1000 \bullet 200}{20{,}000 + 468{,}800} = \frac{200{,}000}{488{,}800} = 0.4 \text{ min}$$

5.2 The vertical tank provides the best filtering action because the equipment dead time is 20% of the residence time per Equation 5-3a, which means the filter time constant is 80% of the residence time.

5.3 The static mixer has the smallest magnitude of equipment dead time because it has the smallest volume.

5.4 The pond or lagoon has the largest magnitude of equipment dead time because it has the largest volume.

Chapter 6

6.1 Equation 6-2c is used to calculate the allowable reagent error. The maximum control valve stick-slip must be half of this allowable error per Equation 6-2d.

From Figure 6-2a:

$$E_r = 80 \bullet \frac{B}{A} = 80 \bullet \frac{0.1}{0.8} = 10\%$$

$$S_s = 0.5 \bullet E_r = 5\%$$

From Figure 6-2b:

$$E_r = 80 \bullet \frac{B}{A} = 80 \bullet \frac{0.1}{0.8} = 1\%$$

$$S_s = 0.5 \bullet E_r = 0.5\%$$

6.2 The reagent error is doubled, and the *p*H error is increased by a factor of 20.

6.3 If the leak is caused by a stuck stem, a valve stem position indicator in the control room could be used, but valve stem position transmitters and indicators are rarely installed. If the leak is caused by valve seat corrosion or erosion, the leak is difficult to detect and can result in a significant waste of both reagents. Monitoring the *p*H controller output would show a larger than normal base reagent demand. If the influent flow is stopped and the set point is above the acid reagent *p*H, the base reagent valve will remain open. An acid and base reagent system should have flowmeters to detect leaks.

6.4 Equations 6-4b and 6-4c are used to calculate that the signal portions are 79.2% and 19.8% for the acid and base valves, respectively. One of the valve actions must be reversed because the reagents have opposite effects on the *p***H** measurement. If the acid valve is reversed, the resulting split range works as indicated in the following table, which shows a split-range gap of 1% at the transition point.

$$S_1 = 99 \bullet \frac{X_1}{X_1 + X_2} = 99 \bullet \frac{1000}{1000 + 250} = 99 \bullet 0.8 = 79.2\%$$

$$S_2 = 100 - G - S_1 = 100 - 1 - 79.2 = 19.8\%$$

Controller Output (%)	Acid Valve Position (%)	Base Valve Position (%)
0	100	0
39.6	50	0
79.2	0	0
80.2	0	0
90.1	0	50
100	0	100

6.5 The valve dead time from the deadband is approximately equal to the valve deadband divided by the rate of change of the valve signal. For a 5% valve deadband and a 0.1% per second signal rate of change of controller output, the dead time is 50 s.

Chapter 7

7.1 The advantages of a properly installed caustic reagent dilution system are:

- Larger control trim size that is less likely to plug
- Smaller reagent transportation time delay
- Greater injection velocity
- Greater injection Reynolds number
- Lower freezing point
- Lower corrosion rate for steels

7.2 A high buffer capacity means that wash water or contaminant coatings on the electrodes would not change the *p***H** of the test solution.

7.3 From Figure 7-4a, the residence time must be 20 times the dissolution time or 640 min for the dry lime in Figure 7-4b.

7.4 The reagent could be diluted to increase the volumetric reagent flow rate, or the reagent pipe could be purged with a gas to prevent backfilling.

Chapter 8

8.1 The electrode time constant will increase. Most of this time constant becomes loop dead time because it is small compared to the vessel time constant, and the peak error is reached in one dead time, so the time to the first peak is increased.

8.2 At best, the nonlinear controller gain matches the *p***H** process gain at three points on the titration curve. The nonlinear controller has three points on the titration curve. The nonlinear controller has three additional settings (lower and upper breakpoints and notch gain) that may require some field adjustment due to the inadequacy of the straight-line approximation. The linear reagent demand controller eliminates the nonlinear effects of the titration curve and restores the full potential of the controller to that dictated by the loop dead-time-to-time-constant ratio. The linear reagent demand controller is tuned like any other conventional linear controller.

8.3 If the correction for an influent *p***H** disturbance must start before a time equal to one loop dead time has transpired, *p***H** feedforward control should be added.

8.4 The acid error below 0 *p***H** corresponds to a large reagent error due to the flatness of the titration curve. The error worsens with time, so compensating for the error is impossible. The *p***H** measurement will be higher than the actual *p***H,** so the feedforward signal will be less than required. While the error is in the safe direction, the performance of *p***H** feedforward control will probably not be greater than that obtained by the flow feedforward that is already incorporated into a linear reagent demand controller. The additional cost of a *p***H** feedforward controller is probably not justified.

8.5 The reagent should be added to the circulation pump suction to form a fast inline feedback loop. The circulation flow is normally held constant, and the *p***H** disturbance from the column sump is normally slow compared to the inline loop dead time due to the large equipment time constant of the sump volume. Adding reagent to the sump would add the dead time of the unagitated sump to the *p***H** control loop. Adding reagent to the spray entering the top of the column would add the transportation delay from the top to the sump and the equipment dead time of the sump to the *p***H** control loop.

8.6 The adaptive algorithm is not forced to continually compensate for an operating point nonlinearity and can be designed to concentrate on a time-varying nonlinearity. If the titration curve is fixed, the linear reagent demand controller compensates for the operating point nonlinearity. The justification for an adaptive controller is to compensate for a time-varying nonlinearity caused by a changing titration curve, uncertainties in the identified curve, a variable time constant, or variable dead time.

Appendix C Control Valve Positioners

C.1 History [1]

Historically, pneumatic positioners were high-gain proportional-only controllers with a relay to provide amplification and volumetric flow as needed. Because the error for a proportional-only controller is inversely proportional to the controller gain, the error was small and readily corrected by the process controller. The tests for valve response were step changes of 10% or more, not revealing deadband and resolution. It was not recognized that most changes in controller output, particularly for load disturbances, are successive small steps. Also, because there was no readback signal of the actuator shaft position unless a special device was installed, visual recognition of shaft movement required large signal changes. Consequently, many positioners did not respond well to small signal changes. For example, the 86.5% response time for a step change less than 0.2% could be 50 s for a step in the same direction and 100 s for a step in the opposite direction. Not understood is that this dramatic increase in response time can cause a constant amplitude oscillation similar to a limit cycle or that a temporary fix is to greatly increase the integral time in the process proportional-integral-derivative (PID) controller. If the PID controller has external-reset feedback, enabling the feature and coupling it with fast, accurate readback of valve position can enable increasing the process PID gain, giving tighter control.

Spool positioners were offered as less-expensive positioners. These positioners often could not respond to changes less than 2%. Also, on-off rotary valves were offered that were attractively priced. High capacity, low leakage, and low price became popular. The linkages and connections created a deadband from lost motion (backlash) of 5% or more, and the seal for tight shutoff created a resolution limitation from friction of 5% or more. Because the readback to the positioner was the actuator shaft position, the lack of response of the valve stem and the ball or disk itself was not known by the positioner. The consequences were seen in the process variable most quantifiable by a sensitive low-noise flow measurement.

Electronic and digital positioners came on the scene, but the diagnostics based on the actuator shaft or stem rather than the closure member did not address the prevalent problem of on-off valves posing as throttle valves. Because instrument and valve specifications have fields for capacity and leakage and not for response requirements, and because budgets emphasize cost reduction, this and other problems related to deadband, resolution, and response time persist.

Integral action was added to positioners, not realizing that the necessitated lower valve position controller gain increased response time, created limit cycles from lost motion and resolution, and increased limit cycle periods. Because some diagnostics flag an offset as undesirable, integral action is often turned on as a default, and an integral deadband is set just larger than the resolution to suppress limit cycles. This also made the valve travel gain more consistent. Not recognized is that overshoot in a positioner response and differences in travel gain with a small remaining offset from the positioner gain maximized are corrected by the process controller. The positioner's job is to provide a fast-immediate response to the demands of the process controller. Turning on integral action and creating a slower response time can also cause a violation of the cascade rule for fast loops. The positioner should be viewed as a secondary loop that will cause oscillations if it is not at least five times faster than the process loop that can be considered to be the primary loop in this cascade control. Many advances in positioner design described in Section C.7 deal with these issues.

C.2 Deadband [1]

Deadband is the change in signal required upon a reversal of signal direction to get the valve to start to respond. Deadband can be the result of lost motion and resolution. If the deadband is much larger than the resolution, the cause is most likely lost motion (backlash) shown in Figure C-2a or a series of 0.2% step changes in the signal.

Lost motion often results from gaps in linkages and imperceptible looseness in a shaft-to-stem and stem-to-closure member (e.g., ball, disk, and plug) connections aggravated

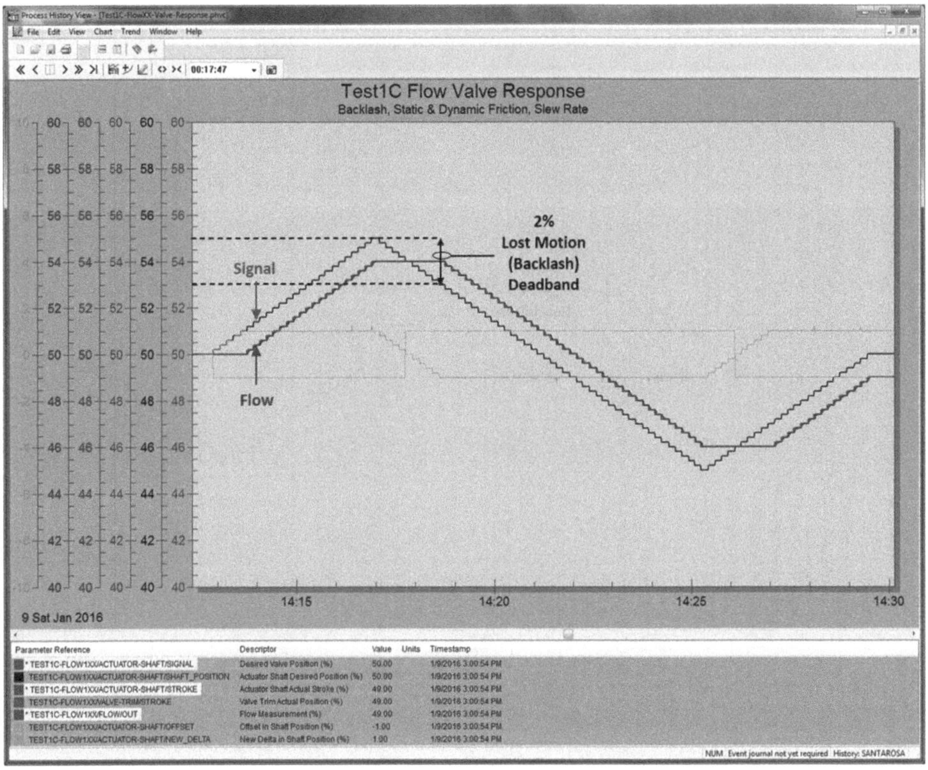

Figure C-2a. Open-loop test result for 2% lost motion.

by friction and a low gain positioner. The linkages, such as link-arm and scotch-yoke, and pinned for rotary valves tend to have the greatest lost motion. Rotary valves that have actuators with a rotary motion, splined shaft-to-stem connections, and integrally cast stem-to-closure members have the least lost motion, as do sliding stem valves with a good shaft-to-stem clamp connection to a vertical actuator. The distinction of whether the deadband is the result of lost motion or resolution is important for addressing and fixing the cause in valve design and mitigating the consequences by using features such as external-reset feedback and understanding the impact of tuning on limit cycles as detailed in Chapter 8, "Control System." particularly the counterintuitive reduction in lost motion oscillation amplitude by an increase in PID gain.

C.3 Resolution [1]

Resolution is the smallest change in signal in the same direction as the previous change in signal after a valve has stopped moving that results in a new valve movement in the same direction. The response to larger changes is a series of steps with a generally incomplete total response because of quantification by steps. Figure C-3a shows the 1% steps in valve position from the resolution that results from a series of 0.2% step changes in the signal. Diaphragm actuators can have a resolution of 0.1%,

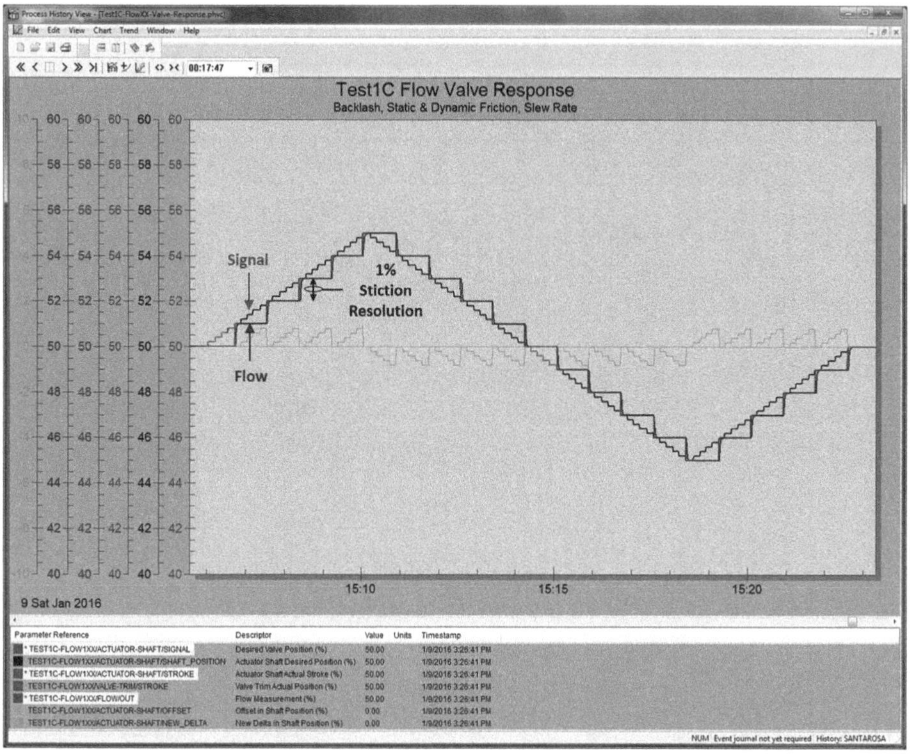

Figure C-3a. Open-loop test result for 1% resolution.

whereas piston actuators tend to have a sensitivity of 1% that can get worse from the wear and tear of piston seals. A large increase in response time of some positioners for step sizes below 0.2% can be translated to a resolution of 0.2%. Spool positioners may have a resolution of only 2%.

The steps result when there is a quantification of response from a rack-and-pinion or digital actuator or static friction to dynamic (e.g., sliding) friction ratio greater than 1. Static friction opposes the start of motion, and dynamic friction opposes the continuation of motion. The ratio is most often greater than 1, resulting in stick-slip (frequently termed *stiction*), increasing for greater friction. Friction is always present. Some sources of greater friction are packing, a bend or loss of smoothness In the sliding stem surface, a closure member rubbing against a seat or a seal, and an inadequately sized actuator. Valves with graphite packing and low leakage rates tend to have the greatest friction. Friction increases when packing is tightened and the valve is operating near the closed position (e.g., <20% open). Most response times from valve suppliers are for loose packing and midrange valve positions (e.g., 50% open). The actuator size quoted to reduce valve cost may result in a poorer-than-possible resolution, particularly near the closed position when pressure drops can be the highest. Practitioners can ask for the actuators to be sized to handle 150% or more of the maximum pressure drop to improve valve response.

C.4 86% Response Time (T_{86}) [1]

The 86% response time, the interval between the initiation of an input signal step change and the moment that the response of a dynamic reaches 86.5% of its full steady-state value, is used to compare the response of control valves. The step response time includes the dead time before the dynamic response. It is common for the step response time to vary, sometimes significantly, depending on many factors such as positioner type, actuator type and selection, valve friction, step size and direction, and reversals. The proportion of the step response time that is dead time, which limits performance, also varies because of some of these same factors. It is common for the step response time to be much larger for small changes in the input signal, especially signal reversal, and much larger than normal. It is also common for the step response time to be much larger for large changes in the input signal where the rate of position change is limited by the positioner flow coefficients. With proper tuning, smart positioners can generally provide more consistent step response times for small step changes. However, using low-flow relays in positioners to reduce air consumption results in a longer step response time. Except for special applications like compressor surge control, the small to medium step size response is more important to process control because the changes in PID output from one execution to another are usually quite small. When the valve step response time is a significant portion of the loop response time, a longer step response time for small or large steps can cause issues with the control loop performance. These issues can be mitigated by such techniques as scheduled tuning and external-reset feedback combined with accurate, fast valve position readback. Still, the best solution is to specify valves that meet the required step response time for specific step size ranges.

C.5 Testing [1]

After a positioner is tuned for maximum gain with minimal or no reset action, the response time should be measured for 0.2%, 2%, and 20% step changes in signal in both directions for the entire operating range with a goal of achieving a response time of 2 s or less for all cases, including those near the closed position. If integral action is used, an integral deadband should be set to stop oscillations. The travel gain improved by integral action, particularly for small steps, is less important than the response time made faster by more gain action because a fast response is more important than preventing position overshoot. For large actuators or compressor surge control, a volume booster on the positioner output(s) with a bypass valve slightly open may be used to meet the response time requirement for large step changes.

Testing should indicate the actual closure member (e.g., ball, disk, plug) position. The actuator shaft position used for position readback is not good enough for many types of valves. There can be lost motion from backlash and shaft windup for rotary valves and poor sensor readback connections for all valves. For these cases, a sensitive low-noise flow measurement may be needed to measure installed response, which has some advantages in terms of the actuator size being able to deal with operating conditions (e.g., fouling, pressure drop, temperature).

Also, the packing for high-temperature applications can have abnormally high friction at room temperature, which will negatively affect the valve performance. A more accurate performance test of the valve system in these applications may require that the valve be tested near the operation temperature if possible. In some cases, a modified packing setup (e.g., less torque on the packing gland follower bolts) at room temperature can emulate the packing friction at high temperatures.

C.6 Volume Boosters [1]

Using volume boosters instead of positioners, advocated for fast loops decades ago, can cause a severe safety risk. Positioners are required for piston actuators. It is not commonly recognized that using a volume booster instead of a positioner on a diaphragm actuator can cause an open butterfly valve to slam shut because of unbalanced forces on the disk and positive feedback from the diaphragm flexure slightly changing the gas volume and high outlet port sensitivity of the volume booster. If a volume booster is needed, it is put on the positioner output with the booster bypass valve slightly open to stop high-frequency oscillations, and the whole assembly (exemplified in Figure C-6a) is thoroughly tested. It may be necessary to increase the airline and actuator connection sizes and use a high-capacity air filter regulator.

C.7 Advances in Digital Positioner Capabilities [1]

Advances in digital positioners may improve valve response time, precision, and linearity to meet process control improvement objectives. It is important that plant Electrical and Instrument (E&I) design and maintenance engineers and technicians know how to enable these features and adjust the parameters needed, especially when technical support is not available (e.g., in the middle of the night). It is also essential that plant employees understand how particular features can help or inadvertently hurt the process PID performance.

Positioners may offer different parameters for opening and closing and for small and large signal changes, including a larger gain to start motion and smarter forms of

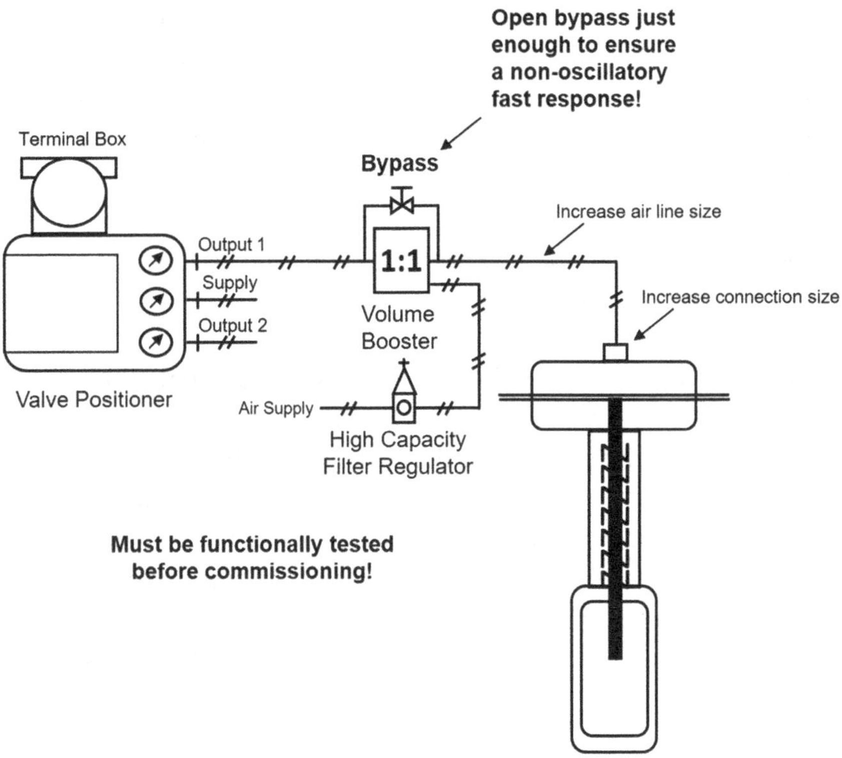

Figure C-6a. Volume booster on the positioner output for a diaphragm actuator.

derivative action to stop motion at the right position. Typically, positioners have an exhaust rate greater than the fill rate, making moving to the air-fail positioner faster but the response time slower for movement away from the air-fail position. Also, a positioner response time is generally slower for small signal changes because of sensitivity limitations and slower for large signal changes because of the slew rate limit. Without these advances, the PID should be tuned for the slowest response unless the it has external-reset feedback enabled with a fast, accurate readback.

Positioners may offer lost motion and resolution compensation. For example, a step is added for a change in signal direction that is equal to the deadband (the lost motion), and when motion stops, there is a kick equal to the resolution to start motion. The step and kick must not be larger than the associated lost motion or resolution, or a continual oscillation will develop. Because lost motion and resolution are often variable, changing with time in service, position, and process operating conditions, accurately setting the step and kick is challenging. A lead-lag may also be offered to help get through lost motion and resolution.

Positioners may offer the option to convert an inherent equal percentage flow characteristic into an inherent linear flow characteristic or vice versa. An equal percentage characteristic helps compensate for process gain changes in the temperature loops. However, the installed flow characteristic may be quite different than the inherent flow characteristic, particularly for ratios of valve pressure drop to total system pressure drop less than 0.5. Positioners may offer the ability to do a piecewise signal characterization. However, the availability of more process information and the need for better visibility and maintainability via an operator interface often result in the signal characterization being done in the PID configuration. The PID may be able to adapt the installed flow characteristic to deal with fouling.

Positioners may start to move the valve for an increasing signal at 0% and then move the valve to 15% for a 0.6% signal to ensure the first 15% is quickly bypassed because it is a range without flow. This can also be done via signal characterization and, in some PIDs, by increasing the reset action by a factor of 16 between a low output limit and an anti-reset windup limit. Some positioners may turn off the air supply to the actuator when the signal is less than 0.5% and go to full air supply to the actuator when the signal is greater than 99.5%. This prevents the PID output from having to overdrive and underdrive the positioner signal to fully open and fully close the positioner, respectfully.

C.8 Stiction [1]

Stiction is a common term used in the literature and workplace for the stick-slip behavior seen in a control valve response. Stick occurs when the valve is stationary and the change in actuating signal is insufficient to overcome static friction. When the change in the actuating signal is high enough to overcome the static friction, the valve suddenly moves because the dynamic friction is less than the static friction. Thee position after movement is determined by the balancing of actuating and frictional forces and the stiffness of the actuator system, and the actuating signal. If the actuating signal is continuously increasing and the net actuating force is always greater than the dynamic friction, once the valve is in motion, the valve's motion may be continuous after the initial step. On the other hand, if the net actuating force is lower than the dynamic friction at any time, the motion will stop and the stick cycle will repeat in the extreme, resulting in a staircase response. The degree of slip increases with the ratio of static friction to dynamic friction. Although the focus is on control valves, many valve response aspects apply to other final control elements, such as dampers and guide vanes.

ANSI/ISA-75.25.01-2000 (R2010), *Test Procedure for Control Valve Response Measurement from Step Inputs*, defines *resolution* as the "smallest step increment of input signal in one direction for which movement of the output is observed. Resolution is expressed as percentage of input span."

According to the definition, resolution is the tendency of a control valve to move in finite steps when responding to step changes in the input signal applied in the same direction. This happens when the control valve sticks in place, having stopped moving after the previous step change.

Note that this resolution behavior can be created by the impact of friction, as well as other phenomena, and that the measure of resolution from this standard is independent of its source.

The resolution metric is useful because it is definitive for testing, and the limit cycle that results from both resolution and stiction are similar in terms of general shape and the effect of process gain and tuning settings (described in Chapter 8). and in terms of possible solutions, such as the kick described in Section C.7, "Advances in Digital Positioner Capabilities," and by external-reset feedback as noted in Chapter 8. The limit cycle amplitude is proportional to the process gain, and the limit cycle period is faster for more aggressive tuning. Fast self-regulating loops (e.g., flow) have a square wave process variable (PV) response, and integrating loops have a triangular PV response. The undershoot from a slip less than stick and the overshoot from a slip greater than stick may be observable in an accurate low-noise flow measurement. Data-driven stiction models have been developed that help identify the undershoot and overshoot [2]. If the valve signal continues to ramp fast enough to keep the valve moving, there is no additional stick-slip. It is also important to distinguish the effect of hysteresis and lost motion and the patterns of the closed-loop oscillations in the PV and PID output for loops with small time constants (e.g., flow) and with a near or true integrating response [1].

Sliding and dynamic friction are too complex and variable to be completely predicted by first-principle models. The impact of stiction on PV is affected by the current to pneumatic converter (I/P); the pneumatic signal and air supply line size, length, and connection; actuator sensitivity; the piston actuator O-rings; the positioner sensitivity and tuning; possible booster design and setup; the stem packing type and tightening; the ratio of the static to dynamic friction for the packing and other internal components; the internal seal or seat or cage; fouling; and the installed flow characteristic. The static friction from graphoil packing can be 7 to 10 times greater than Teflon packing. The static friction near the closed position is possibly several times larger than at the mid-throttling range because of seat and seal friction, particularly for valves designed for tight shutoff. A "stiffer" actuator system, for example, a higher spring rate and a higher bias pressure in double-acting pneumatic actuators reduce the slip amount. Some actuator sizing methods do not consider static friction could increment the total thrust requirement by 50%. For this reason and the benefit of better resolution and the ability to deal with inevitable unknowns, actuator sizes are increased by at least 50%.

Before a change in signal direction, movement stops momentarily. The observed resolution or stiction is 50% to 100% of the observed deadband after subtracting lost motion, such as backlash and shaft windup that are affected by positioner-to-shaft connection, linkages, and shaft-to-stem connection. Valves designed for on-off service have particularly severe stiction and lost motion. If the valve signal moves in the original direction to the point of slip, the observed increase in deadband is almost twice the resolution or stick. Most citations in the literature show the increase in deadband as equal to the resolution or stick. Some do not note that it is necessary to subtract the lost motion from the deadband. Note that the official definition of *hysteresis* is a bowing of the response curves for increasing and decreasing signals that does not include a deadband. A common source of hysteresis is the flexure of an actuator diaphragm and spring.

If there are two or more integrators because of integrating action in the process or in multiple PIDs (e.g., in cascade control) and a positioner without integral deadband, limit cycles can develop from lost motion. Greater proportional and derivative action in the process PID and positioner can reduce the dead time and limit cycle period from resolution (stiction) and lost motion.

Data-driven stiction models must be updated periodically, and parameters must be stored as a function of valve position and operating conditions. If these models are accurate and up to date, they could be used to provide a more intelligent adaptive kick-in controller output to start motion. There is an opportunity for hybrid models to take advantage of some first principles to deal with changing conditions proactively. An accurate, fast readback of the actual valve position is needed for modeling and external-reset feedback. A rotary valve without a splined shaft-to-stem connection and with an integrally cast stem in a ball or disk suffer from twisting action, which leads to shaft windup and 8% or more error in the readback position if it is a tight shutoff valve. An error can also develop in the positioner feedback mechanism in connection to the actuator shaft. The update of an auxiliary HART variable for readback is typically not fast enough. A fast, precise, low-noise flow measurement is often needed because of readback limitations.

An unusually slow step response time for small signal reversals to the control valve for a fast, self-regulating, or integrating process can cause a "stick-slip" cycle identical to that caused by stiction. In fact, the two issues can both contribute to the stick-slip cycle.

The resolution metric serves a useful purpose by enabling a straightforward test and the consequential recognition of the tuning and the complete control valve design needed to achieve better performance. Actuator teeth and digital resolution also contribute to resolution limits. Using resolution and other definitive and easily measurable metrics, such as valve response time, deadband, travel gain, and process gain provide guidance for better valve applications without getting into unnecessary complications. For test results showing the impact of tuning and external-reset feedback on the lost motion from backlash, poor positioner sensitivity, static friction/positioner or I/P gain, analog-to-digital bit resolution, and resolution from stiction, see Reference 2. The proposed Annex A for ANSI/ISA-TR75.25.02-2000 (R2010) Control Valve Response from Step Inputs describes how these and other metrics affect performance and concludes with suggested minimum and maximum test step sizes and starting points.

References

[1] ANSI/ISA-TR5.9-2023, *PID Algorithms and Performance* Technical Report (Research Triangle Park, NC: ISA [International Society of Automation], 2023).

[2] Gregory McMillan and Pierce Wu, "How to Specify Valves and Positioners That Do Not Compromise Control," *Control* (March 2016).

Appendix D Review of Algebra with Logarithms

$$\log(m \bullet n) = \log(m) + \log(n)$$

$$\log(m / n) = \log(m) - \log(n)$$

$$\log(10^{p}) = p \bullet \log(10) = p \quad \text{antilog}(p) = p$$

$$\log(m \bullet 10^{p}) = \log(m) + p$$

Therefore, to convert from hydrogen activity to *p***H**:

$$p\mathbf{H} = -\log\ (a_{H})$$

$$p\mathbf{H} = -\log\ (m \bullet 10^{-p})$$

$$p\mathbf{H} = -\log\ (3 \bullet 10^{-4})$$

$$p\mathbf{H} = -[\log\ (m) - p]$$

For example, if the hydrogen activity is 0.0003:

$$pH = -\log\ (3 \bullet 10^{-4})$$

$$pH = -[\log\ (3) - 4] = 3.523$$

And to convert back from *p***H** to hydrogen activity:

$$a_H = -\text{antilog}\ (pH)$$

$$a_H = -\text{antilog}\ [-(\log\ (m) - p)]$$

$$a_H = m \bullet 10^{-p}$$

For example, if the *p***H** is 3.523:

$$a_H = -\text{antilog}\ [-(0.477 - 4)]$$

$$a_H = -\text{antilog}\ [-(\log\ (3) - 4)]$$

$$a_H = 3 \bullet 10^{-4}$$

Appendix E
Enhanced PID

E.1 History [1, 2]

Analyzers are a large source of dead time because of the sample transportation delay, analyzer cycle time, and multiplexor reporting time. The best location of an at-line analyzer is in the field as close as possible to the sample point to minimize the sample transportation delay and the changes in sample temperature. The location of the analyzer in an analyzer house is chosen to provide a better environment for the analyzer and its maintenance, but it is often not realized that the impact of the additional transportation delay on the total loop dead time can be significant. Also, because the analyzer result is at the end of the cycle time, the analysis dead time is one-half times the dead time. If the analyzer serves several unit operations, the dead time from multiplexing is significant. If the analysis is done in a lab, there is also a reporting time delay that can be huge.

Some proportional-integral-derivative (PID) controllers attempt to deal with this large, discrete update dead time by using an execution rate after a set-point change that equals the total dead time from the analyzer. However, this creates an additional dead time for unmeasured disturbances that is one-half the dead time from the analyzer.

If a PID is allowed to execute at a normal rate of X seconds when there is an update, the derivative mode considers the entire change from the last analyzer result to have occurred in X seconds, causing a kick or jump in the PID output from the derivative mode, so large derivative action is not used despite being generally beneficial on a near-integrating, true integrating, and runaway process.

For wireless measurements, the dead time is one-half the update rate plus any communication delays. Although the consequential dead time is much smaller than from an analyzer, it can be a concern for fast loops.

The enhanced PID does not execute unless a set point is changed or a process variable update is detected beyond the measurement noise or repeatability limit, indicating a new analyzer or wireless result. The positive feedback filter for integral action shown in Chapter 8, "Control System," Figure 8-12a, uses the elapsed time from the last update to compute an exponential result for the integral mode contribution from the updated filter input, that is, the new external-reset feedback signal. The derivative mode uses the elapsed time from the last update to compute the derivative mode contribution.

E.2 Performance [1, 2]

The enhanced PID provides more robust control, including protection against failures of the manipulated variable (e.g., a stuck valve or frozen secondary loop process variable) and the controlled variable (e.g., an analyzer or wireless measurement) because external-reset feedback is used and the input to the onetime execution of the integrator is the actual manipulated variable.

The greatest robustness and increase in performance occur for a self-regulating process when the analyzer or wireless dead time exceeds the 63% response time of the process (e.g., the process dead time plus process time constant). In this case, the PID gain can be increased to be the inverse of the maximum open-loop self-regulating process gain. The reset time can be as small as the maximum dead time without the analyzer or wireless device. Also, an increase in the dead time from the analyzer or wireless measurement does not require retuning. Offline analyzers could be used with extremely large and variable reporting times, but the ability to reject an unmeasured disturbance deteriorates with increases in the time between analyzer results and wireless updates.

For integrating processes, robustness is increased, and it may be possible to increase the PID gain slightly. Still, oscillations may start if the wireless and analyzer dead time approaches one-half the arrest time or the original loop dead time. In this case, the arrest time should be increased and the PID gain decreased accordingly.

E.3 Static Mixer Wireless pH Control Examples [3]

The enhanced PID controller dramatically improves the response of cascade loops by enabling a secondary loop immediate response to changes in its set point. Because the static mixer process dead time (6 s due to injection delay) and process time constant (6 s due to electrode lag) is much less than the default update rate (60 s), the enhanced PID controller gain can be increased. With the secondary flow and the primary *p***H** loop controller gain equal to the inverse of the process gain, the response to *p***H** set-point changes and load disturbances is impressive. In contrast, the traditional PID controller gains for both loops were a factor of 10 lower to prevent oscillations. If the default update rate of the secondary loop is not about four times faster than the primary loop, the enhanced PID controller gain must be detuned to stabilize the loops because the oscillations from the secondary loop can be amplified by the primary loop. For a primary *p***H** loop default update rate of 60 s, the secondary flow default update rate must be 16 s or faster.

For a 60 s *p***H** update time and load upsets, the enhanced PID controller reduced the Integrated Absolute Error by 40% in Figure E-3b. Because both the flow and *p***H** loop have a process dead time that is not significantly less than the process time constant, the peak error of the enhanced PID controller and traditional loop for a load disturbance is the full open-loop error (error with the loop in manual). For a 60 s *p***H** update time and set-point changes, the enhanced PID controller reduced the rise time by 60%, reduced the settling time by 84%, and eliminated the overshoot that was 40% of the set-point change in Figure E-3a. The enhanced PID controller eliminates the large upset caused by a failure and restoration of a pH measurement signal seen in Figure E-3c.

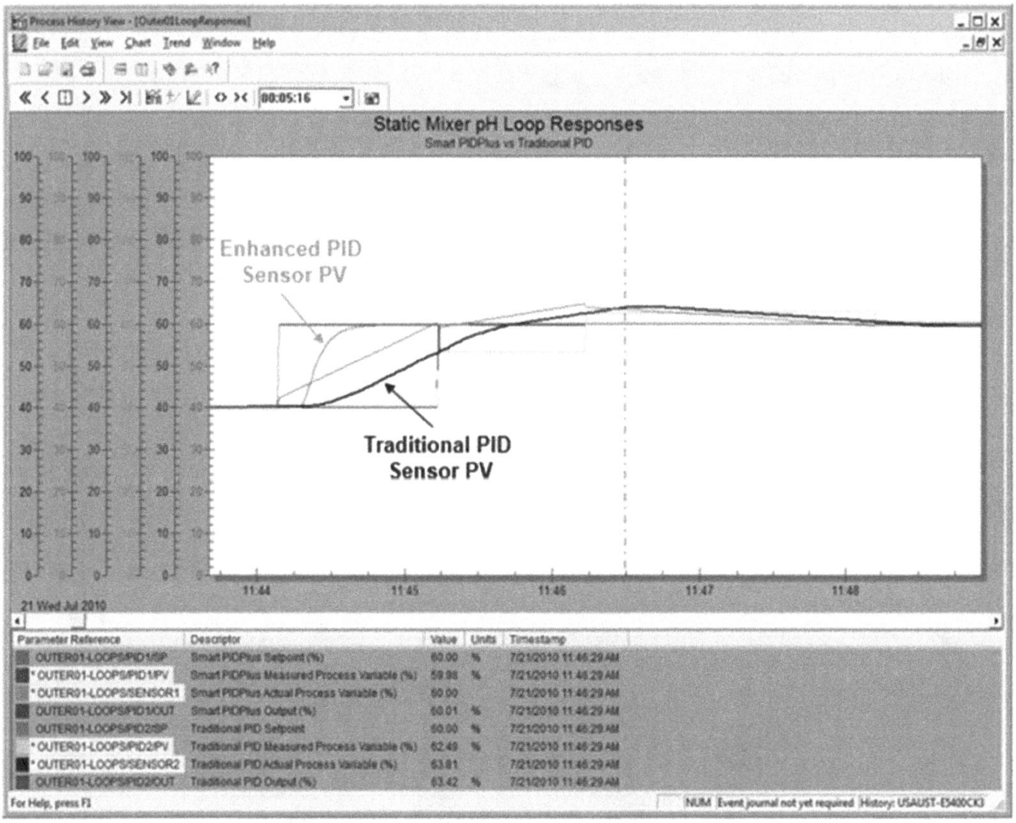

Figure E-3a. *p*H set-point response of enhanced and traditional PID (update time = 60 s) controllers.

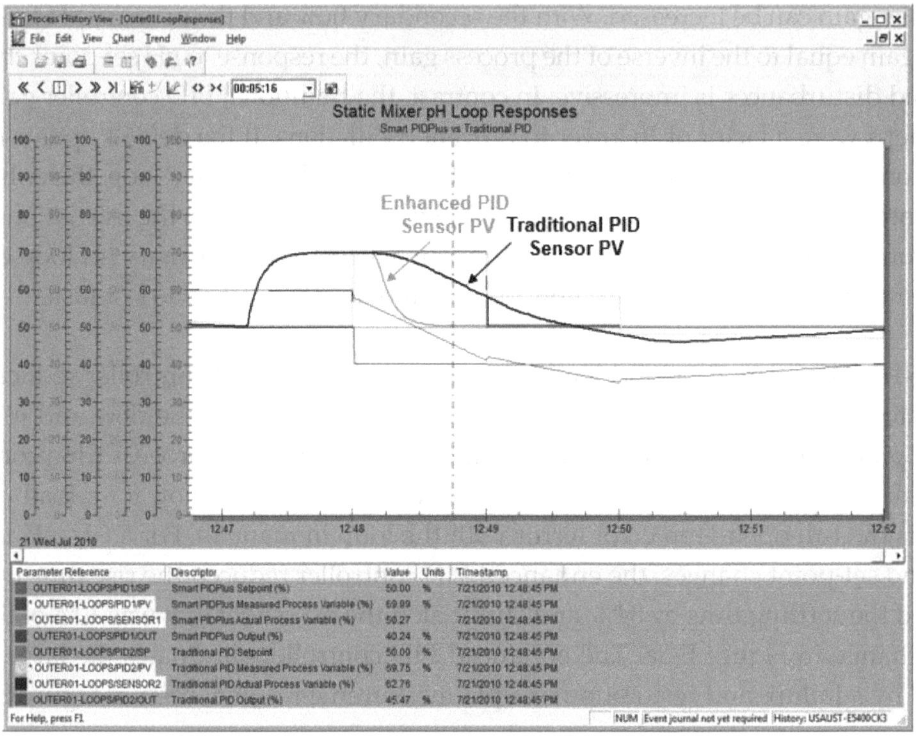

Figure E-3b. *p*H load response of enhanced and traditional PID (update time = 60 s) controllers.

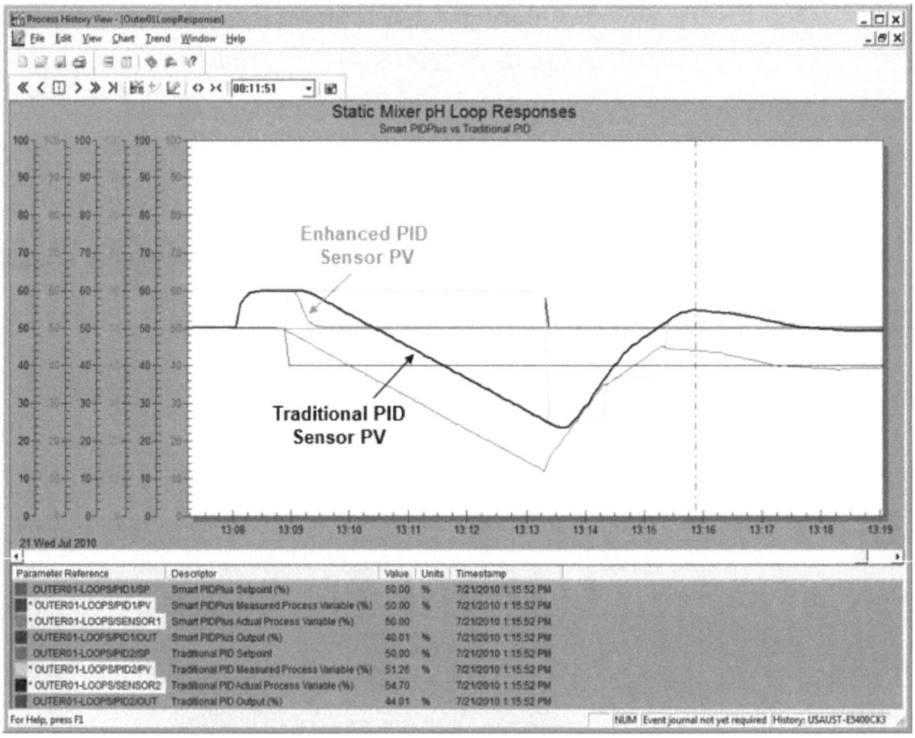

Figure E-3c. *p*H failure response of enhanced and traditional PID (update time = 60 s) controllers.

References

[1] ISA-TR5.9-2023, *Proportional-integral-derivative (PID) Algorithms and Performance* (Research Triangle Park, NC: ISA [International Society of Automation], 2023).

[2] Gregory K. McMillan, "Wireless: Overcoming Challenges in PID Control & Analyzer Applications," *InTech* 57, no. 4 (July/August 2010).

[3] Gregory K. McMillan, Christopher Stuart, Rehman Fazeem, Zachary Sample, and Timothy Schieffer, *New Directions in Bioprocess Modeling and Control,* 2nd ed. (Research Triangle Park, NC: ISA [International Society of Automation], 2020).

Appendix F First-Principle Process Relationships

F.1 General Implications

First-principle relationships can define process causes and effects that can lead to improved controller tuning and performance by selecting better tuning rules and process variables for scheduling tuning settings. They also affect the choice of control valve trim and the feedforward design. Understanding these relationships does not require a degree in chemical engineering but presumes some understanding of common terms (e.g., heat transfer coefficient and area), relationships (e.g., ideal gas law), and physical concepts (e.g., conservation of mass and energy).

Equations have been developed from first-principle relationships for the process gains, dead times, and time constants of volumes with various degrees of mixing. The results show that for well-mixed volumes with negligible injection delays, the effect of flow cancels out for the controller gain if one of the following methods is used: lambda self-regulating rule where lambda is set equal to the dead time, or the reaction curve method. The effect of flow also cancels out for the reset time in addition to the controller gain if the process is treated as a "near-integrator" and the lambda integrating tuning rule is used. This is because the flow rate cancels out in the computation of the ratio of process gain to time constant that is the "near-integrator" gain. This ratio and "near-integrator gain" are inversely proportional to the process holdup mass (e.g., liquid mass). For temperature control, however, the effect of changes in liquid mass cancels out because a change in level increases the heat transfer surface area covered. Several authors have mistakenly tried to schedule controller tuning based on liquid level for reactor temperature control. One author reported being bewildered by its failure. This is not the case for gas pressure control. The equations show that liquid level has a profound effect on the process integrating gain for vessel pressure control because it changes the vapor space volume without any competing effect. To summarize, the integrator gain for composition and gas pressure is inversely proportional to liquid level (liquid mass). For temperature, the effect of level cancels out unless the level is above or below the heat transfer surface area, which is unusual but can occur at the beginning or end of a batch when coils instead of a jacket are used for heat transfer. For temperature, the integrator gain is nearly always proportional to the overall heat transfer coefficient that is a function of mixing, process composition, and fouling or frosting.

The equations also show that if the transport delay for flow injection is large compared to the time constant, which does occur for reagent injection in dip tubes for *p***H** control, then the controller gain will be proportional to flow. Note that *p***H** control is a class of concentration control.

To control temperature and concentration in a pipe, the process dead time and process gain are both inversely proportional to flow and the process time constant is essentially zero, which makes the actuator, sensor, transmitter, or signal filter time lag the largest time constant in the loop. Thus, the largest automation system lag determines the dead time to time constant ratio. For a static mixer, there is some mixing, and the process time constant is inversely proportional to flow but is usually quite small compared to other lags in the loop. The controller gain is generally proportional to flow for both cases.

Finally, the preceding discussion has implications as to whether a flow feedforward multiplier or summer should be used and whether a linear or equal percentage trim should be used. A flow feedforward multiplier and equal percentage trim, which both have a gain proportional to flow, can help compensate for a process gain that is inversely proportional to flow provided the process time constant is not also inversely proportional to flow. This is generally the case for temperature and concentration control of plug flow volumes (pipelines, static mixers, and heat exchangers). For well-mixed volumes, feed-forward summers and an installed linear characteristic for valves are generally best. For control valves, this corresponds to a linear trim when the available pressure drop is much larger than the system pressure drop or critical pressure drop so the installed flow characteristic is close to the inherent flow characteristic.

The results are also useful for determining the dead time to time constant ratio, which has a profound effect on the tuning factors used and the performance of dead time compensation, which has been discussed on the *Control* magazine Control Talk Blog (https://www.controlglobal.com/blogs/controltalkblog/) and the Control Talk column "Voices" (https://www.controlglobal.com/voices/) sites.

F.2 Batch Implications

Plug flow volumes can always be considered as continuous because the volume is completely full and anything entering will be discharged after a transportation delay.

A back-mixed volume is partially full. If the liquid discharge flow is zero, this volume can be considered to be in the batch mode rather than the continuous mode. If the flows are all sequenced and charged based on time and/or totals, the vessel operation can be considered to be *pure batch*. If reactor flows are ratioed and manipulated by a control loop, the vessel operation can be classified as *fed-batch*.

Level has an integrating response whether in the batch or continuous mode. In the batch mode, it is a zero load integrator in that all the feed flows must be zero for the level to stop rising. Level has a one-sided integrating response in the batch mode because the level can only rise and not drop. This type of response causes overshoot for any controller with reset action. Proportional plus derivative (PD) controllers with a zero or negative bias can be tuned for zero overshoot.

The temperature response of a back-mixed volume remains self-regulating even for zero liquid discharge flow unless the liquid level is above or below the heat transfer surfaces. However, the temperature response some self-regulation and behaves more like a "near-integrator."

The concentration response of a back-mixed volume becomes integrating for a zero discharge flow. This is not obvious because the discharge flow cancels out of the differential equation from applying the multiplicative rule of integration in transitioning the derivative from the rate of accumulation of component mass to the rate of accumulation of component concentration. The multiplicative rule of integration says that the integration of the product of the derivatives of two variables (e.g., energy and mass rate of change) can be expressed as the sum of the two derivatives. The effect of zero discharge flow is more recognizable if we consider the case of a zero reaction rate so that the process time constant is simply the residence time (liquid mass divided by the liquid feed flow rate). The increase in mass for a fixed feed rate over the residence time is the feed rate multiplied by the residence time. The result is an increase in mass equal to the existing mass. This doubling of mass doubles the residence time and therefore the process time constant. Consequently, the process never reaches a steady state because the process time constant is constantly increasing as the level is rising for a zero discharge flow.

For the case of zero reaction, the integrator is a zero load integrator because the component feed must be zero for the concentration of the component to stop rising. The concentration here has a one-sided integrating response in that the concentration can only increase and not decrease. This would also be the case for reaction products where there is only a forward reaction (no reverse or side reactions). As with level, overshoot is a problem unless PD controllers are used. Alternately, the controlled variable can be translated to a rate of change of concentration as noted applying model-predictive control (MPC) for bioreactor biomass and product concentration.

Gas pressure is an integrator regardless of liquid discharge flow as long as the pressure in the vessel has a negligible effect on vent flow, which is the case for large or critical pressure drops. If this is not the case, the gas pressure response becomes self-regulating, but for large volumes and small vent flows it behaves like a "near-integrator."

F.3 Results

These equations are developed in the "Back-Mixed Volumes" and "Plug Flow Volumes" sections in this appendix. They are numbered according to the point of derivation so the user can better locate and understand their origination. The nomenclature with engineering units is summarized at the end of the appendix.

The integrating process gain (K_{ip}) for the control of liquid level by manipulating a flow:

$$K_{ip} = 1 / (\rho_o \bullet A_o) \tag{F-4d}$$

The integrating process gain (K_{ip}) for the control of pressure per gas law with a specific gas constant (R_g) by a flow:

$$K_{ip} = [(R_g \bullet T_g) / V_g] \tag{F-5d}$$

To manipulate jacket temperature to control outlet temperature, the main process time constant (τ_p) is (positive feedback if heat of feed and reaction exceeds the product of heat transfer coefficient and area):

$$\tau_p = (C_p \bullet M_o) / \left[C_p \bullet F_f - \Delta Q_r / \Delta T_o + U \bullet A\right] \tag{F-6g}$$

To manipulate jacket temperature to control outlet temperature, the process gain (K_p) is:

$$K_p = (U \bullet A) / [C_p \bullet F_f - \Delta Q_r / \Delta T_o + U \bullet A] \quad \text{(F-6h)}$$

To manipulate jacket temperature to control outlet temperature, the near-integrating process gain (K_{nip}) is:

$$K_{nip} = (U \bullet A) / (C_p \bullet M_o) \quad \text{(F-6i)}$$

To manipulate feed temperature to control outlet temperature, the process gain (K_p) is:

$$K_p = (C_p \bullet F_f) / [C_p \bullet F_f - \Delta Q_r / \Delta T_o + U \bullet A] \quad \text{(F-6j)}$$

To manipulate feed flow to control outlet temperature, the process gain (K_p) is:

$$K_p = (C_p \bullet T_f) / [C_p \bullet F_f - \Delta Q_r / \Delta T_o + U \bullet A] \quad \text{(F-6k)}$$

To manipulate jacket temperature, the additional small secondary process time constant associated with the heat capacity and mass of the jacket wall is:

$$\tau_{p2} = (C_w \bullet M_w) / [U \bullet A] \quad \text{(F-6l)}$$

The process dead time (θ_p) from the turnover time for temperature and concentration control in a well-mixed volume is:

$$\theta_p = (M_o / \rho_o) / [(F_f + F_a + F_r) / \rho_o + F_v / \rho_v] \quad \text{(F-6m)}$$

The process dead time (θ_p) from injection delay for concentration control is:

$$\theta_p = V_1 / (F_1 / \rho_1) \quad \text{(F-6n)}$$

To manipulate feed flow to control reactant concentration (X_{Ao}), the main process time constant (τ_p) is:

$$\tau_p = M_o / (R_x + F_f) \quad \text{(F-7g)}$$

To manipulate feed flow to control reactant concentration (X_{Ao}), the process gain (K_p) is:

$$K_p = X_{Af} / (R_x + F_f) \quad \text{(F-7h)}$$

To manipulate feed flow to control reactant concentration, the near-integrating process gain (K_{nip}) is:

$$K_{nip} = X_{Af} / M_o \tag{F-7i}$$

To manipulate feed concentration to control reactant concentration, the process gain (K_p) is:

$$K_p = F_f / (R_x + F_f) \tag{F-7j}$$

For plug flow volumes where different streams are being combined, the process gain (K_p) for controlling the temperature of the mixture (T_f) by manipulating the flow is:

$$K_p = dT_f / dF_1 = T_1 / \Sigma F_i \tag{F-9a}$$

For plug flow volumes where different streams are being combined, the process gain (K_p) for controlling the composition of component A (X_{Af}) in the mixture by manipulating the flow is:

$$K_p = dX_{Af} / dF_1 = X_{A1} / \Sigma F_i \tag{F-9b}$$

For plug flow volumes where different streams are being combined, the process dead time for controlling the temperature or composition of the mixture by manipulating the flow is:

$$\theta_p = V_1 / (F_1 / \rho_1) + V_p / \Sigma(F_i / \rho_i) \tag{F-9c}$$

The process time constant is essentially zero for true plug flow. For a static mixer there is some back mixing; the residence time in Equation F-9c is split between a dead time and time constant per Equations F-9d and F-9e.

$$\theta_p = V_1 / (F_1 / \rho_1) + x \bullet V_p / \Sigma(F_i / \rho_i) \tag{F-9d}$$

$$\tau_p = (1 - x) \bullet V_p / \Sigma(F_i / \rho_i) \tag{F-9e}$$

F.4 Derivations

There are three types of processes: self-regulating, integrating, and runaway as shown in Figures F-4a, F-4b, and F-4c, respectively. A self-regulating process will decelerate to a new steady state. An integrating process will continually ramp. A runaway process will accelerate until it hits a relief or interlock setting.

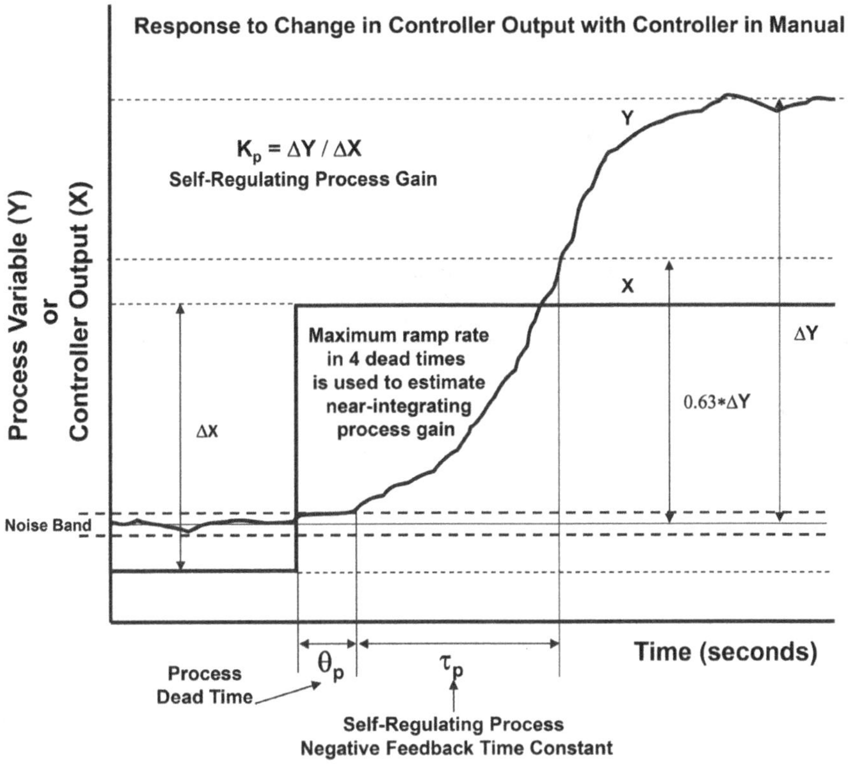

Figure F-4a. Self-regulating (negative feedback) process.

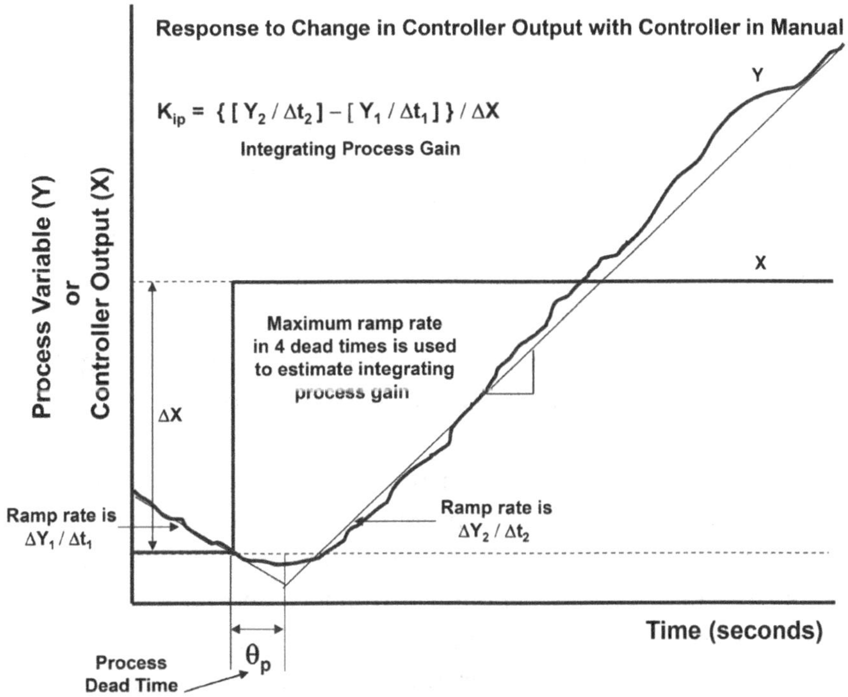

Figure F-4b. Integrating (zero feedback) process.

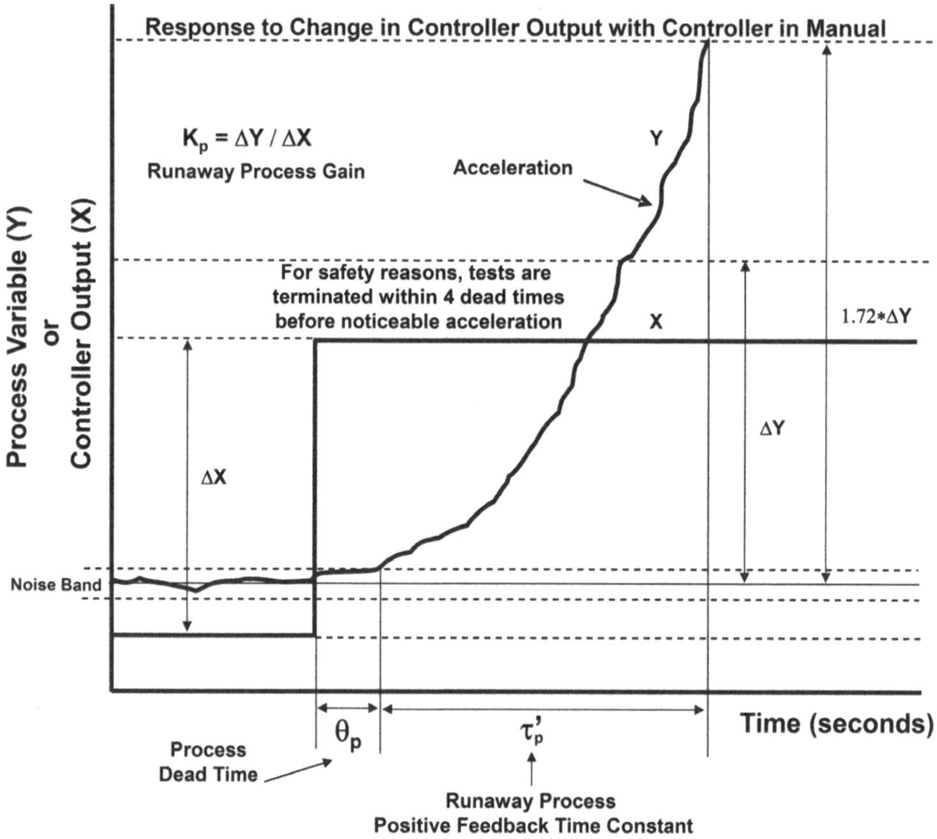

Figure F-4c. Runaway (positive feedback) process.

More than 90% of the processes are self-regulating. However, many of the continuous and fed-batch processes in the chemical industry with the greatest direct economic benefits behave and can be best treated as "near-integrating" processes. The classic integrating process is a pure batch or level process. Fewer than 1% of the processes are runaway. When these exist, understanding the runaway response is critical in terms of safety and control because of the propensity to accelerate and reach a point of no return. Runaway responses are almost exclusively associated with highly exothermic reactors used in plastics and specialty chemical production.

This appendix develops the equations for process dynamics for back-mixed volumes and plug flow volumes. The back-mixed volumes section applies to volumes whenever an agitator pumping rate, an eductor or recirculation liquid flow rate, or a gas evolution or sparge rate produces enough turbulence and back mixing to make the mixture more uniform in the axial in addition to the radial direction of the volume. Gas volumes generally have enough turbulence and fast enough gas dispersion rates to be treated as a back-mixed volume. The equations therefore hold relatively well for an evaporator and a single distillation stage due to turbulence from the vapor flow. The plug volume section applies to static mixers, pipelines, coil inlets, and jacket inlets where the turbulence from pipe fittings or internal mixing elements creates enough radial mixing to make the mixture uniform over the cross section of the pipe or nozzle inlet but provides little axial mixing.

F.5 Back-Mixed Volumes

For a back-mixed volume, the process gains and time constants can be readily identified if the ordinary differential equations for the rate of accumulation of energy or material in the volume are set up so that the process output of interest (Y) is on the right side with a unity coefficient. From this simple generic form, we can identify the process time constant (τ_p) as the coefficient of the derivative of the process output (dY/dt) and the process gain (K_p) as the coefficient of the process input (X). The process output (Y) and net input (X) can be viewed as the controlled variable and manipulated variable biased by a load, respectively. Many other terms can exist, but these are not shown in the following equations. These missing terms can be categorized as disturbances.

If the sign of the unity coefficient of the process output on the right side is negative (Equation F-1a), the process has negative feedback. As the process output changes, the negative feedback slows down and eventually halts the excursion of the process output at its new steady state when it balances out the effect of the process input and the disturbances.

$$\tau_p \bullet dY / dt = K_p \bullet X - Y \tag{F-1a}$$

Integrating this ordinary differential equation mathematically provides the time response of a change in the process output (ΔY) for a step change in the process input (ΔX). The step occurs at $t = 0$.

$$\Delta Y = K_p \bullet \left(1 - e^{-t/\tau_p}\right) \bullet \Delta X \tag{F-1b}$$

If the process output does not appear on the right side (Equation F-2a), there is no process feedback. As the process output changes, there is no feedback to slow it down or speed it up so it continues to ramp. There is no steady state. The ramping will stop only when X is zero, which occurs when the manipulated variable is equal to the load.

$$dY / dt = K_{ip} \bullet X \tag{F-2a}$$

$$\Delta Y = K_{ip} \bullet t \bullet \Delta X \tag{F-2b}$$

Often in the more important loops for concentration, pressure, and temperature control of large volumes, the time constant in Equation F-1a is so large that the time to reach steady state is beyond the time frame of interest. Because these loops with small dead time to time constant ratios should be tuned with integrating process-tuning rules as discussed in Chapter 8, the controller only sees the first part of the excursion before the inflection point and deceleration by negative process feedback. In this case, the response is best characterized by a near-integrating process gain calculated per Equation F-2c.

$$K_{nip} = K_p \,/\, \tau_p \tag{F-2c}$$

If the sign of the unity coefficient of the process output on the right side is positive (Equation F-3a), the process has positive feedback. As the process output changes, the positive feedback speeds up the excursion unless disturbances counteract the effect of the process input and output.

$$\tau_p{}' \bullet dY \,/\, dt = k_p \bullet X + Y \tag{F-3a}$$

$$\Delta Y = K_p \bullet (e^{t/\tau_p{}'} - 1) \bullet \Delta X \tag{F-3b}$$

Consider a mixed volume with a jacket and vapor space. There are liquid reactant feeds, gas feeds (sparged through the liquid and added directly to the vapor space), an outlet liquid flow, a vent gas flow, and a jacket coolant flow. Usually, there are multiple components of interest. For example, consider liquid or gas acid and base reagent or reactant components (a, b) to produce primary and secondary liquid or gas products (c, d, e). Consider also that there are typically water and nitrogen gas components (w, n).

The ordinary differential equation for accumulating liquid mass as shown in Equation F-4a includes inlet flows added directly to the liquid volume (ΣF_i), vapor flow rates from evaporation and vaporization (ΣF_v), and an outlet liquid flow rate (F_o). The liquid level depends on the density and cross-section area of the liquid. Equation F-4a can then be reformulated to Equation F-4b to include the process variable of interest, liquid level (L_l), in the derivative.

$$dM_o \,/\, dt = \Sigma F_i - \Sigma F_v - F_o \tag{F-4a}$$

$$d(\rho_o \bullet A_o \bullet L_o) \,/\, dt = \Sigma F_i - \Sigma F_v - F_o \tag{F-4b}$$

If we consider the density (ρ_o) to be a weak function of composition and therefore constant like the cross-sectional area (A_o), we can take these terms outside the derivative and divide through to get an equation for level (L_o) in the form of Equation F-2a. Now it is evident that the integrating process gain (K_{ip}) for manipulating a flow in or flow out is simply the inverse of the product of the liquid density and cross-section area (Equation F-4d).

$$dL_o / dt = [1/(\rho_o \bullet A_o)] \bullet [\Sigma F_i - \Sigma F_v - F_o] \quad \text{(F-4c)}$$

$$K_{ip} = 1/(\rho_o \bullet A_o) \quad \text{(F-4d)}$$

The ordinary differential equation for accumulating gas mass as shown in Equation F-5a includes inlet flows added directly to the gas volume (ΣF_i); vapor flow rates from gas sparging, evolution, and vaporization (ΣF_v); and an exit gas flow rate (F_g). Equations of state such as the ideal gas law can be used to express this relationship for a given composition using a specific gas constant (R_g) that includes a compressibility factor. Equation F-5a can then be reformulated to Equation F-5b to include the process variable of interest, gas pressure (P_g), in the derivative.

$$dM_g / dt = \Sigma F_i + \Sigma F_v - F_g \quad \text{(F-5a)}$$

$$d[(P_g \bullet V_g)/(R_g \bullet T_g)] / dt = \Sigma F_i + \Sigma F_v - F_g \quad \text{(F-5b)}$$

If we consider changes in the gas volume (V_g) and gas temperature (T_g) to be much slower than changes in the gas pressure (P_g) and therefore relatively constant during the integration step, we can take these terms outside the derivative and divide through to get an equation for pressure in the form of Equation F-2a. Now it is evident that the integrating process gain (K_{ip}) for manipulating flows in or flow out is simply the product of the specific gas coefficient (R_g) and the absolute gas temperature divided by the gas volume (Equation F-5d). This assumes a change in pressure does not significantly change the gas flow out of the volume, which is usually the case for a pressure drop across the vent valve that is large or critical.

$$dP_g / dt = [(R_g \bullet T_g)/V_g] \bullet [\Sigma F_i + \Sigma F_v - F_g] \quad \text{(F-5c)}$$

$$K_{ip} = [(R_g \bullet T_g)/V_g] \quad \text{(F-5d)}$$

The ordinary differential equation (ODE) for accumulating energy as shown in Equation F-6a includes the effects of feed temperature, heat of reaction as a function of temperature, heat of vaporization, and heat transfer to the jacket. If we consider the specific heat capacity to be relatively constant and use the multiplicative rule of integration, we can express the differential equation in the generic form of Equation F-1a in terms of temperature to show the process feedback in Equation F-6f, the final form of the ODE where the discharge flow cancels out. The relative magnitude of the terms in the denominator of Equation F-6g determines the feedback sign.

$$\begin{aligned} dQ_o / dt = C_p \bullet \Sigma(F_i \bullet T_i) - C_p \bullet F_o \bullet T_o + (\Delta Q_r / \Delta T_o) \bullet T_o \\ + H_x \bullet R_x - H_v \bullet F_v - U \bullet A \bullet (T_o - T_j) \end{aligned} \quad \text{(F-6a)}$$

$$dQ_o / dt = d(C_p \bullet M_o \bullet T_o) / dt = C_p \bullet (dM_o / dt) \bullet T_o + C_p \bullet M_o \bullet (dT_o \bullet dt) \quad \text{(F-6b)}$$

$$F_f = \Sigma F_i \quad \text{(F-6c)}$$

$$T_f = \Sigma(F_i \bullet T_i) / \Sigma F_i \quad \text{(F-6d)}$$

$$C_p \bullet (dM_o / dt) \bullet T_o = C_p \bullet (F_f - F_o) \bullet T_o \quad \text{(F-6e)}$$

$$\begin{aligned} & C_p \bullet M_o \bullet (dT_o / dt) \\ & = C_p \bullet F_f \bullet T_f + H_x \bullet R_x - H_v \bullet F_v + U \bullet A \bullet T_j - [C_p \bullet F_f - \Delta Q_r / \Delta T_o + U \bullet A] \bullet T_o \end{aligned} \quad \text{(F-6f)}$$

To manipulate jacket temperature to control outlet temperature, the main process time constant (τ_p) is (positive feedback if heat of feed and reaction exceeds the product of heat transfer coefficient and area):

$$\tau_p = (C_p \bullet M_o) / [C_p \bullet F_f - \Delta Q_r / \Delta T_o + U \bullet A] \quad \text{(F-6g)}$$

To manipulate jacket temperature to control outlet temperature, the process gain (K_p) is:

$$K_p = (U \bullet A) / [C_p \bullet F_f - \Delta Q_r / \Delta T_o + U \bullet A] \quad \text{(F-6h)}$$

To manipulate jacket temperature to control outlet temperature, the near-integrating process gain (K_{nip}) is:

$$K_{nip} = (U \bullet A) / (C_p \bullet M_o) \quad \text{(F-6i)}$$

To manipulate feed temperature to control outlet temperature, the process gain (K_p) is:

$$K_p = (C_p \bullet F_f) / [C_p \bullet F_f - \Delta Q_r / \Delta T_o + U \bullet A] \quad \text{(F-6j)}$$

To manipulate feed flow to control outlet temperature, the process gain (K_p) is:

$$K_p = (C_p \bullet T_f) / [C_p \bullet F_f - \Delta Q_r / \Delta T_o + U \bullet A] \quad \text{(F-6k)}$$

To manipulate jacket temperature, the additional small secondary process time constant associated with the heat capacity and mass of the jacket wall is:

$$\tau_p = (C_w \bullet M_w) / [U \bullet A] \tag{F-6l}$$

Any change in the temperature at the heat transfer surfaces or the feed inlet must be dispersed and back mixed into the volume. This process dead time (θ_p) is the turnover time that can be approximated as the liquid inventory divided by the sum of the feed flow rate (F_f), agitator pumping rate (F_a), recirculation flow rate (F_r), and vapor evolution rate or vapor bubble rate (F_v). Because this turnover time is computed in terms of volumetric flow rates, the liquid mass and the mass flow rates are divided by their respective densities as shown in Equation F-6m.

$$\theta_p = (M_o / \rho_o) / [(F_f + F_a + F_r) / \rho_o + F_v / \rho_v] \tag{F-6m}$$

If there is an injector (dip tube or sparger ring) volume, a change in composition at the nozzle must propagate by plug flow to the discharge points of the dip tube or sparger ring. The dead time for a feed flow (F_1) is the injector volume (V_1) divided by the injection mass flow (F_1) divided by its respective density (ρ_1).

$$\theta_p = V_1 / (F_1 / \rho_1) \tag{F-6n}$$

The ODE for accumulating liquid reactant mass (M_A) as shown in Equation F-7a includes the effects of feeds (F_i) with a reactant mass fraction (X_{Ai}), reaction rate (R_x), and outlet flow (F_o). The feeds can be from raw material, intermediate products, recycle streams, or multistage reactors. If we use the multiplicative rule of integration, we can express the differential equation in the generic form of Equation F-1a in terms of concentration to show the process feedback in Equation F-7f, the final form of the ODE.

$$dM_A / dt = \Sigma(F_i \bullet X_{Ai}) - (R_x + F_o) \bullet X_{Ao} \tag{F-7a}$$

$$dM_A / dt = d(M_o \bullet X_{Ao}) / dt = (dM_o / dt) \bullet X_{Ao} + M_o \bullet (dX_{Ao} / dt) \tag{F-7b}$$

$$F_f = \Sigma F_i \tag{F-7c}$$

$$X_{Af} = \Sigma(F_i \bullet X_{Ai}) / \Sigma F_i \tag{F-7d}$$

$$(dM_o / dt) \bullet X_{Ao} = (F_f - F_o) \bullet X_{Ao} \tag{F-7e}$$

$$M_o \bullet (dX_{Ao} / dt) = F_f \bullet X_{Af} - (R_x + F_f) \bullet X_{Ao} \tag{F-7f}$$

To manipulate feed flow to control reactant concentration (X_{Ao}), the main process time constant (τ_p) is:

$$\tau_p = M_o / (R_x + F_f) \tag{F-7g}$$

To manipulate feed flow to control reactant concentration (X_{Ao}), the process gain (K_p) is:

$$K_p = X_{Af} / (R_x + F_f) \tag{F-7h}$$

To manipulate feed flow to control reactant concentration, the near-integrator gain (K_{nip}) is:

$$K_{nip} = X_{Af} / M_o \tag{F-7i}$$

To manipulate feed concentration to control reactant concentration, the process gain (K_p) is:

$$K_p = F_f / (R_x + F_f) \tag{F-7j}$$

The process dead times from turnover time and from feed injection are the same as computed in the section for temperature control (Equations F-6m and F-6n).

F.6 Plug Flow Volumes

For plug flow volumes where different streams are being combined, the process gain for controlling the temperature (T_f) or composition (X_{Af}) of the mixture (often a feed to downstream equipment) can be computed by taking the derivative of Equations F-6d and F-7d with respect to the manipulated flow for stream 1 (F_1) to give Equations F-9a and F-9b, respectively. In both cases, the process gain is inversely proportional to total flow (ΣF_i).

$$K_p = dT_f / dF_1 = T_1 / \Sigma F_i \tag{F-9a}$$

$$K_p = dX_{Af} / dF_1 = X_{A1} / \Sigma F_i \tag{F-9b}$$

The process dead time for manipulating a flow for stream 1 (F_1) is the sum of the injection delay for stream 1 and the piping delay from the point of injection to the point of temperature or composition measurement. For plug flow, the residence time, which is the second expression in Equation F-9c, becomes dead time.

$$\theta_p = V_1 / (F_1 / \rho_1) + V_p / \Sigma(F_1 / \rho_1) \tag{F-9c}$$

The process time constant is essentially zero for true plug flow. For a static mixer there is some back mixing; the residence time in Equation F-9c is split between a dead time and a time constant per Equations F-9d and F-9e.

$$\theta_p = V_1 / (F_1 / \rho_1) + x \bullet V_p / \Sigma(F_i / \rho_i) \tag{F-9d}$$

$$\tau_p = (1 - x) \bullet V_p / \Sigma(F_i / \rho_i) \tag{F-9e}$$

It is obvious from the preceding discussion that both the process gain and dead time are inversely proportional to total flow.

F.7 Controller Tuning

The implication of the results can be best seen if lambda is set equal to the total loop deadtime (θ_o), resulting in Equation F-10a for the proportional-integral-derivative (PID) gain. If the open-loop time constant (τ_o) is large compared to the dead time, the process response is termed *near-integrating* and the ratio of the open-loop self-regulating process gain (K_o) to the time constant is used to approximate the open-loop integrating process gain (K_i) in Equation 5.10b. This PID gain is about half of the PID gain estimated by the Ziegler-Nichols reaction curve method.

$$K_c = 0.5 \bullet \frac{\tau_o}{K_o \bullet \theta_o} \tag{F-10a}$$

$$K_c = 0.5 \bullet \frac{1}{K_i \bullet \theta_o} \tag{F-10b}$$

$$K_o = K_v \bullet K_p \bullet K_m \tag{F-10c}$$

The open-loop time constant (τ_o) in the numerator of Equation F-10a is the largest time constant in the loop wherever it occurs. Hopefully, the process is mixed well enough and the instrumentation is fast enough that the largest time constant is in the process ($\tau_o = \tau_P$) and not the automation system. A large time constant in the process slows down the disturbance and is desirable. A large time constant in the measurement and final element is detrimental because it slows down the ability of the controller to see and react to disturbance, respectively.

The open-loop self-regulating process gain (K_o) in the denominator is dimension-less (Δ%PV/Δ%OUT). The open-loop gain is the product of the manipulated variable gain (K_v); the process gain (K_p); the slope of the plot of the process variable versus the manipulated variable (e.g., flow) and the measurement gain (K_m). For a loop that manipulates a control valve or variable speed drive, the manipulated variable gain is the slope of the valve's or variable speed drive's installed flow characteristic. For linear installed flow characteristics, the manipulated variable gain is the valve or variable speed drive maximum flow capacity divided by 100%. For the primary loop of a cascade control system, the manipulated variable gain is the secondary loop set-point span divided by 100%. For temperature and composition control and particularly pH control, the process variable is often plotted versus the ratio of the manipulated flow to feed flow. For this case, the process gain from the slope of the plot must be divided by the feed flow as seen in the process gain equations for the material balance. The measurement gain is 100% divided by the process variable measurement span. Thus, changes in calibration span affect the computed controller gain. A decrease in measurement span may increase measurement accuracy, but the controller gain may need to be decreased to ensure robustness and a less oscilla-tory response.

Finally, the dead time (θ_o) in the denominator is really the total loop dead time, which is the sum of the process dead time (θ_p) plus all the small time lags and delays in the loop. While the names *open-loop time constant* (τ_o), *open-loop self-regulating process gain* (K_o), and *total loop dead time* (θ_o) for the parameters in Equation F-10a are more definitive, most of the control literature uses the terms *process time constant*, *process gain*, and *process dead time* indiscriminately.

F.8 Nomenclature

Process Parameters

A_o = cross-sectional area of liquid level (m^2)
A = heat transfer surface area (m^2)
C_p = heat capacity of process (kJ/kg • °C)
C_w = heat capacity of wall of heat transfer surface (kJ/kg • °C)
F_a = agitator pumping rate (kg/s)
F_f = total feed flow (kg/s)
F_g = gas flow (kg/s)
F_i = feed stream *i* flow (kg/s)
F_o = vessel outlet flow (kg/s)
F_r = recirculation flow (kg/s)
F_v = vaporization rate (kg/s)
H_v = heat of vaporization (kJ/kg)
H_x = heat of reaction (kJ/kg)
L_o = liquid level (m)
M_A = component A mass (kg)

M_g = gas mass (kg)
M_o = liquid mass (kg)
M_w = mass of wall of heat transfer surface (kg)
P_g = gas pressure (kPa)
T_f = total feed temperature (°C)
T_g = gas temperature (°K)
T_i = feed stream *i* temperature (°C)
T_o = vessel outlet temperature (°C)
t = time (s)
Q_o = total heat of liquid (kJ)
Q_r = heat from reaction (kJ)
R_x = reaction rate (kg/s)
R_g = specific gas constant including compressibility factor ((kPa • m^3)/(°K • kg per mole))
ρ_g = gas density (kg/m^3)
ρ_i = stream *i* density (kg/m^3)
ρ_o = liquid density (kg/m^3)
ρ_v = density of vapor (kg/m^3)
U = overall heat transfer coefficient (kJ/m^2 • °C)
V_g = gas volume (m^3)
V_i = injection (e.g., dip tube or sparger ring) volume (m^3)
V_p = piping volume (m^3)
x = fraction of volume that is plug flow
X_{Af} = total feed component A concentration (mass fraction)
X_{Ai} = feed stream *i* component A concentration (mass fraction)
X_{Ao} = vessel outlet component A concentration (mass fraction)

Generic Terms

X = process input (manipulated variable) (eu)
Y = process output (controlled variable) (eu)

Dynamic Parameters

K_c = PID controller gain (dimensionless)
K_i = open-loop integrating process gain (PV percent per second per CO percent) (1/seconds)
K_{ip} = integrating process gain (PV engineering units/MV engineering units)
K_m = measurement gain (PV percent / PV engineering units)
K_o = open-loop self-regulating process gain (PV percent / CO percent) (dimensionless)

K_p = process gain (PV engineering units / MV engineering units)
K_{nip} = near-integrating process gain (PV engineering units per second / MV engineering units)
K_v = valve or VSD gain (slope of installed flow characteristic) (flow engineering units/CO percent)
τ_p = negative feedback process time constant (seconds)
τ_p' = positive feedback process time constant (seconds)
τ_o = open-loop time constant (primary time constant) (seconds)
θ_p = process dead time (seconds)
θ_o = total loop dead time (seconds)

Appendix G
Gas Pressure Dynamics

The interactive time constants of multiple gas volumes (e.g., vessels, columns, and headers) in series separated by flow resistances (e.g., valves and fittings) can be estimated by using Equations G-1 through G-4 for turbulent and sonic flow [1].

$$\tau_i = \frac{C_i}{\sum_i \frac{1}{R_i}} \tag{G-1}$$

$$C_i = \frac{V_i}{P_a} \tag{G-2}$$

For turbulent flow:

$$R_i = \frac{2 \bullet \Delta P_i}{F_i} \tag{G-3}$$

For sonic flow:

$$R_i = \frac{P_i}{F_i} \tag{G-4}$$

where

C_i = capacitance of the gas volume i (scf/psi)
F_i = gas flow through resistance i (scfm)

P_a = atmospheric pressure at resistance i (psi)
P_i = inlet pressure at resistance i (psi)
ΔP_i = pressure drop across resistance i (psi)
R_i = resistance to the gas flow into or out of volume i (psi/scfm)
V_i = gas volume i (scf)
τ_i = interactive time constant of gas volume i (min)

If the intermediate flow resistance pressure drops for turbulent nonsonic flow between volumes in a gas system are less than 2% of the system inlet and outlet flow resistance pressure drops, then all the individual volumes can be summed and substituted into Equation G-2 to give a single process time constant for the gas system. The universal gas sizing equation can be used to estimate the C_v of the fitting or valve necessary to prevent creating multiple time constants. If the pressure drops between volumes are not negligible (flow coefficient of internal resistance does not satisfy Equation G-5), then the system should be partitioned into two major volumes and Appendix I used in conjunction with the equations here to estimate the equivalent two noninteractive time constants for two volumes in series. The sum of the noninteractive time constants should be then used to estimate the process dead time.

$$C_v > \frac{F}{P \bullet \sqrt{\frac{0.02 \bullet \Delta P}{P}}} \bullet \sqrt{\frac{G \bullet T}{520}} \tag{G-5}$$

where

C_v = gas flow coefficient of internal resistance (scfm/psi)
F = gas flow through the system (scfm)
G = gas specific gravity in the system (dimensionless)
P = pressure at the system inlet (psi)
ΔP = pressure drop at the system outlet (psi)
T = absolute temperature in the system (°R)

Reference

[1] Peter Harriott, *Process Control* (New York: McGraw-Hill, 1964).

Appendix H Charge Balance to Model *p*H

H.1 Introduction

Aqueous systems are composed of ions in various degrees of dissociation in water. These contributions come from salts, acids, and bases.

Salts

$$B_{\varepsilon}^{+\lambda} A_{\beta}^{-\omega} \leftrightarrow \varepsilon B^{+\lambda} + \beta A^{-\omega} \tag{H-1a}$$

Acids

$$H_{\omega} A^{-\omega} \underset{pK_{a1}}{\leftrightarrow} H^{+} + H_{\omega-1} A^{-1} \underset{pK_{a2}}{\leftrightarrow} H^{+} + H_{\omega-2} A^{-2} \underset{pK_{a3}}{\leftrightarrow} \ldots \tag{H-1b}$$

Bases

$$B^{+\lambda}(OH)_{\lambda} \underset{pK_{b1}}{\leftrightarrow} OH^{-1} + B(OH)_{\lambda-1}^{+1} \underset{pK_{b2}}{\leftrightarrow} OH^{-1} + B(OH)_{\lambda-2}^{+2} \underset{pK_{b3}}{\leftrightarrow} \ldots \tag{H-1c}$$

H.2 Definitions

pH

$$pH = -\log_{10}[H^+] \tag{H-2a}$$

pK_a

$$pK_a = -\log_{10} K_a \tag{H-2b}$$

K_w

$$H_2O \underset{K_w}{\leftrightarrow} H^+ + OH^- \ \ and \ \ K_w = K_b \bullet K_a = [OH^+] \bullet [H^+] \tag{H-2c}$$

For a system composed of these species, there are three fundamental relationships available:

1. Principle of electroneutrality in a charge balance
2. Conservation of mass in a species balance
3. Equilibrium relationships

The way these relationships come together will depend on the valence of the ions considered. The derivations for monovalent through trivalent ions follow.

H.3 Monovalent Ion

The monovalent ion can enter the system through an acid (if an anion), a base (if a cation), or a salt. Consider the dissociation of an anion in an acid:

$$[HA] \overset{K_{a1}}{\leftrightarrow} [A^{-1}] + [H^{+1}] \ \ and \ \ K_{a1} = \frac{[H^+] \bullet [A^{-1}]}{[HA]} \tag{H-3a}$$

In addition, there may be any number of salts contributing the anion to the system as:

$$B_{\varepsilon}^{+\lambda} A_{\beta}^{-1} \rightarrow \varepsilon B^{+\lambda} + \beta A^{-1} \tag{H-3b}$$

We can then write a species balance about the system at a constant volume as:

$$[HA] + [A^{-1}] = X_a + \sum_{j=1}^{n} \beta_j \bullet X_j = E \tag{H-3c}$$

where X_a and X_j are initial concentrations of acid and salt, respectively. The β_j is the stoichiometric coefficient associated with the anion for the jth salt. The right side of this equation is a constant and can be noted as E.

$$[HA]+[A^{-1}]=E \tag{H-3d}$$

Taking the equilibrium expression and rearranging:

$$\frac{[H^{+1}]\bullet[A^{-1}]}{K_{a_1}}=[HA] \tag{H-3e}$$

Substituting Equation H-3d into Equation H-3e:

$$\frac{[H^{+1}]\bullet[A^{-1}]}{K_{a_1}}+[A^{-1}]=E \tag{H-3f}$$

Solving for $[A^{-1}]$:

$$[A^{-1}]=\frac{E}{\left(\frac{[H^{+1}]}{K_{a_1}}+1\right)} \tag{H-3g}$$

Using the definition of pK_a and pH:

$$[A^{-1}]=\frac{E}{\left(10^{-pH}\bullet 10^{pK_{a1}}+1\right)}=\frac{E}{\left(10^{pK_{a1}-pH}+1\right)} \tag{H-3h}$$

For a base, K_b can be used instead:

$$B^{+1}(OH)_1 \underset{pK_{b1}}{\leftrightarrow} OH^{-1}+B^{+} \tag{H-3i}$$

$$[B(OH)]+[B^{+1}]=X_b+\sum_{j=1}^{n}\varepsilon_j \bullet X_j=E \tag{H-3j}$$

and:

$$K_b=\frac{[OH^{-}]\bullet[B^{+}]}{B(OH)} \tag{H-3k}$$

$$[B^{+}]=\frac{E}{\left(\frac{[OH^{-}]}{K_{b_1}}+1\right)} \tag{H-3l}$$

For $\frac{[OH^{-}]}{K_{b_1}}$:

$$[OH^{-}]\bullet[H^{+}]=K_w \rightarrow [OH^{-}]=\frac{K_w}{[H^{+}]} \tag{H-3m}$$

and with:

$$K_w = K_a \bullet K_b \rightarrow K_b = \frac{K_w}{K_a} \tag{H-3n}$$

then:

$$\frac{[OH^-]}{K_{b_1}} = \underbrace{\frac{K_w}{[H^+]\bullet K_{b_1}}}_{Substituting\,Eq.H-3n} = \frac{\overbrace{K_w}}{[H^+]\bullet \frac{K_w}{K_a}} = \frac{K_a}{\underbrace{[H^+]}_{Simplifying}} = 10^{pH-pKa_1} \tag{H-3o}$$

By introducing the sign s, which is positive for bases and negative for acids, we can define P_1:

$$P_1 = 10^{s\bullet(pH-pKa_1)} \tag{H-3p}$$

The charge balance now becomes (with respect to the acid form with no base and salts):

$$0 = \sum_{i=1}^{n} \pm X_i \bullet \left[B_i^{\pm X_i}\right] \tag{H-3q}$$

$$0 = -[A^-] - [OH^-] + \sum_{j=1}^{l} \lambda_j \bullet \beta_j \bullet [B^{+\lambda}] + [H^+] \tag{H-3r}$$

$$0 = \underbrace{\frac{E}{P_1+1}}_{M_1} + [OH^-] + \underbrace{\sum_{j=1}^{l} \lambda_j \bullet \beta_j \bullet [B^{+\lambda}] + [H^+]}_{C_s} \tag{H-3s}$$

$$0 = -M_1 - [OH^-] + C_S + [H^+] \tag{H-3t}$$

Using Equation H-3m:

$$0 = -M_1 - \frac{K_w}{[H^+]} + C_S + [H^+] \tag{H-3u}$$

Substituting definitions for pK_w and pH:

$$0 = -M_1 - 10^{pH-pK_w} + 10^{-pH} + C_S \tag{H-3v}$$

If a base is used instead, the sign of M_1 will change, giving:

$$M_1 = s\bullet \frac{X_a + \sum_{j=1}^{n} \beta j \bullet X_j}{P_1+1} = s\bullet \frac{E}{P_1+1} \tag{H-3w}$$

H.4 Divalent Ion

As before, the divalent ion can enter the system through an acid (if an anion), a base (if a cation), or a salt. For a divalent acid:

$$[H_2A]\overset{K_{a1}}{\leftrightarrow}[HA^{-1}]+[H^{+1}]\ and\ K_{a_1}=\frac{[H^+]\bullet[HA^-]}{[H_2A]} \tag{H-4a}$$

$$[HA^{-1}]\overset{K_{a2}}{\leftrightarrow}[H^+]+[A^{-2}]\ and\ K_{a_2}=\frac{[H^+]\bullet[A^{-2}]}{[HA^-]} \tag{H-4b}$$

And for a salt that completely dissociates:

$$B_{\varepsilon}^{+\lambda}A_{\beta}^{-2}\rightarrow\varepsilon B^{+\lambda}+\beta A^{-2} \tag{H-4c}$$

The species balance becomes:

$$[H_2O]+[HA^{-1}]+[A^{-2}]=X_a+\sum_{j=1}^{n}\beta_j\bullet X_j=E \tag{H-4d}$$

Substituting equilibrium expressions:

$$\frac{[H^+]\bullet[HA^{-1}]}{K_{a_1}}+\frac{[H^+]\bullet[A^{-2}]}{K_{a_2}}+[A^{-2}]=E \tag{H-4e}$$

$$\frac{[H^+]^2\bullet[A^{-2}]}{K_{a_1}\bullet K_{a_2}}+\frac{[H^+]\bullet[A^{-2}]}{K_{a_2}}+[A^{-2}]=E \tag{H-4f}$$

Solving for A^{-2}:

$$[A^{-2}]=\frac{E}{1+\frac{[H^+]}{K_{a_2}}\bullet\left(1+\frac{[H^+]}{K_{a_1}}\right)} \tag{H-4g}$$

The charge balance becomes:

$$0=C_S-[HA^-]-2\bullet[A^{-2}]+[H^+]-[OH^-] \tag{H-4h}$$

with:

$$[HA^-]=\frac{[H^+]\bullet[A^{-2}]}{K_{a_2}} \tag{H-4i}$$

The charge balance becomes:

$$0 = C_S - \frac{[H^+] \bullet [A^{-2}]}{K_{a2}} - 2 \bullet [A^{-2}] + [H^+] - [OH^-] \tag{H-4j}$$

Isolating $[A^{-2}]$:

$$0 = C_S - [A^{-2}] \bullet \left(2 + \frac{[H^+]}{K_{a_2}}\right) + [H^+] - [OH^-] \tag{H-4k}$$

Substituting for $[A^{-2}]$:

$$0 = C_S - \left(2 + \frac{[H^+]}{K_{a_2}}\right) \bullet \frac{E}{1 + \frac{[H^+]}{K_{a_2}} \bullet \left(1 + \frac{[H^+]}{K_{a_1}}\right)} + [H^+] - [OH^-] \tag{H-4l}$$

Casting into terms of P_1 and P_2:

$$0 = C_S - \underbrace{(2 + P_2) \bullet \frac{E}{1 + P_2 \bullet (1 + P_1)}}_{M_2} + [H^+] - [OH^-] \tag{H-4m}$$

Note that a similar approach from the monovalent case can be used to show the derivation for bases where the sign of $M2$ will change, leading to the final expression:

$$M_2 = s \bullet (2 + P_2) \bullet \frac{\left(X_a + \sum_{j=1}^{n} \beta_j \bullet X_j\right)}{1 + P_2 \bullet (1 + P_1)} = s \bullet (2 + P_2) \bullet \frac{E}{1 + P_2 \bullet (1 + P_1)} \tag{H-4n}$$

H.5 Trivalent Ion

For the trivalent acid:

$$[H_3A] \overset{K_{a1}}{\leftrightarrow} [H_2A^{-1}] + [H^{+1}] \overset{K_{a2}}{\leftrightarrow} [H^{+1}] + [HA^{-1}] \overset{K_{a3}}{\leftrightarrow} [H^{+1}] + [A^{-3}] \tag{H-5a}$$

$$K_{a_1} = \frac{[H_2A^-] \bullet [H^+]}{[H_3A]} \quad K_{a_2} = \frac{[HA^{-2}] \bullet [H^+]}{[H_2A^-]} \quad K_{a_3} = \frac{[A^{-3}] \bullet [H^+]}{[HA^{-2}]} \tag{H-5b}$$

The species balance becomes:

$$[H_3A] + [H_2A^-] + [HA^{-2}] + [A^{-3}] = X_a + \sum_{j=1}^{n} \beta_j \bullet X_j = E \tag{H-5c}$$

Substituting for equilibrium as before:

$$\frac{[A^{-3}]\bullet[H^+]^3}{K_{a1}\bullet K_{a2}\bullet K_{a3}}+\frac{[A^{-3}]\bullet[H^+]^2}{K_{a2}\bullet K_{a3}}+\frac{[A^{-3}]\bullet[H^+]}{K_{a3}}+[A^{-3}]=E \quad \text{(H-5d)}$$

Solving for $[A^{-3}]$:

$$[A^{-3}]=\frac{E}{\left(\frac{[H^+]^3}{K_{a1}\bullet K_{a2}\bullet K_{a3}}+\frac{[H^+]^2}{K_{a2}\bullet K_{a3}}+\frac{[H^+]}{K_{a3}}+1\right)} \quad \text{(H-5e)}$$

The charge balance in this instance becomes:

$$0=C_S+[H^+]-[OH^-]-[H_2A^-]-2\bullet[HA^{-2}]-3\bullet[A^{-3}] \quad \text{(H-5f)}$$

Substituting for equilibrium expressions:

$$0=C_S+[H^+]-[OH^-]-\frac{[A^{-3}]\bullet[H^+]^2}{K_{a2}\bullet K_{a3}}-2\bullet\frac{[A^{-3}]\bullet[H^+]}{K_{a3}}-3\bullet[A^{-3}] \quad \text{(H-5g)}$$

Isolating $[A^{-3}]$:

$$0=C_S+[H^+]-[OH^-]-[A^{-3}]\bullet\left(\frac{[H^+]^2}{K_{a2}\bullet K_{a3}}+2\bullet\frac{[H^+]}{K_{a3}}+3\right) \quad \text{(H-5h)}$$

Substituting for $[A^{-3}]$:

$$0=C_S+[H^+]-[OH^-]-\frac{E}{\left(\frac{[H^+]^3}{K_{a1}\bullet K_{a2}\bullet K_{a3}}+\frac{[H^+]^2}{K_{a2}\bullet K_{a3}}+\frac{[H^+]}{K_{a3}}+1\right)}\bullet\left(\frac{[H^+]^2}{K_{a2}\bullet K_{a3}}+2\bullet\frac{[H^+]}{K_{a3}}+3\right) \quad \text{(H-5i)}$$

Substituting for P_1, P_2, and P_3:

$$0=C_S+[H^+]-[OH^-]-\frac{3+P_3\bullet(2+P_2)}{1+P_3\bullet(1+P_2\bullet(1+P_1))}\bullet E \quad \text{(H-5j)}$$

Note that a similar approach from the monovalent case can be used to show the derivation for bases where the sign of $M3$ will change, leading to the final expression:

$$M_3=s\bullet\left(X_a+\sum_{j=1}^{n}\beta_j\bullet X_j\right)\bullet\frac{3+P_3\bullet(2+P_2)}{1+P_3\bullet(1+P_2\bullet(1+P_1))}=s\bullet E\frac{3+P_3\bullet(2+P_2)}{1+P_3\bullet(1+P_2\bullet(1+P_1))} \quad \text{(H-5k)}$$

H.6 Temperature

Note that equilibrium constants (pK_a values, including K_w) will be functions of temperature. Many sources in the literature have a variety of data.

H.7 Activity

As the concentrations of ionic species in the system increase, the ideality of the system will diminish. For a general equilibrium equation, activity coefficients can be included:

$$K_a' = \gamma_{H^+} \bullet [H^+] \bullet \gamma_{A^-} \bullet \frac{[A^-]}{[HA] \bullet \gamma_{HA}} = K_a \bullet \frac{\gamma_{H^+} \bullet \gamma_{A^-}}{\gamma_{HA}} \qquad \text{(H-7a)}$$

However, for relatively dilute systems ($1 \leq 0.1$), activity can be used to correct nonidealities using the Debye-Hückel equation at 25°C with water as the solvent:

$$\log(\gamma_{H^+}) = -0.5115 \bullet \frac{I^{0.5}}{1 + 2.962 \bullet I^{0.5}} \qquad \text{(H-7b)}$$

where I is the ionic strength. Ionic strength can be calculated as:

$$I = 0.5 \bullet \sum m_i \bullet Z_i^2 \qquad \text{(H-7c)}$$

To estimate the activity of other species at low ionic strength ($1 \leq 0.1$), the general Debye-Hückel equation is of the form:

$$\log(\gamma_i) = \frac{\phi \bullet Z_i^2 \bullet I^{0.5}}{1 + \theta \bullet \text{Å} \bullet I^{0.5}} \qquad \text{(H-7d)}$$

where Å depends upon the ion.

For H^+, Å $= 9$ [1] $\varphi = 0.5115$ *and* $\theta = 0.3291$.

For molarity and water 25 °C [2].

For K^+, Cl^-, Å $= 3$.

For CO_3^{-2}, Å $= 4.5$ [2].

H.8 Nomenclature

β = stoichiometric coefficient associated with an anion
ε = stoichiometric coefficient associated with a cation
λ = positive charge associated with a cation
ω = negative charge associated with an anion
B = cation (such as Ca^{2+})
A = anion (such as Cl^-)
C_S = contribution of salts to a charge balance
E = molar concentration of an acid or base and its common ions to a system

I = ionic strength (see Equation H-7c)
K_a = acid dissociation constant
P_i = *p***H** factor of a particular dissociation i
s = sign of a particular contribution; defined as –1 for acids and +1 for bases
pH = negative of the base 10 log of hydronium ion concentration
pK_a = negative of the base 10 log of the acid dissociation constant
pK_w = negative of the base 10 log of the water dissociation constant
M_i = overall contribution of the ith component to the charge balance
m_i = molal concentration of the ith species (moles/(kg of solvent))
X = molar concentration of acid, base, or salt in solution (mol/L)
H = hydrogen
K_w = equilibirum constant for the autoionization of water
K_b = *base* dissociation constant
m_i = molality of species i
Z_i = charge of species i
γ_i = activity coefficient of a given species i
φ = Debye-Hückel temperature parameter 1
θ = Debye-Hückel temperature parameter 2
Å = *ion size paramater*

References

[1] Jacob Kielland, "Individual Activity Coefficients of Ions in Aqueous Solutions," *Journal of American Chemical Society* 59 (September 1937): 1675–1678.

[2] Roger G. Bates, *Determination of pH: Theory and Practice*, 2nd ed. (New York: Wiley, 1973).

Appendix I
Interactive to Noninteractive Time Constant Conversion

Interactive time constants can be partitioned into two subsystems in series with their capacitance and resistance components to calculate the equivalent noninteractive time constants. The quadratic equation can then be used to find the two noninteractive time constants for two capacitances that have an inlet and outlet resistance and are separated by a resistance [1]. Interactive time constants exist when the flow or transfer of mass or heat into, within, and out of a system depend on the operating conditions as driving forces (e.g., pressure drop for flow, concentration differences for mass transfer, and temperature difference for heat transfer). For gas pressure systems, the flow resistance of valves and fittings into, within, and out of the volumes are important. For thermal systems such as vessel coils and jackets, heat exchanger tubes, and thermowells, the resistance of convective heat transfer of the inside and outside surfaces and the resistance of conductive heat transfer within the surfaces are important. The capacitances offer the retention of energy or material in a subsystem, and the resistances oppose the transfer of material and energy between the subsystems.

I.1 Equivalent Noninteractive Time Constants

$$\tau_{n1} = \frac{2 \bullet A}{B - \sqrt{B^2 - 4 \bullet A \bullet C}} \tag{I-1a}$$

$$\tau_{n2} = \frac{2 \bullet A}{B + \sqrt{B^2 - 4 \bullet A \bullet C}} \tag{I-1b}$$

$$A = R_1 \bullet C_1 \bullet R_2 \bullet C_2 \bullet R_3 \tag{I-1c}$$

$$B = R_1 \bullet C_1 \bullet R_3 + R_1 \bullet C_1 \bullet R_2 + R_2 \bullet C_2 \bullet R_3 + R_1 \bullet C_2 \bullet R_3 \tag{I-1d}$$

$$C = R_1 + R_2 + R_3 \tag{I-1e}$$

where

A = first term of quadratic equation
B = second term of quadratic equation
C = third term of quadratic equation
C_1 = capacitance of section 1
C_2 = capacitance of section 2
R_1 = resistance at inlet to section 1
R_2 = resistance between sections 1 and 2
R_3 = resistance at outlet of section 2
τ_{n1} = noninteractive time constant 1 (minutes)
τ_{n2} = noninteractive time constant 2 (minutes)

I.2 Original Interactive Time Constants

The following equations can be used to compute the interactive time constants for a system partitioned into two sections with an inlet resistance (R_1) and outlet resistance (R_3) separated by a resistance (R_2). The capacitances and resistances for gas pressure systems, including engineering units, are defined in Appendix G.

$$\tau_{i1} = \frac{C_1}{\frac{1}{R_1} + \frac{1}{R_2}} \tag{I-2a}$$

$$\tau_{i2} = \frac{C_2}{\frac{1}{R_2} + \frac{1}{R_3}} \tag{I-2b}$$

where

C_1 = capacitance of section 1
C_2 = capacitance of section 2
R_1 = resistance at inlet to section 1
R_2 = resistance between sections 1 and 2
R_3 = resistance at outlet of section 2
τ_{i1} = interactive time constant 1 (minutes)
τ_{i2} = interactive time constant 2 (minutes)

Reference

[1] Peter Harriot, *Process Control* (New York: McGraw-Hill, 1964).

Appendix J
Jacket and Coil Temperature Control

The heating and cooling system is critical for reactors because temperature plays such a huge role in determining the reaction rate and selectivity (formation of desired product). ***The limit to what a valve position controller (VPC) can do to increase the production rate depends on the capability of the heating and cooling system. The system can also be the source of discontinuities and nonlinearities. Most reactor temperature control problems beyond tuning can be traced back to deficiencies in the reactor coil or jacket temperature system design.***

Heating is required for endothermic reactions (reactions that consume energy), liquids that are being vaporized, or bringing a reactor up to operating temperature. Cooling is needed for exothermic reactions (reactions that produce heat), vapors that are being condensed, and bringing a reactor down to operating temperature. Often both heating and cooling are required. Split-range control is used to go back and forth between heating and cooling. In split-range control for an exothermic reactor with a fail-open cooling valve, as the temperature controller output increases from 0% to the split-range point, the coolant valve goes from wide open to completely closed. As the temperature controller output increases from the split-range point to 100%, the heating valve goes from closed to wide open. The split-range point is traditionally set at 50% but would be better set so that the change in temperature for a change in flow is about the same for each valve. Although cooling and heating may be done by coils as well as jackets, many of the control considerations are the same. Jacketed reactors will be used for the purposes of discussion and illustration. Cascade temperature control is the most prevalent strategy in which the primary reactor temperature proportional-integral-derivative (PID) controller provides a set point to a secondary jacket temperature PID for throttling hot and cold fluids (Figures J-1a and J-1b). For highly exothermic reactors, boiler feedwater is added under level control and the reactor temperature PID output is the jacket outlet steam pressure PID set point. These jacket control schemes are suitable for batch as well as continuous operation.

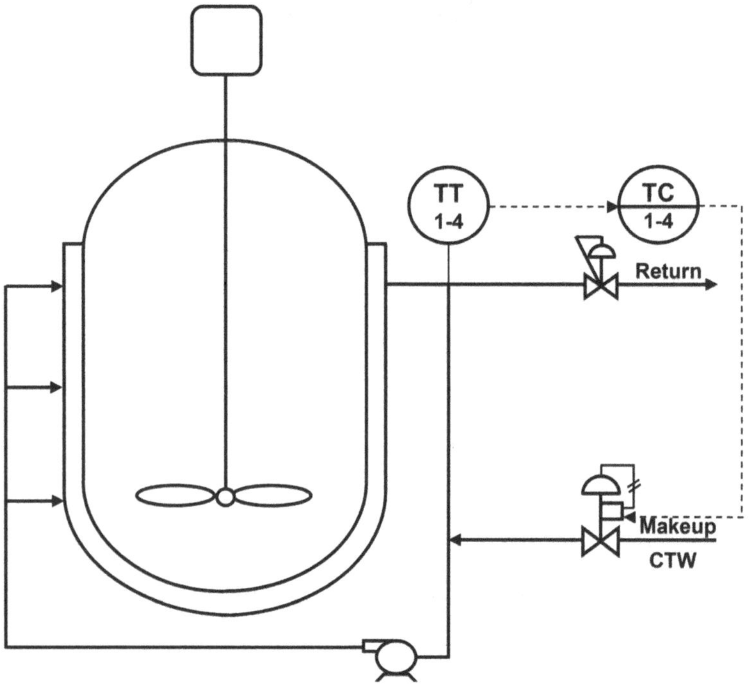

Figure J-1a. Jacket-outlet temperature control offers moderation of mixing and phase discontinuities from the attenuation by jacket volume.

Using cascade control, where the reactor temperature controller output is the set point of a coil or jacket temperature loop, offers considerable performance improvements. The jacket temperature controller can correct for disturbances to the jacket before they affect the reactor temperature. The jacket temperature loop also isolates process and valve gain nonlinearity from the reactor loop. For a negligible increase in the heat of reaction compared to the heat transfer capability, the process gain for the reactor temperature loop approaches unity.

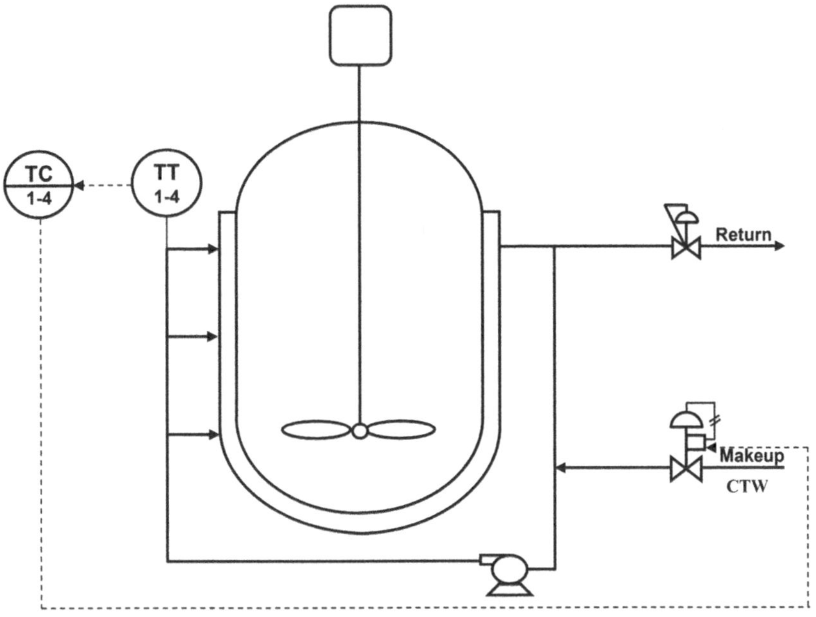

Figure J-1b. Jacket-inlet offers faster correction of jacket inlet disturbances such as coolant temperature.

Coils generally offer a faster temperature response than a jacket by decreasing in the volume and increasing in the velocity. Both of these work to decrease the process dead time, which is the coil volume divided by the utility flow rate. The increase in velocity increases the heat transfer coefficient, but this is partially offset by a decrease in the surface area. An increase in the product of the heat transfer coefficient and surface area (UA) will decrease the secondary process lag in the thermal response. The transition in split-range operation is faster, which is useful for a valid transition between hot and cold utility streams but can be problematic for inadvertent transitions from stick-slip and an integrating response in the process or controller.

Whether the secondary loop uses coil or jacket inlet or outlet temperature is often a matter of tradition for a particular company or process industry. The dynamic response of the cascade control system to reactor disturbances such as feed and reaction rate are the same for jacket inlet and outlet temperature control. ***The coil or jacket inlet temperature control will correct for changes in cooling or heating utility supply temperature and pressure sooner than the transportation delay through the jacket.*** The jacket loop process dead time is also less by the amount of this delay, enabling a faster reset time setting and faster correction of valve nonlinearities. ***At the coil or jacket inlet, the mixing of the recirculation with the hot or cold makeup flow may be incomplete and the discontinuity in the transition from hot to cold more abrupt. Locating*** the temperature sensor on the jacket outlet offers time for mixing and volume for smoothing transitions. Less measurement temperature noise can translate to a higher controller gain and less overreaction to the discontinuity at a split-range transition.

The difference between the reactor and the jacket outlet temperature (approach temperature) can provide an inferential measurement of the heat transfer coefficient (U) for a constant jacket circulation flow, a given production rate, and the heat transfer area (A). The approach temperature increases as UA decreases. For residence time control, the increase in level will increase the heat transfer area covered by reactants, offsetting the increase in heat release with the increase in production rate and eliminating the need for production rate and level correction. For fed-batch reactors and continuous reactors without residence time control, a correction for level is needed to compute U from the UA.

For the coil or jacket temperature control to rapidly adjust cooling and heating for disturbances to the jacket and set-point changes from the reactor temperature control, the coil or jacket PID response must be fast and achieved by a fast sensor and fast tuning.

A tight-fitting sensor bottomed in a thermowell with a high thermal conductivity metal and tapered tip provides a fast measurement. Spring loading can ensure the sensor sheath is bottomed. The clearance between the sheath outside diameter and the thermowell inside diameter must be minimized and the fluid velocity maximized. While grounded thermocouple sensors are a few seconds faster than resistance temperature detector (RTD) sensors, the difference is insignificant compared to the effect of thermowell design and fluid velocity. The thermowell design details are also important for reactor temperature. The greater sensitivity and lower drift of the RTD is important for jacket and reactor temperature measurements. The lower drift reduces maintenance, and the higher sensitivity provides faster recognition. Using RTDs also facilitates more accurate online heat transfer computation for process diagnostics and inferential measurements of reaction rates. Thermocouples are preferred for temperatures above 400°C where RTD insulation resistance and sensor integrity become problematic.

A high controller gain and a small reset time provide fast tuning. The jacket temperature controller is a self-regulating process with a maximum PID gain that is about half of the open-loop time constant divided by the product of the open-loop gain and dead time as detailed in the section in Chapter 8 on controller tuning. The minimum reset setting is about four times the loop dead time for the jacket temperature PID controller.

Most temperature loops tend to develop an oscillation across the split-range point in making the transition between heating and cooling. The transition creates a discontinuity from the change in utility fluid, the increase in friction, and a loss of sensitivity in the near each valve's closed position. The worst case is often the transition between steam and coolant due to the huge difference in temperatures and the creation of bubbles in the coolant or water droplets in the steam. Steam valves tend to have higher seating and sealing friction due to the higher temperatures and pressures. Hot and cold liquids and tempered water systems offer a much smoother transition than steam and coolant, and they reduce nonlinearities.

Steam does not provide uniform heating in coils or a jacket. Steam typically collects in the top of the jacket and condensate collects in the bottom. Hot spots can develop around inlets. Thermal shock and steam hammer can damage glass-lined vessels. The time required to drive steam completely out of the jacket before cooling water is introduced causes a considerable delay in the control system. Improper trap design or operation can cause condensate buildup.

Adding hot water instead of steam directly into the jacket provides a more uniform heat distribution, a dramatically smoother transition between heating and cooling, and a more efficient and maintainable system. For rapid heating, using direct steam injection heaters and pressurized water can provide hot water temperatures well above 100°C (Figure J-2). If the injection heater has hundreds of small orifices, the bubbles are extremely small and are rapidly and quietly mixed into the water. Variable orifice steam injection heaters are not as quiet, and the mixing is not as complete. Using jacket outlet temperature reduces the possibility of bubbles hitting the temperature sensor.

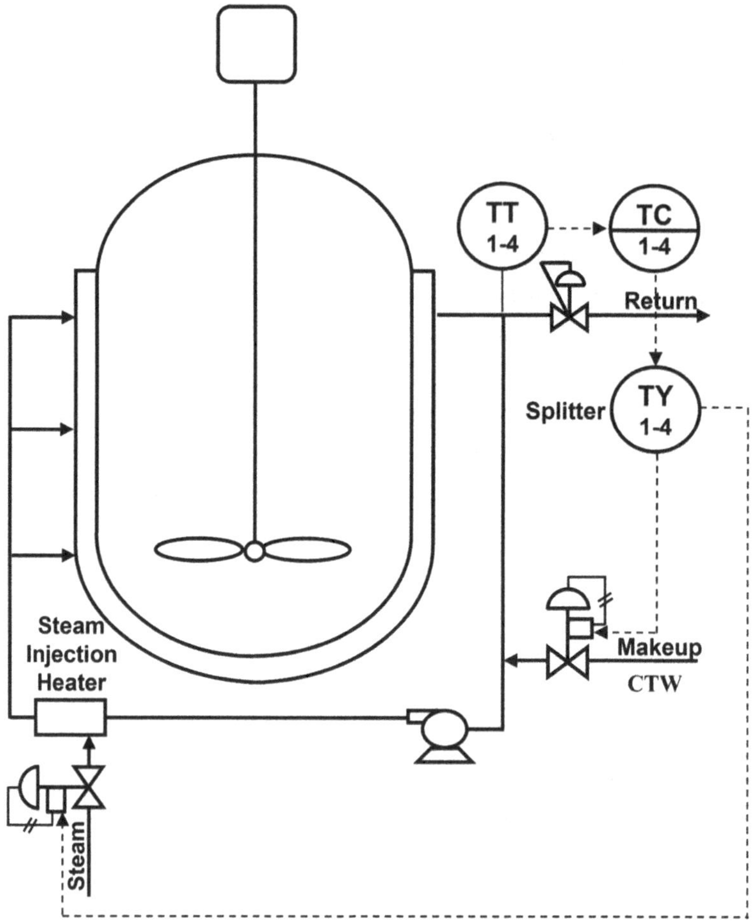

Figure J-2. Using steam injection heaters to create hot water offers rapid heating and tight control of coil or jacket temperature with smoother split-ranged transitions.

Appendix K PID Forms and Conversion of Tuning Settings

K.1 Conversion of Settings to Correct Units and Form

The equations in this book are based on the Standard form with specific tuning setting units where the controller gain is dimensionless, the reset time is seconds (seconds per repeat), and the rate time is seconds. Proportional-integral-derivative (PID) tuning settings must first be checked for units and, if necessary, undergo a units conversion before being used.

1. Convert proportional band (%) to controller gain (% / %) (dimensionless). Gain = 100 % / proportional band
2. Convert reset setting in repeats per minute to reset time in seconds per repeat. Seconds per repeat = 60 / repeats per minute
3. Convert rate time in minutes to rate time in seconds. Seconds = 60 • minutes

4. After the tuning setting units are verified, use the Series and Parallel forms to convert tuning settings to Standard form tuning settings per the following equations.

   The primed tuning settings are for the Series form, the double-primed tuning settings are for the Parallel form, and the unprimed tuning settings are for the Standard form used in this book.

To convert from Series to Standard form controller gain:

$$K_c = \frac{T_i' + T_d'}{T_i'} \bullet K_c' \tag{K-1a}$$

To convert from Series to Standard form reset (integral) time:

$$T_i = \frac{T_i' + T_d'}{T_i'} \bullet T_i' = T_i' + T_d' \tag{K-1b}$$

To convert from Series to Standard form rate time:

$$T_d = \frac{T_i'}{T_i' + T_d'} \bullet T_d' \tag{K-1c}$$

Note that if the rate time is zero, the Series form (Figure K-1a) and the Standard form (Figure K-1c) settings are identical. When using the Standard form, if the rate time is greater than 1/4 the reset time, the response can become oscillatory. If the rate time exceeds the reset time, the response can become unstable from a reversal of action from these modes. The Series form inherently prevents this instability by increasing the effective reset time as the rate time is increased.

We can convert from the Standard form to the Series form using the following equations if the reset time is equal to or greater than four times the rate time ($T_i \geq 4 \bullet T_d$).

$$K_c' = \frac{K_c}{2} \bullet \left[1 + \left(1 - 4 \bullet \frac{T_d}{T_i}\right)^{0.5}\right] \tag{K-1d}$$

$$T_i' = \frac{T_i}{2} \bullet \left[1 + \left(1 - 4 \bullet \frac{T_d}{T_i}\right)^{0.5}\right] \tag{K-1e}$$

$$T_d' = \frac{T_i}{2} \bullet \left[1 - \left(1 - 4 \bullet \frac{T_d}{T_i}\right)^{0.5}\right] \tag{K-1f}$$

The Parallel form (Figure K-1b) is used in some distributed control systems (DCSs) and programmable logic controllers (PLCs) to isolate the proportional mode tuning setting from the other modes. The gain setting does not affect the contribution from the integral and derivative modes and is sometimes called *noninteracting*. We can convert from the Parallel form to the Standard form by using the following equations.

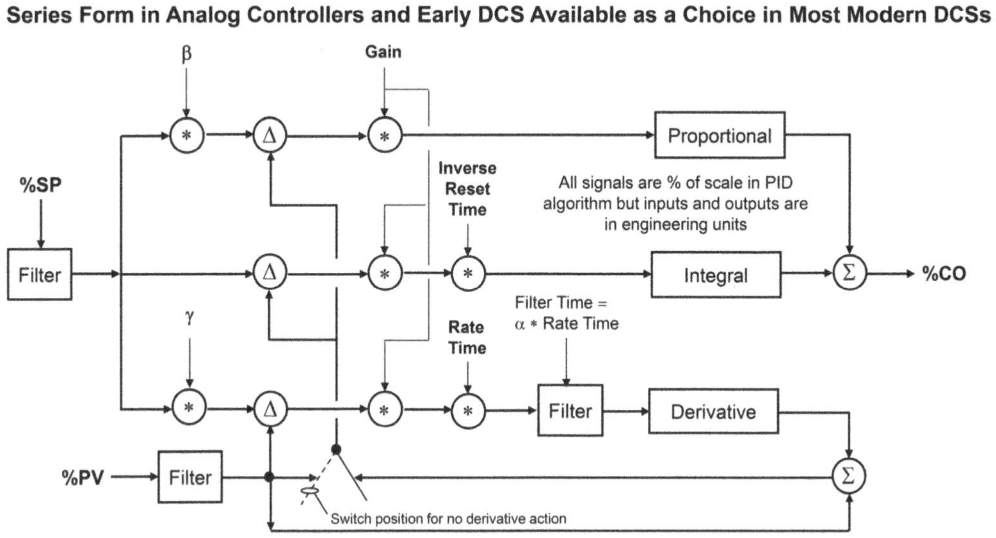

Figure K-1a. Series form has derivative mode computed in series with other modes.

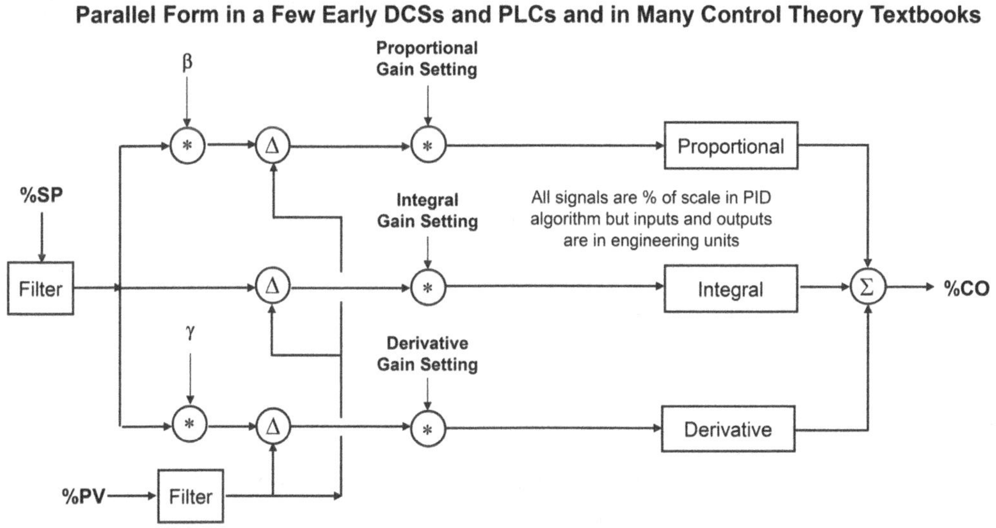

Figure K-1b. Parallel form has independent mode gain settings.

$$K_c = K_c'' \tag{K-1g}$$

$$T_i = K_c'' \bullet T_c'' \tag{K-1h}$$

$$T_d = \frac{T_d''}{K_c''} \tag{K-1i}$$

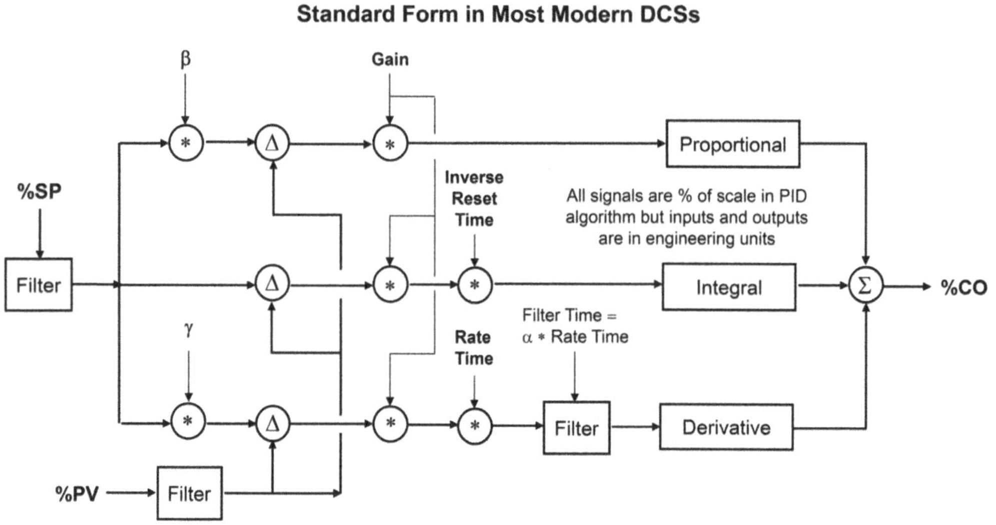

Figure K-1c. Standard form has modes in parallel with proportional gain factor.

We can convert from the Standard form to the Parallel form by using the following equations:

$$K_c'' = K_c \tag{K-1j}$$

$$T_i'' = \frac{T_i}{K_c} \tag{K-1k}$$

$$T_d'' = K_c \bullet T_d \tag{K-1l}$$

where

K_c = controller gain for *Standard* form (%/%) (dimensionless)
K_c' = controller gain for *Series* form (%/%) (dimensionless)
K_c'' = controller gain for *Parallel* form (%/%) (dimensionless)
T_i = integral time (reset time) for *Standard* form (seconds)
T_i' = integral time (reset time) for *Series* form (seconds)
T_i'' = integral time (reset time) for *Parallel* form (seconds)
T_d = derivative time (rate time) for *Standard* form (reset time) (seconds)
T_d' = derivative time (rate time) for *Series* form (seconds)
T_d'' = derivative time (rate time) for *Parallel* form (seconds)

K.2 Difference Equations for Standard and Series Form

The difference equations for a reverse-acting Standard form PID controller with rate limiting where the filter time on the derivative mode is a fraction α of the rate time are as follows:

$$\%CO_n = P_n + I_n + D_n + \%CO_i \tag{K-2a}$$

$$P_n = K_c \bullet (\beta \bullet \%SP_n - \%PV_n) \tag{K-2b}$$

$$I_n = \frac{K_c}{T_i} \bullet (\%SP_n - \%PV_n) \bullet \Delta t_x + I_{n\text{-}1} \tag{K-2c}$$

$$D_n = \frac{K_c \bullet T_d \bullet [\gamma \bullet (\%SP_n - \%SP_{n-1}) - (\%PV_n - \%PV_{n-1})] + \alpha \bullet T_d \bullet D_{n-1}}{\alpha \bullet T_d + \Delta t_x} \tag{K-2d}$$

The difference equations for a reverse-acting Series form PID controller with rate limiting where the filter time on the derivative mode is a fraction α of the rate time are as follows:

$$\%CO_n = PD'_n + ID'_n + \%CO_i \tag{K-2e}$$

$$PD'_n = K'_c \bullet (\beta \bullet \%SP_n - \%PV_n - D'_n) \tag{K-2f}$$

$$ID'_n = \frac{K'_c}{T_i} \bullet (\%SP_n - \%PV_n - D'_n) \bullet \Delta t_x + ID'_{n\text{-}1} \tag{K-2g}$$

$$D'_n = \frac{K'_c \bullet T'_d \bullet [\gamma \bullet (\%SP_n - \%SP_{n-1}) - (\%PV_n - \%PV_{n-1})] + \alpha \bullet T'_n \bullet D'_{n-1}}{\alpha \bullet T'_d + \Delta t_x} \tag{K-2h}$$

If the structure used does not have integral action, the term $\%CO_i$ is an adjustable bias and the integral terms I_n and ID'_n are zero (Equations K-2c and K-2g are not used).

K.3 Positive Feedback Implementation of the Integral Mode

The positive feedback implementation of the integral mode as shown in Figure K-3a effectively yields Equations K-2a through K-2d as shown in the following derivation for a proportional-integral (PI) controller using Laplace transforms. Instead of an integrator, a filter whose input is the controller output and whose output is added to the contribution from the proportional mode is used to provide integral action in the positive feedback implementation. The filter time is the integral time setting. When external-reset feedback is enabled, the input to the filter is switched from the controller output to the external-reset feedback signal.

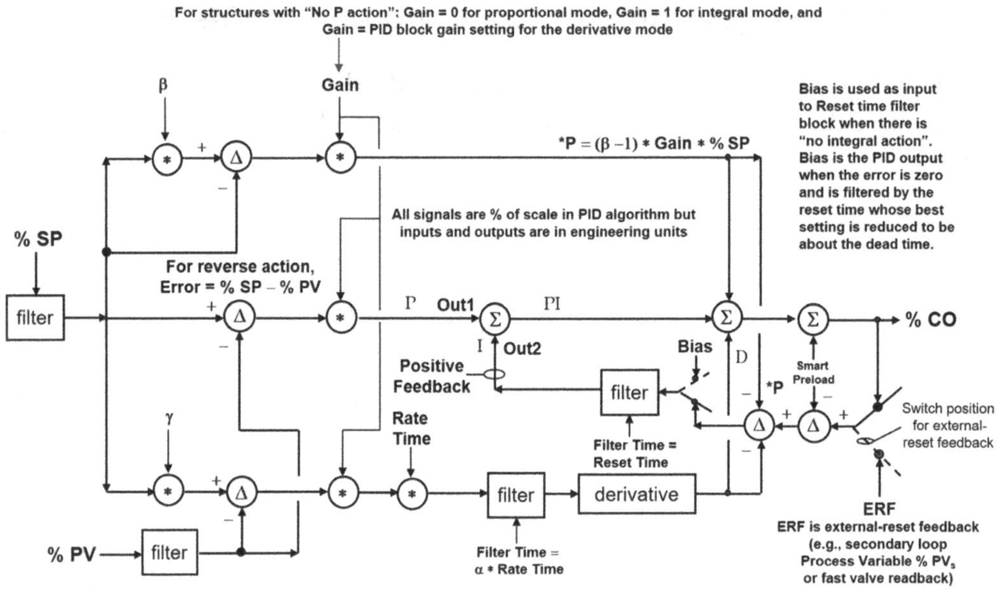

Figure K-3a. Standard form with positive feedback implementation of integral mode and external-reset feedback.

Using the positive feedback implementation of the integral mode is the basis for the enhanced PID controller developed for analyzers and wireless devices (described in Appendix C) that is effective for these and other devices with discontinuous updates especially when the update time is much longer than the process 63% response time.

$$O(s) = K_c \bullet E(s) + \frac{1}{1 + T_i \bullet s} \bullet O(s) \tag{K-3a}$$

$$O(s) \bullet \left(1 - \frac{1}{1 + T_i \bullet s}\right) = K_c \bullet E(s) \tag{K-3b}$$

$$O(s) \bullet \left(\frac{T_i \bullet s}{1 + T_i \bullet s}\right) = K_c \bullet E(s) \tag{K-3c}$$

$$\frac{O(s)}{E(s)} = K_c \bullet \left(\frac{1 + T_i \bullet s}{T_i \bullet s}\right) = K_c + \frac{K_c}{T_i \bullet s} \tag{K-3d}$$

where

$\%CO_i$ = controller output at the transition from manual (MAN) or remote output (ROUT) modes (%)

$\%CO_n$ = controller output for current scan n (%)

$\%CO_{n-1}$ = controller output for last scan n–1 (%)

$\%PV_n$ = process variable for current scan n (%)

$\%PV_{n\text{-}1}$ = process variable for last scan n–1 (%)
$\%SP_n$ = set point for current scan n (%)
$\%SP_{n\text{-}1}$ = set point for last scan n–1 (%)
K_c = controller gain for *Standard* form (% / %) (dimensionless)
K_c' = controller gain for *Series* form (% / %) (dimensionless)
T_i = integral time (reset time) for *Standard* form (seconds)
T_i' = integral time (reset time) for *Series* form (seconds)
T_d = derivative time (rate time) for *Standard* form (reset time) (seconds)
T_d' = derivative time (rate time) for *Series* form (seconds)
P_n = proportional mode contribution for *Standard* form for current scan n (%)
$P_{n\text{-}1}$ = proportional mode contribution for *Standard* form for last scan n–1 (%)
PD_n' = proportional-derivative mode contribution for *Series* form for current scan n (%)
PD_{n-1}' = proportional-derivative mode contribution for *Series* form for last scan n–1 (%)
I_n = integral mode contribution for *Standard* form for current scan n (%)
$I_{n\text{-}1}$ = integral mode contribution for *Standard* form for last scan n–1 (%)
ID_n' = integral-derivative mode contribution for *Series* form for current scan n (%)
ID_{n-1}' = integral-derivative mode contribution for *Series* form for last scan n–1 (%)
Δt_x = execution time of PID block (seconds)
α = alpha factor for derivative mode filter (fraction of rate time) (0–1) (dimensionless)
β = beta set-point weight factor for proportional mode (0–1) (dimensionless)
γ = gamma set-point weight factor for derivative mode (0–1) (dimensionless)
E(s) = Laplace transform of controller error (%)
O(s) = Laplace transform of controller output (%)

Appendix L
Liquid Mixing Dynamics

The dead time and time constant of tanks, static mixers, and pipelines for liquid composition control can be estimated by using Equations L-1 and L-2 for an estimated or computed equivalent agitation flow (F_a). For static mixers and pipelines, the agitation flow for back mixing is minimal and rarely exceeds 25% of the feed flow (F_f). These volumes are characterized as plug flow, and most of the residence time (volume divided by feed flow) becomes dead time. The lack of a process time constant translates to no slowing down of a fast disturbance. The saving grace is that the residence time of these systems is relatively short (e.g., 3 s), so the dead time introduced is minimal and may be less than the electrode time constant or signal filter that provides some smoothing and prevents the control loop from becoming dead time dominant.

$$\theta_p = 0.5 \bullet \frac{V}{F_f + F_r + F_a} \tag{L-1}$$

$$\tau_p = \frac{V}{F_f} - \theta_p \tag{L-2}$$

For axial agitation where liquid height is between 50% and 150% of vessel diameter:

$$F_a = 7.48 \bullet N_q \bullet N_s \bullet D_t^3 \tag{L-3a}$$

$$N_q = \frac{0.4}{\left[\frac{D_i}{D_t}\right]^{0.55}} \quad \text{(L-3b)}$$

The volume cancels out in the ratio of the process dead time to the 63% response time (dead time plus time constant), which is important for the minimum peak error if the recirculation plus agitation flow is much larger than the feed flow.

$$\frac{\theta_p}{\tau_p + \theta_p} = 0.5 \bullet \frac{F_f}{F_f + F_r + F_a} \quad \text{(L-4)}$$

where

D_i = impeller diameter for axial (back) mixing (feet)
D_t = tank diameter for axial (back) mixing (feet)
N_s = impeller speed for axial (back) mixing (revolutions per minute)
N_s = impeller discharge coefficient for axial (back) mixing (gallons per minute per revolutions per minute • ft^3)
F_a = agitation flow for axial (back) mixing (gallons per minute)
F_f = total feed flow (gallons per minute)
F_r = recirculation flow (gallons per minute)
V = liquid volume (gallons)
θ_p = process dead time (minutes)
τ_p = process time constant (minutes)

For self-regulating processes, the peak error is proportional to the dead time to time constant ratio for a dead time much smaller than the time constant. If the recirculation flow and agitation flow are much greater than the feed flow, the volume cancels out in the ratio of Equation L-2 to Equation L-1, giving Equation L-4 showing that the size of the recirculation and agitation flow relative to feed flow is important. Because the integrated (accumulated) error is proportional to the ratio of the dead time squared to time constant, the volume does not cancel out and will increase the amount of off-spec product. The recirculation flow may tend to provide more radial mixing than axial agitation unless the entry point is vertical instead of side entry. An eductor can magnify the recirculation flow and improve the mixing pattern.

For pipelines, there is no recirculation flow and the agitation flow for back mixing is about 20% of the feed flow. For static mixers and laminar flow, the back mixing is less, resulting in more of the residence time becoming dead time and making the volume more plug flow with only radial mixing (no axial mixing). Plug flow is desirable for reactions because the residence time is constant for each infinitesimal volume of reactants [1]. Plug flow is undesirable for inline systems where reaction time is not a consideration (e.g., static mixers for blending, stock consistency, and neutralization) because there is no smoothing of axial fluctuations.

Equations L-1 and L-2 lead to some interesting conclusions about using well-mixed tanks for composition control. If the tank volume is increased but the feed, recirculation, and agitation flows are kept constant, the peak error is the same but the integrated (accumulated) error is increased in proportion to the increase in tank volume. If the tank volume is made exceptionally large, the dead time may be so large that the minimum reset time setting may cause a reset cycle. If the recirculation and agitation flows were increased in proportion to the tank volume, the peak error would be reduced and the integrated error would be about the same as before the increase in volume and recirculation and agitation flows. The increased tank volume will slow down upstream disturbances and attenuate upstream oscillations for downstream control systems. Thus, exceptionally large tanks improve composition control performance if the loops are upstream or downstream but not on the tank, unless the agitation flow is scaled up with the volume. If the large tank is upstream of the loop, the reagent or coolant usage will be reduced when there are short-term fluctuations in reagent or cooling demand.

Reference

[1] T. Bor, "The Static Mixer as a Chemical Reactor," *British Chemical Engineering* 16, no. 7 (1971), 610.

Appendix M
Modeling *p*H Systems in Digital Twin

M.1 The Right Solution

The great need for dynamic models for every aspect of a *p*H system application is not commonly recognized. The failure to meet objectives is greater than in most other common applications. Furthermore, the failure can be drastic, and the solution may not be known. Often the problem persists for years because of a lack of fundamental understanding of the critical roles of equipment, piping, and automation system design in meeting the many challenges posed by the exceptional nonlinearity and sensitivity of the *p*H process. Fixing the problems can cost hundreds of thousands of dollars and take months, even if the solution is found.

Solutions can be found to failures avoided by using a dynamic model in the project definition stage before estimates are done and the design starts. The model must use an accurate titration curve; all the delays and lags in the equipment, piping, and automation system; and the installed flow characteristic, resolution, and deadband for all control valves and variable-speed drives (VSDs). The model can start as a standalone dynamic simulation integrated with a projected control system that is eventually migrated into a digital twin to finalize the control system design and the testing, training, maintenance, and continuous improvement parameters. The model and the test case figures for each section can be found online.

For a much more extensive understanding of the capability and critical role of the dynamic model in process control (detailed in Chapter 9, "Digital Twin") to enable exploration, discovery, prototyping, testing, training, sustainability, and continuous improvement to maximize performance and minimize lifecycle costs, see References 1 through 8. Dynamic models are more critical in *p***H** systems because of the many challenges and misunderstandings. The right solution involves getting the right model to get the right curve, staging, equipment and piping, valves, electrodes, and automation solution. Consult the best practices at the end of each chapter in this book to help achieve the best model and best total system design and implementation.

M.2 The Right Curve

The critical first step is to set up the charge balance inputs that provide a process *p***H** that matches the lab titration curves for the spectrum of expected operating conditions, particularly compositions and temperatures. Conjugate salts can greatly moderate the slope of a strong acid and strong base curve, as detailed in Appendix H. It is extremely important to add carbonic acid to the charge balance to account for its absorption into the process fluid due to the exposure of the process fluid to air.

M.3 The Right Staging

The general rule of thumb is that a stage is required for every 2 *p***H** units in the maximum deviation of the feed *p***H** from the pH set point. For a 7 *p***H** set point and a feed below 3 *p***H** or above 11 *p***H**, three stages of neutralization are required. The equipment for a stage can be either a vessel or a static mixer. The final stage is generally a vessel. The vessel should be well mixed in most cases. A volume is generally required downstream of any static mixer to smooth out oscillations. Often the static mixer is considered for the first stage because the titration curve is rather flat (process gain low), resulting in smaller oscillations. All stages must have precise control valves or VSDs and minimized reagent injection delays.

The following sections will first address the effect of equipment, piping, valve, and sensor dynamics on the last stage, which is the most challenging stage because of the steepness of the titration curve and the most important in terms of meeting objectives and the effect on downstream processes. We will then follow up with test cases to show the relative performance of the last stage for different equipment and automation systems, including the benefits of linear reagent demand and flow ratio control using valves and flowmeters with exceptional precision and rangeability to possibly eliminate a stage of neutralization.

M.4 The Right Equipment and Piping

The best way to deal with the many common problems and identify and justify the exceptional equipment and piping design often needed for *p*H systems is to simulate the dynamic response of innovative versus conventional designs.

The most common problem is using conventional dip tube designs with a minimum dip tube diameter of 1 in to withstand the turbulence from agitation and a length that extends down at least halfway into the liquid volume to the agitator blades. The dip tube volume for a large vessel is usually more than a gallon. Considering reagent flow rates are often less than 1 gal/h, particularly in the latter stages where *p*H control is more important, it can take hours for the reagent to push out the backfill of vessel fluid in the dip tube at the start of control. Also, when the reagent addition is stopped, the reagent will migrate from the dip tube filled with reagent for hours after the reagent valve is closed. The dip tube can be shortened, but this introduces a mixing delay and noise from poor mixing at the point of addition.

The best solution is injecting reagent into a recirculation line right before the entry point into the vessel. The recirculation line also adds a mixing flow rate to the agitator pumping rate that reduces the turnover time, which is the mixing delay. If an eductor is used, the effective mixing flow rate is three to five times the recirculation rate. The reagent valve is also close coupled to an isolation valve mounted on the nozzle to the recirculation line so the injection delay is negligible.

A similar benefit from reducing reagent injection delay can be achieved by injecting reagent into the feed stream. However, the feed flow is usually much less than the recirculation flow, and feeding the reagent through the drip tube adds a larger and more variable reagent injection delay and, therefore, turnover time. Also, when the feed stops, the injection delay becomes disastrously large because of the large dip tube volume resulting from its large diameter to prevent movement from agitation and a long length to get close to the agitator.

M.5 The Right Valves

A resolution limitation in the reagent control valve or VSD will cause a limit cycle that is amplified by the high process gain of a steep titration curve, as detailed in Equation 8-10dd. A decrease in the proportional-integral-derivative (PID) gain and an increase in the PID reset time increases the limit cycle period, as detailed in Equation 8-10ee. A dynamic model is essential to show the limit cycle for a given resolution limitation and the impact of the process gain of PID tuning. Faster periods are more effectively filtered by volumes, as shown in Equation 8-10pp, so there is some benefit of a secondary inline *p***H** loop in the recirculation line with the vessel *p***H** as the primary loop in a cascade control system setup. In this case, the reagent is injected into the inlet of the static mixer with *p***H** sensors about 20 pipe diameters downstream of the mixer outlet. The mixer is near the reentry point of the recirculation line into the vessel to minimize transportation delay.

The best resolution available in a well-designed pneumatic actuated throttling valve is about 0.1%. The more common resolution is 0.5%, and it can be as large as 2% for piston actuators. For on-off valves designed for tight shutoff posing as throttling valves, the resolution could be as bad as 8%.

M.6 The Right Electrodes

Most users are unaware of how slow the electrode response is from the start for various designs and, additionally, from deterioration due to time and operating conditions. Often the suppliers do not furnish information, and if they do, it is for ideal conditions for a new electrode stating a response time that does not include the percentage of the final response. The user should ask the supplier for the 86% response time (T86), which is the time to reach 86% of the final response for different directions, velocities, and operating conditions. For example, the T86 may be 10 times larger for a velocity less than 0.1 ft/s compared to 4 ft/s, twice as large for decreases versus increases in *p***H**, and four times larger for a strong acid and strong base system. This T86 metric is useful in including the delay in the start of the response and the greater-than-expected slowing down of the response. T86 is used in quantifying the valve speed of response in the ISA-TR75.25 technical report on valve response testing and is advocated as a common metric for many types of sensors.

The difference in response time can be quite significant for different electrode designs. For example, the T86 of a new flat glass electrode can be 40 s, whereas the T86 of a new spherical glass electrode can be 4 s assuming a velocity of 1 ft/s past the electrodes. Even more dramatic is the deterioration of even the fastest electrodes because of aging or fouling of the glass measurement electrode, causing a T86 of 400 s or more.

M.7 The Right Automation System

The most prominent opportunities for control strategies involve using signal characterization, adaptive tuning, and intelligent ratio control. The first step is a piecewise linear fit via a signal characterizer block to compute the x-axis of the titration curve, which is a ratio of reagent flow to influent flow. Note that the lab curve has an implied ratio of reagent volume added to sample volume that can be converted to a reagent-to-feed-flow ratio taking into account any differences in lab-versus-field reagent concentration and temperature.

The controlled variable and set point are transformed from *p*H to the x-axis of the curve, often simply scaled at 0% to 100% reagent demand. The number of line segments is increased in the control region to better include the process nonlinearity where it matters the most. The *p*H controller becomes a reagent demand controller with most of the titration curve nonlinearity effect on process gain eliminated. The greater remaining nonlinearity, known as the *hidden factor* for all composition control loops, is inversely proportional to the feed flow revealed in the first-principle ordinary differential equations for a material balance detailed in Appendix F. This can be deduced from the need to divide the reagent flow by the feed flow to get the x-axis of the titration curve. Fortunately, for a well-mixed vessel, the increase in the process gain for a decrease in feed flow is compensated for by the increase in residence time and, therefore, the process time constant. For injecting reagent into a recirculation line, the hidden factor has a negligible effect on the best PID gain. However, reagent injected into a feed line will show an increase in reagent injection delay for a significant decrease in feed flow into a large dip tube that will require a decrease in the PID gain and increase in reset time set for the best load response for low feed flows. For equipment with little to no back mixing, such as the static mixer, the decrease in feed flow increases the process dead time but has a negligible effect on the process time constant, therefore necessitating a detuning of the PID for low feed flows. For these cases, adaptive tuning can schedule the PID tuning settings as a function of feed flow.

A signal characterizer is also used to provide intelligent ratio control. The distance between the feed *p*H to a point on the titration curve, just short of the *p*H set point to allow for curve and flow measurement errors (e.g., 6.8 *p*H for a 7 *p*H set point), is used to compute the change in the ratio of reagent flow to feed flow. This change in ratio is then used to update the reagent flow controller set point via a ratio controller. The ratio controller output, which is the ratio set point multiplied by the feed flow, becomes the reagent flow controller set point. The *p*H or linear reagent demand controller provides a bias correction of the ratio controller output. The ratio set point and actual ratio resulting from the bias correction must be visible to operations, maintenance, and engineering support. The operator should be able to set the ratio manually if there is a problem with the *p*H or reagent demand controller. It is critical that the feed and reagent flow controllers have the high accuracy and rangeability best achieved by Coriolis meters, which also provide an inferred composition measurement from an extremely accurate density measurement. For large feed flows, less costly magmeters can be used with some sacrifice of accuracy and rangeability. Vortex and differential head meters should be avoided. If precise reagent valves and recirculation line reagent injection are used, the combination of intelligent ratio control and reagent demand control based on accurate titration curves can possibly eliminate the need for one state of neutralization.

M.8 The Right Model

Having the blocks and objects to simulate the titration curve and the equipment, piping, valve, VSD, sensor, and automation system dynamics greatly simplifies achieving the right model to enable the best system design and sustained performance. The focus can be on selecting and parameterizing the blocks and objects and on connecting streams and signals.

The transportation delay of piping, dip tubes, and static mixers can be simulated by an object whose transportation delay is a function of flow rate and volume, like a conveyor object. For a reagent dip tube, an additional effect occurs when the reagent flow is zero: bidirectional equilibration flow between the vessel and dip tube contents is driven by composition differences. The equilibration flow of reagent out of the dip tube is gradually superseded by an equilibration flow of vessel fluid into the dip tube. The static mixer typically has a volume that delivers about a 6 s transportation delay at normal flow. Transportation delay must be included for piping between the vessel, electrodes, and analyzers.

A backlash-stiction block can simulate the lost motion deadband and resolution limitation in a control valve or VSD. The sensor lag can be simulated with a filter block with the filter time constant set to the 63% response time that goes through a rate limit block to match the slowing down of the valve or VSD response, making the 86% response time more than twice the filter time constant.

A ratio block provides ratio control with a bias that is scaled to provide a negative and positive feedback correction (e.g., –50% to 50% of the ratio scale). An enhanced PID block with a two degrees of freedom (2DOF) structure and external-reset feedback (e.g., dynamic reset limit) is used for all controllers to maximize the load response capability. A simplified relay method autotuner block (e.g., Rough N Ready Tuner block) is used to provide the starting point after a relatively quick, robust test. External reset protects against unnecessary split-range point crossings and helps to deal with slow or suspended responses of valves, VSDs, sensors, or secondary controllers in cascade control. Rate limits can be installed to prevent unnecessary split-range point crossings and enable a smooth, gradual optimization with a fast getaway for abnormal conditions by a valve position controller. All these additional capabilities are achieved without having to retune the controller if there is a fast and accurate update of the actual manipulated variable (e.g., valve position, VSD speed, secondary loop process variable).

The 2DOF structure enables tuning the PID for maximum load disturbance rejections and setting the set-point weight limits for proportional and derivative action to achieve the desired set-point response.

If wireless measurements are used, a block to simulate the default update rate and exception reporting is used to include the additional dead time introduced into the measurement by the wireless transmitter. There is also an analyzer block to simulate the cycle time and analysis time of online analyzers. The transportation delay of the sample must be included along with the multiplex time that occurs when one analyzer is used for multiple streams.

A future value block can be used to compute the rate of change of the *p*H measurement, giving fast updates to a future value prediction with a time interval setting that provides a good signal-to-noise ratio. The future value can be used to prevent Resource Conservation and Recovery Act *p*H violations and predict batch end points. The rate of change signal can be used for diagnostics and provide batch slopes for profile control.

References

[1] Gregory K. McMillan, Christopher Stuart, Rehman Fazeem, Zachary Sample, and Timothy Schieffer, *New Directions in Bioprocess Modeling and Control,* 2nd ed. (Research Triangle Park, NC: ISA [International Society of Automation], 2020).

[2] Gregory K. McMillan and Mark S. Sowell, "Virtual Control of Real pH," *Control* (November 2007).

[3] Gregory K. McMillan, Roger D. Reedy, and John P Moulis, "Virtual Plant Provides Real Insights," *Chemical Processing* (October 2008).

[4] Gregory K. McMillan and Martin E. Berutti, "Virtual Plant Virtuosity," *Control* (August 2017).

[5] Gregory K. McMillan and P. Hunter Vegas, *Process/Industrial Instrumentation and Controls Handbook,* 6th ed. (New York: McGraw-Hill Education, 2019).

[6] Christopher Stuart, Rehman Fazeem, and Gregory K. McMillan, "Improve pH Control," *Chemical Processing* (March 2017).

[7] Gregory K. McMillan and Christopher Stuart, "Simulation Breeds Innovation," *Control* (November 2019).

[8] Gregory K. McMillan, "Simulation Enhances Career and System Performance," *Control* (August 2020).

Index

Printed in the USA/Agawam, MA
September 17, 2026

910661.230